"十二五"职业教育国家规划教材
经全国职业教育教材审定委员会审定

21世纪高等院校
移动开发人才培养规划教材

Android移动开发项目式教程（第2版）

谢景明 主编　钟闰禄 陈长辉 冯敬益 王志球 副主编

Mobile Internet Devolpment in Android
(2nd Edition)

人民邮电出版社
北京

图书在版编目（CIP）数据

Android移动开发项目式教程 / 谢景明主编. -- 2版
. -- 北京 : 人民邮电出版社, 2015.7(2023.1重印)
21世纪高等院校移动开发人才培养规划教材
ISBN 978-7-115-36062-5

Ⅰ. ①A… Ⅱ. ①谢… Ⅲ. ①移动终端—应用程序—程序设计—高等学校—教材 Ⅳ. ①TN929.53

中国版本图书馆CIP数据核字(2014)第137067号

内 容 提 要

本书内容主要包括 7 部分，第 1 部分讲解搭建 Android 开发环境的方法，第 2 部分讲解在 Android 上开发基本界面的方法，第 3 部分讲解利用 SQLite 存储、管理数据的方法，第 4 部分讲解开发多媒体播放器的方法，第 5 部分讲解开发手机相机的方法，第 6 部分讲解开发地图应用的方法，第 7 部分讲解通过网络连接获取天气预报信息的方法。内容较好地涵盖 Android 的界面开发、数据存储、多媒体开发、百度地图开发、网络应用、多线程、XML 解析、JSON 解析等技术，通过丰富的实例来详细介绍它们的技术原理和使用方法。

本书以简单易懂的项目为主线进行讲解，对实现项目所需的知识点进行全面的介绍，并给出了任务实现的详细操作步骤。全书由浅入深、实例生动、易学易用，可以满足不同层次读者的需求。

◆ 主　　编　谢景明
副 主 编　钟闰禄　陈长辉　冯敬益　王志球
责任编辑　王　威
责任印制　杨林杰

◆ 人民邮电出版社出版发行　　北京市丰台区成寿寺路 11 号
邮编　100164　　电子邮件　315@ptpress.com.cn
网址　http://www.ptpress.com.cn
北京七彩京通数码快印有限公司印刷

◆ 开本：787×1092　1/16
印张：18.75　　　　2015 年 7 月第 2 版
字数：494 千字　　　　2023 年 1 月北京第 15 次印刷

定价：42.00 元

读者服务热线：(010)81055256　印装质量热线：(010)81055316
反盗版热线：(010)81055315
广告经营许可证：京东市监广登字20170147号

前　言

本教材的修订是广东省教育科研“十二五”规划2012年度研究项目立项课题（立项编号为2012JK173成果之一）。

Android 操作系统是一款由谷歌（Google）和开放手机联盟共同开发并发展的移动设备操作系统，已经成为世界上最流行的手机操作系统。它易学、易用、功能强大，极大地降低了开发嵌入式应用程序的难度，使程序开发的效率大大提高，已经广泛应用于移动电话、机顶盒、平板电脑等嵌入式消费类电子设备的开发。学习Android开发技术，可以帮助Java程序员进一步提升自身的专业能力，更好地适应移动互联网时代下的技术发展。目前，我国很多院校的计算机相关专业，都将“Android 程序设计”作为一门重要的专业课程。为了帮助院校的教师能够比较全面、系统地讲授这门课程，使学生能够熟练地使用Android进行软件开发，特编写这本《Android移动开发项目式教程（第2版）》。

本书的体系结构是按照项目式的写法来编写的，根据实际项目对Android的常见技术要求，组织了7个难度循序渐进的独立项目，并将每一个项目划分为较为独立的任务，以“任务分析—相关知识—任务实施”这一思路，将Android技术的知识融入具体任务的实现中。在内容编写方面，我们注意难点分散、循序渐进；在文字叙述方面，我们注意言简意赅、重点突出；在实例选取方面，我们注意要求实用性强、针对性强。

在开始学习每一个项目之前，建议读者先对项目的任务有个完整的了解。通过运行书上所附带的程序，对程序的功能有一个直观感受，然后逐个去了解实现任务所需要的知识，掌握基本的概念。在每个项目的最后，仿照本书的程序重新编写一遍，最后独立地完成每章的实训项目，以达到温故知新的目的。

本书每个项目都附有实践性较强的实训，可以供学生上机操作时使用。本书配备了PPT课件、源代码、习题答案、教学大纲、课程设计等丰富的教学资源，任课教师可到人民邮电出版社教学服务与资源网（www.ptpedu.com.cn）免费下载使用。本书的参考学时为90学时，其中实践环节为40学时，各部分的参考学时参见下面的学时分配表。

项　目	课程内容	学时分配	
		讲　授	实　训
项目一	建立Android开发环境	4	2
项目二	开发标准身高计算器	8	7
项目三	开发手机通讯录	11	9
项目四	开发多媒体播放器	8	7
项目五	开发手机相机	7	6

续表

项　目	课程内容	学时分配	
		讲　授	实　训
项目六	开发地图应用	7	5
项目七	开发天气预报程序	5	4
课时总计		50	40

本书由广州番禺职业技术学院的谢景明博士担任主编，钟闰禄、陈长辉、冯敬益、王志球任副主编。由于 Android 技术发展日新月异，而本人水平有限，书中难免存在错误和不妥之处，敬请广大读者批评指正。

编　者

2015 年 4 月

目 录 CONTENTS

项目一 建立 Android 开发环境 1

1.1 背景知识 1
一、移动应用开发技术 1
二、典型移动应用案例 4
三、Android 的发展历史 5
1.2 安装 Sun JDK 6
一、任务分析 6
二、相关知识 6
三、任务实施 7
1.3 安装 Android SDK 12
一、任务分析 12
二、相关知识 12
三、任务实施 12
1.4 安装 Eclipse ADT 17
一、任务分析 17
二、相关知识 17
三、任务实施 19
1.5 测试开发环境 23
一、任务分析 23
二、相关知识 23
三、任务实施 34
1.6 实训项目 38
一、建立 Android 开发环境 38
二、开发运行一个简单的 Android 程序 38

项目二 开发标准身高计算器 39

2.1 背景知识 39
一、常见的手机硬件参数知识 39
二、Android 的像素单位 41
2.2 开发输入界面 41
一、任务分析 41
二、相关知识 42
三、任务实施 58
2.3 进行事件处理 62
一、任务分析 62
二、相关知识 62
三、任务实施 71
2.4 显示计算结果 72
一、任务分析 72
二、相关知识 72
三、任务实施 77
2.5 发布到手机 77
一、任务分析 77
二、任务实施 77
2.6 完整项目实施 79
2.7 实训项目 83
一、用户登录界面 83
二、调查问卷程序 83

项目三　开发手机通讯录　84

3.1 Android 的数据存储技术 84

一、使用 SharedPreferences 存储数据 85

二、文件存储数据 86

三、SQLite 数据库存储数据 88

四、使用 ContentProvider 对外共享数据 89

五、Internet 网络存储数据 89

3.2 添加联系人记录 90

一、任务分析 90

二、相关知识 90

三、任务实施 97

3.3 修改联系人记录 102

一、任务分析 102

二、相关知识 102

三、任务实施 108

3.4 查找号码记录 110

一、任务分析 110

二、相关知识 110

三、任务实施 118

3.5 查看联系人记录 120

一、任务分析 120

二、任务实施 120

3.6 删除号码记录 122

一、任务分析 122

二、相关知识 122

三、任务实施 122

3.7 对外共享数据 123

一、任务分析 123

二、相关知识 123

三、任务实施 126

3.8 设计主界面 127

一、任务分析 127

二、任务实施 127

3.9 完整项目实施 129

3.10 实训项目 147

一、手机通讯录的改进 147

二、我的移动日记 148

三、英语题库系统 148

项目四　开发多媒体播放器　149

4.1 开发多媒体播放界面 149

一、任务分析 149

二、相关知识 149

三、任务实施 154

4.2 播放音乐 156

一、任务分析 156

二、相关知识 157

三、任务实施 164

4.3 播放视频 167

一、任务分析 167

二、相关知识 167

三、任务实施 169

4.4 管理多媒体文件 170

一、任务分析 170

二、相关知识 170

三、任务实施 178

4.5 多线程开发 183
一、任务分析 183
二、相关知识 183
三、任务实施 189
4.6 后台服务 Service 190
一、任务分析 190
二、相关知识 190
三、任务实施 193
4.7 完整项目实施 195
4.8 实训项目 206
一. 开发多媒体播放器 206

项目五 开发手机相机 207

5.1 相机打开界面 207
一、任务分析 207
二、相关知识 208
三、任务实施 211
5.2 相机拍照控制 212
一、任务分析 212
二、相关知识 212
三、任务实施 217
5.3 照片保存和预览 219
一、任务分析 219
二、相关知识 219
三、任务实施 220
5.4 照片浏览 221
一、任务分析 221
二、相关知识 222
三、任务实施 232
5.5 完整项目实施 235
5.6 实训项目 242
一、实现手机录像功能 242
二、对手机晃动进行检测 243

项目六 开发地图应用 244

6.1 显示百度地图 244
一、任务分析 244
二、相关知识 244
三、任务实施 246
6.2 地图基础应用 248
一、任务分析 248
二、相关知识 248
三、任务实施 252
6.3 实现 POI 查询 255
一、任务分析 255
二、相关知识 256
三、任务实施 257
6.4 实现定位 259
一、任务分析 259
二、相关知识 259
三、任务实施 261
6.5 实训项目 263
一、完善百度地图应用程序的开发 263
二、GPS 定位器 263

项目七　开发天气预报程序　264

7.1　获取天气预报信息　264

一、任务分析　264

二、相关知识　264

三、任务实施　278

7.2　下载天气图片　279

一、任务分析　279

二、任务实施　279

7.3　显示天气预报　280

一、任务分析　280

二、相关知识　280

三、任务实施　281

7.4　完整项目实施　282

7.5　扩展项目 JSON 格式接口调用　287

一、任务分析　287

二、相关知识　287

三、项目实施　290

7.6　实训项目　291

一、开发天气预报程序　291

二、开发手机聊天室　292

PART 1

项目一 建立 Android 开发环境

近年来，移动通信技术得到了高速发展，智能手机已逐步普及。随着手机功能的增强，手机程序的种类也越来越多，市场上将手机程序一般分为应用程序和游戏两大类。应用程序有电子书、系统工具、实用软件、多媒体软件、通信辅助软件、网络软件等；游戏类有角色扮演、动作格斗、体育竞技、射击飞行、冒险模拟、棋牌益智等。

Apple 公司 iPhone 和 Google 公司 Android 以及中国移动、Nokia、Samsung 等 IT 界的巨头先后推出手机应用商店，允许企业或程序员将自己开发的手机软件在应用商店出售，用户购买应用程序所支付的费用由应用商店提供者与开发者按照一定的比例进行分成，为软件开发者与用户提供了一个良好的交流平台，使得第三方软件提供者参与手机程序开发的积极性空前高涨。

Android 是主要的手机应用程序开发平台之一。本项目将学习如何利用开源免费软件构建 Android 开发环境，目标是使初学者了解移动应用程序发展情况、不同移动开发技术的主要特点以及 Android 技术的优势，掌握 JDK、SDK、Eclipse，以及 ADT 的安装、配置和使用方法，能够使用 Eclipse 创建和运行简单的 Android 程序。

1.1 背景知识

一、移动应用开发技术

2014 年底，我国手机用户数已经达到了 12.86 亿户，移动电话用户普及率达 94.5 部/百人。手机市场的繁荣促使手机的功能越来越强大，并不断朝着智能化方向发展。近年来中国的智能手机用户增长很快，中国智能手机用户在 2014 年首次超过 5 亿人，由此也引发了对手机应用程序和游戏程序的庞大需求。外部应用程序的丰富程度和易用性，已成为消费者购买某款智能手机的决定因素之一。

中国移动、中国联通、中国电信三家运营商都具有移动网络通信运营资质，现有网络分为 2G、3G 和带宽更高的 4G 网络。4G 是第四代移动电话行动通信标准的简称，该技术集 3G 与 WLAN 于一体，并能够快速传输数据、高质量的音频、视频和图像等。4G 理论上能够以 100Mbps 以上的速度下载，比 4 兆的家用宽带 ADSL 快 25 倍，并能够满足几乎所有用户对于无线服务的要求。除了支持传统的通信业务之外，4G 的无线高速业务也将会在移动终端上产生新的应用，如流媒体点播应用，与物联网、云计算和大数据的综合应用等。

移动应用程序的硬件平台主要是小型嵌入式设备，这些设备具有便携性，但在性能和功能上较 PC 要差，而且缺乏统一的标准，功能、外观和操作方式差别较大。开发移动应用程序的平台主要有 J2ME、Symbian、Windows Mobile、iOS 和 Android，这些技术各有千秋。J2ME、Symbian 和 Windows Mobile 是比较传统的开发技术，已经发展多年，而 iOS、Android、Windows Phone 则是近年来兴起的热门技术。目前，在这些平台上开发的应用程序数量一直在稳定增长，已经有超过百万个手机应用程序在应用商店出售。下面对这些技术做简要介绍。

Android 是 Google 公司于 2007 年 11 月发布的基于 Linux 平台的开源手机操作系统。Android 基于 Linux 技术，由操作系统、中间件、用户界面和应用软件组成。Android 产品线较为丰富，覆盖商务、时尚、娱乐、中低端市场等各种人群。Android 不仅对第三方软件完全开放，可免费提供给开发人员，使之可以对源代码进行修改，而且集成了大量的 Google 应用程序，例如 Google 地图、Gmail 邮箱等，大大增强了 Android 手机的功能。Android 主要支持的开发语言为 Java 和 C/C++。

J2ME 是 Java 2Micro Edition 的缩写，于 1999 年 6 月在 JavaOne 开发者大会上亮相。J2ME 是为机顶盒、移动电话和 PDA 等嵌入式消费电子设备提供的 Java 语言平台，包括虚拟机 KVM 和一系列标准化的 Java API。J2ME 和标准版的 J2SE、企业版的 J2EE 一起构成 Java 技术的三大版本。J2ME 程序的执行方式是字节码解释，性能上会受到一定程度的影响。目前国内该技术已经接近淘汰。

Symbian 公司成立于 1998 年 6 月，是由 Ericsson、Motorola、Nokia 等公司共同持股组成的合资公司。2008 年 6 月 Symbian 公司被 Nokia 全资收购，成为其旗下公司。Symbian 曾经是手机领域中应用范围最广的操作系统之一，提供多个不同版本的人机界面。Symbian 主要支持的开发语言为 C++、Java 和 C。目前国内该技术已经接近淘汰。

Windows Mobile 是由微软公司在 2003 年 6 月发布的，从 7.0 版本后更名为 Windows Phone。在此之前微软的智能终端设备操作系统主要分为 Pocket PC 和 Smartphone 两类。原形为 Windows CE，是 Microsoft 用于 Pocket PC 和 Smartphone 的软件平台。Windows Mobile 的优势在于将熟悉的 Windows 桌面扩展到移动设备中，界面设计、功能应用与 PC 类似，内置有 Microsoft Office、Media Player 等软件。Windows Mobile 主要支持的开发语言为 C#、C++和 VB。现在微软公司力推的是 Windows Phone 平台。

iPhone 是 Apple 公司在 2007 年 1 月举行的 Macworld 上宣布推出的，2007 年 6 月在美国上市。iPhone 使用了众多增强用户体验的领先技术，例如，多触点式触摸屏技术允许用户同时通过多个触点进行操作；基于传感器的隐式输入技术提高了手机的智能水平；全新的用户界面设计技术提高了手机的易用性；手机应用商店提供了源源不断的实用程序。iPhone 将原本普通的手机电话变成了一个集潮流时尚于一体且功能强大的随身工具，引起了手机设计领域的一次新变革。iPhone 的操作系统为 iOS，其主要支持的开发语言为 Objective-C、C、C++、JavaScript。

相比传统的软件发布，手机应用商店为程序开发者提供了更广阔的平台，程序开发者可将手机应用程序发布到网站上，分享自己的作品，并通过下载来和服务商按比例进行收益分配。图 1-1、图 1-2、图 1-3 分别为三家著名的手机应用商店，感兴趣的读者可登录网站浏览或者下载各种类型的手机移动应用程序，从中获取软件设计的灵感。

图 1-1　中国移动的移动应用商城（mm.10086.cn）

图 1-2　机锋软件（http://apk.gfan.com/）

图 1-3　亚马逊应用商店（http://www.amazon.cn/mobile-apps/b?node=146628071）

二、典型移动应用案例

手机的优势在于其不但具有通信、多媒体、支持应用程序等功能，而且还易于携带且便于使用。移动应用主要分为企业应用和个人应用，讲究的都是实用性。进行移动应用开发需要遵循手机的特点，例如屏幕大小、存储空间、供电能力等。下面以移动办公、个人应用和手机游戏等典型应用为例进行介绍。

1．移动办公

移动办公是指办公人员可以随时随地处理与业务有关的事情。要达到这个目标，可通过开发手机上的移动办公软件，实现与企业软件系统或者互联网的连接，获取办公所需的信息，并进行及时处理。移动办公的应用领域非常广泛，主要有：流程审批、行政执法、物流派送、信息查询等。国内许多行业都可以利用移动办公来进一步提高工作效率，例如，商务人士在出差的路途中可以及时地处理单位事务，获得市场信息；工商人员巡查市场时，可通过移动办公系统对商品进行现场监管查询，有效防止假冒伪劣食品的流通等。

移动办公需要将手机、无线网络和企业系统三者有机结合。实现移动办公系统一般需要重点解决两个问题：一是实现客户端软件与企业服务器的无线连接、数据传输等；二是客户端的界面要友好，使用简易，符合手机的操作特点。图 1-4 所示为 Android 移动办公系统界面效果示例。

图 1-4　移动办公系统界面效果示例

2．个人应用

手机出厂自带的软件类型一般较少，往往难以满足用户的个性化需求。个人移动应用系统的主要任务是针对手机自带软件功能不足的问题，开发设计出新的功能，为日常生活中的“衣、食、住、行”提供便利，体现出手机智能化助手的优势。例如，对各种类型的来电进行管理控制，对餐馆、景点、购物场所进行查询、定位。

个人应用中另一类比较重要的应用是移动学习（M-learning）软件。移动学习是通过移动通信、计算机、信息教育等多种技术实现在任何时间、任何地点开展学习的一种新型教育模式。移动学习的主要特征有三个：①自由性大，学习的时间和地点不固定，学习者能够自主安排学习计划；②主动性高，学习通常发生在零散时间或者特定情景下，学习者往往是出于提高稿自身能力或者解决问题的需要而进行学习，这种积极的学习动机更容易产生良好的学习效果；③便捷性强，学习的技术手段更为先进，学习者能够利用移动设备通过无线网络灵活快捷地获取知识。移动学习软件使人们在学习方式上摆脱时间和空间的约束，大大改变了传统的以固定教室为主的教学模式，使知识的传播更为及时和方便，因此可作为课堂教学之外的一种良好补充。如图 1-5 所示为金山词霸 Android 版手机界面。

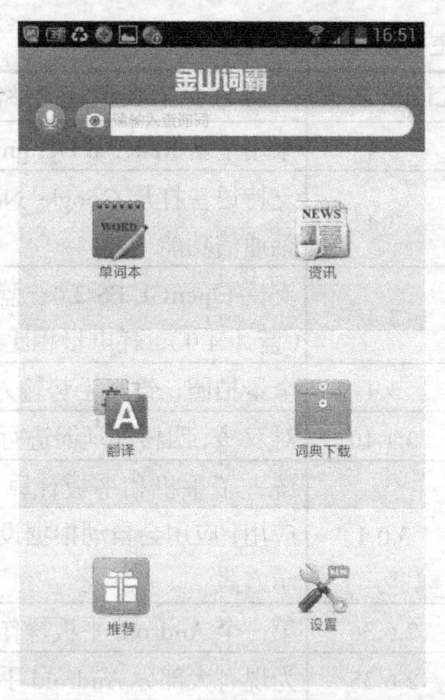

图 1-5　金山词霸 Android 版手机界面

3．手机游戏

近几年随着手机的普及，特别是 3G、4G 网络的完善和智能手机性能的提高，手机游戏逐渐受到了业界的重视，成为移动增值业务的重点发展方向。早期的游戏主要是由手机厂商作为手机附属品提供的，例如“俄罗斯方块”、“贪吃蛇”等，而且这些游戏的画面较为粗糙，规则也很简单。随着手机硬件的提升以及众多开发人员的加入，目前的手机游戏在界面和可玩性上取得了长足的发展，具有较好的娱乐性和交互性。手机游戏的优势是可以在坐车、等人的空闲时间为手机用户提供娱乐休闲。手机游戏按内容属性来分，可以分为角色扮演游戏、动作游戏、策略游戏、格斗游戏等不同类型。

手机游戏的实现方式，可以有单机游戏、网络游戏、蓝牙游戏、模拟器游戏等。很多游戏的创意来自于 PC。如图 1-6 所示为一款 3D 场景的捕鱼达人游戏界面。

图 1-6　3D 场景的捕鱼达人游戏界面

三、Android 的发展历史

Android 操作系统是由 Google 和开放手机联盟共同开发并发展的移动设备操作系统，其最早的版本开始于 2007 年 11 月的 Android 1.0 beta，并且已经更新发布了多个 Android 操作系统版本。伴随着性能、接口的极大改进和功能的丰富，Android 系统的运行变得越来越快、越来越省电，功能越来越丰富，嵌入的 Google 服务也越来越多。从 2009 年 4 月开始，Android 操作系统改用甜点来作为版本代号，并按照大写字母的顺序进行命名（见表 1-1）。在 Android 程序开发中，建议选择 Android 2.1 以上版本。

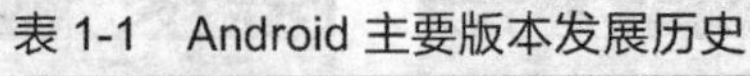

表 1-1　Android 主要版本发展历史

版　　本	发布日期	API 层次	Linux 内核	典型特征举例
5.0 棒棒糖（Lollipop）	2014.6.25	21	3.4	采用全新 Material Design 界面
4.4 奇巧巧克力（KitKAT）	2013.9.3	19	3.4	支持语音打开 Google Now；强大的电话通信功能
4.3 果冻豆（Jelly Bean）	2013.7.24	18	3.4	支持 OpenGL ES 3.0；蓝牙低功耗功能（蓝牙 4.0），耗电量相比之前大幅降低
4.2 果冻豆（Jelly Bean）	2012.10.29	17	3.4	全景拍照，键盘手势输入等
4.1 果冻豆（Jelly Bean）	2012.6.28	16	3.1.10	对双核、四核处理器进行了更好的优化
4.0 冰激凌三明治（Ice Cream Sandwich）	2011.10.19	14–15	3.0.1	统一了手机和平板计算机不同平台的应用，应用会自动根据设备选择最佳显示方式
3.0 蜂巢（Honeycomb）	2011.2.2	11–13	2.6.36	第一个 Android 平板操作系统
2.3 姜饼（Gingerbread）	2010.12.6	9–10	2.6.35	为现有大部分 Android 手机使用的版本
2.2 冻酸奶（Froyo）	2010.5.20	8	2.6.32	支持将软件安装至扩展内存
2.0/2.1 松饼（Eclair）	2009.10.26/2010.1.12	5–7	2.6.29	优化硬件速度，支持蓝牙，内置相机闪光灯和数码变焦
1.6 甜甜圈（Donut）	2009.9.15	4	2.6.29	将相机、摄像机和照片画廊功能合并为一个界面
1.5 纸杯蛋糕（Cupcake）	2009.4.30	3	2.6.27	Android 早期版本，使用不多

1.2　安装 Sun JDK

一、任务分析

本次任务要求完成 JDK 的下载、安装和配置。要完成本次任务，需要思考如下几个问题。

（1）JDK 是什么软件，对于本项目有何作用？

（2）从何处获得正确的 JDK 软件？

（3）JDK 对计算机硬件和操作系统有何要求？

（4）如何安装 JDK？

（5）如何配置 JDK？

二、相关知识

通常而言，进行软件开发并不是从一张白纸开始，往往会利用一些已有的工具。JDK（Java Development Kit）就是为 Java 开发者提供的一组开发工具包，包括了 Java 运行环境（JRE，Java Runtime Environment）、Java 工具和 Java 标准 API 类库。JDK 是进行 Java 程序开发的基础。主流的 JDK 由 Sun 公司（2009 年已经被著名的数据库公司 Oracle 收购）所开发。一些公司和组织也先后推出自己的 JDK，例如 IBM JDK、GNU JDK。JDK 一般有三种版本，此外，还有适用于 Windows、Linux、Solaris 等不同操作系统的版本。

Java 运行环境包含一个 Java 虚拟机（JVM，Java Virtual Machine）和运行 Java 程序所需的类库。其中 Java 虚拟机的主要作用是解释字节码（bytecode），实现 Java 程序的跨平台运行。JRE 一般包含在 JDK 中，也可以独立安装。

常用的 Java 工具有 Javac、Java、JAR、Javap 和 AppletViewer。其中，Javac 是编译器，用于将 Java 源程序（后缀名为.java）转换成字节码文件（后缀名为.class）。Java 是解释器，解释、加载、运行由 Javac 编译得到的字节码文件。JAR 是打包工具，用于将相关的类文件或者资源文件压缩打包成一个 JAR 文件，以便于应用程序的发布。Javap 是反编译器，用于显示字节码文件的含义。AppletViewer 不需要 Web 浏览器就可以调试运行 Applet 小程序。

Java 标准 API 类库（Application Programming Interface）又称应用程序编程接口，通过提供一些预先定义的函数，达到简化开发人员工作的目的。开发人员无需访问源码或理解内部工作机制的细节，通过调用 API 即可实现程序的特定功能。编程语言或二次开发的软硬件环境一般会提供相应的 API。同样，开发 Java 程序是在 Java 标准 API 类库的基础上进行的。JDK 提供了 4 个包：dt.jar 是关于运行环境的类库，主要是 Swing 包；tools.jar 是用 Java 所编写的开发工具的类库，例如 javac.exe、jar.exe 等；rt.jar 包含了 JDK 的基础类库，也就是在 Java 文档帮助中看到的所有类的 Class 文件；htmlconverter.jar 包含了命令转换工具，可将<Applet>标记转换成<Object>和<Embed>标记，是为不支持 Applet 的浏览器或使用 JRE 1.6 以前版本的 IE 浏览器而设计的。

三、任务实施

1．下载 Sun JDK

下载软件一般有两种方法：一种是通过搜索引擎进行搜索，另一种是到开发软件的公司网站进行下载。第一种方法的优点是比较简单，缺点是下载的软件质量没有保证，例如软件不能正常使用、有病毒、版本较旧等。第二种方法的优点是能够保证下载软件的质量，但很多公司是英文网站，会对下载造成一定的困难。下面介绍采用第二种方法——从 Oralce 公司网站下载 Sun JDK 软件。一些初学的开发者往往会觉得到英文网站下载软件很困难，不知道从哪里找到所需要的软件。事实上，网站的下载资源一般都会放在首页的 Downloads 链接上。所以要下载 JDK 软件最好的方法是，先进入 Downloads 界面。

（1）登录 Oracle 公司的网站：http://www.oracle.com，如图 1-7 所示。如果不知道 Oracle 公司的网站，可通过百度搜索引擎搜索获取。主页上有两个地方和软件下载有关：Downloads 和 Top Downloads，其中前者罗列了所有可以下载的资源，后者罗列了网站下载量最多的几种资源。

图 1-7 Oracle 公司主页

（2）将鼠标光标放到“Downloads”按钮上，在显示出的菜单中的“Popular Dounload”下单击“Java for Developers”超级链接，进入到Sun开发者网络页面，如图1-8所示。该页面罗列了Sun提供的各种类型的Java开发工具，例如NetBeans、JavaFx等，并提供下载服务。

图1-8　Sun开发者网络页面

（3）单击“JDK”超级链接，进入JDK下载选择页面，如图1-9所示。在该页面可以选择Windows、Windows x64、Solaris x64、Solaris x86、Linux、Linux x64等多个不同操作系统的版本。本书项目的开发主要是在Windows环境下，所以选择适合于Windows平台的JDK版本。

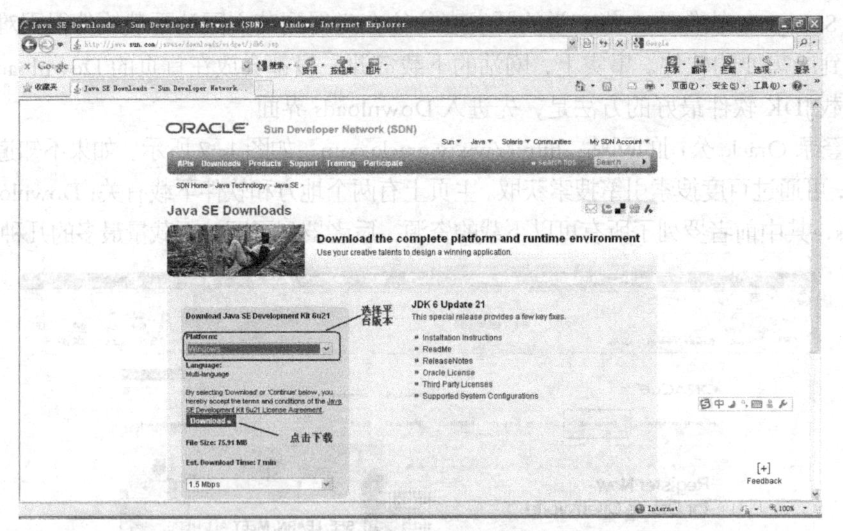

图1-9　JDK下载选择页面

（4）单击【Download】按钮，弹出下载登录页面，如图1-10所示。在该页面中可以有3种操作：一种是如果以前已经注册，可以输入个人账户和密码登录；第二种是通过新创建一个账号进行登录；最后一种是跳过登录步骤。

（5）选择跳过登录步骤，进入JDK文件下载页面，如图1-11所示。该页面列出了所下载的JDK文件名和大小，直接单击文件名，即可将文件下载到本地计算机指定的文件夹中。注意，由于 JDK 的版本更新较快，用户下载的文件名可能和本书不一样，一般情况下不会影响开发环境的建立。

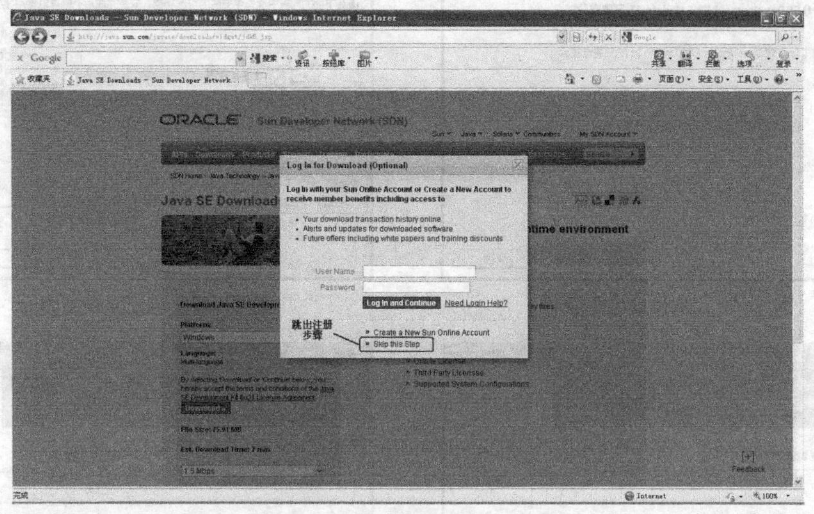

图 1-10　下载登录页面

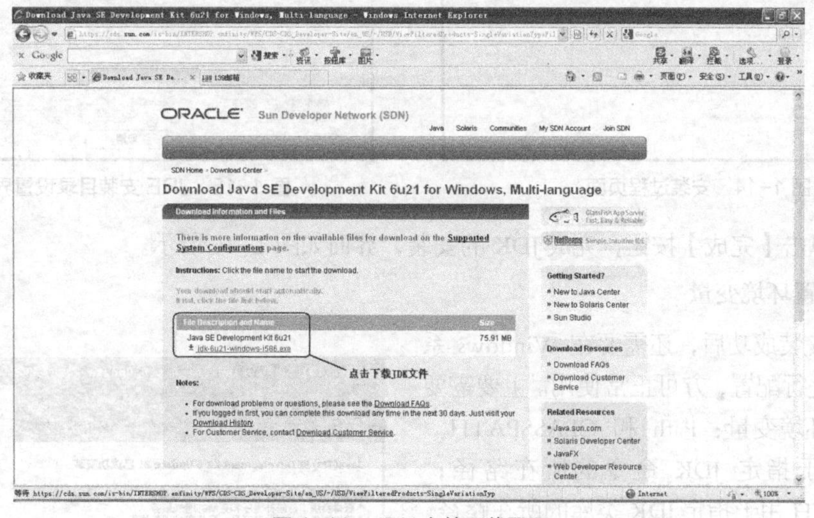

图 1-11　JDK 文件下载页面

2．安装 JDK 软件

JDK 的安装很简单，直接双击所下载的 JDK 文件，即可进行安装。下面是主要的安装步骤。

（1）双击上一节下载的 JDK 文件【jdk-6u21-windows-i586.exe】，进入安装界面，如图 1-12 所示。

（2）单击【下一步】按钮，进入自定义安装界面，如图 1-13 所示。在本界面上可以选择要安装的功能，还可以指定安装的目录。

（3）按照默认配置，单击【下一步】按钮，即进入安装状态，安装过程界面如图 1-14 所示。

（4）安装过程中，弹出 JRE 的安装目录设置界面，如图 1-15 所示，可按照默认路径进行安装。

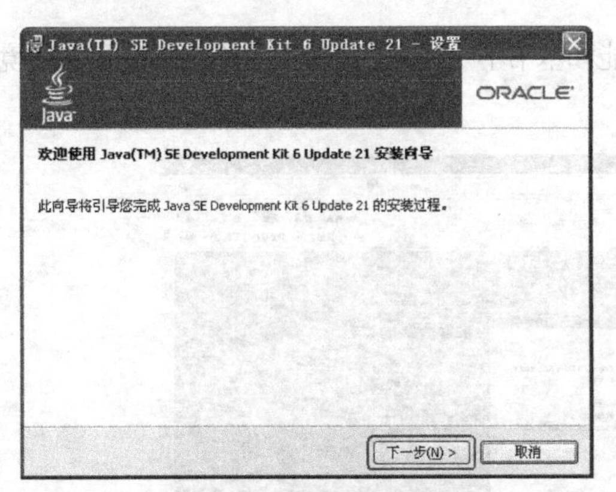

图 1-12　JDK 安装界面

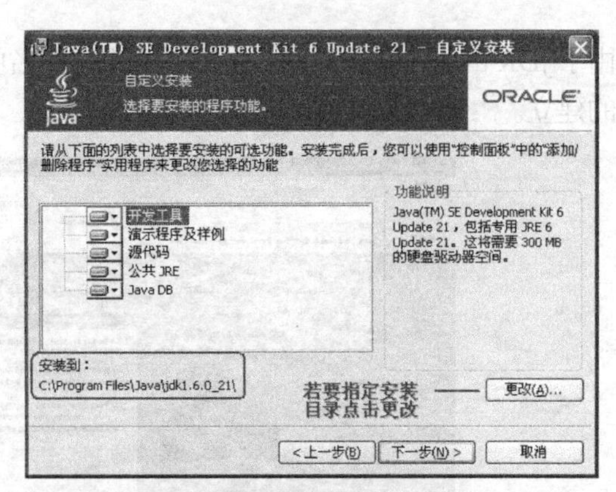

图 1-13　自定义安装界面

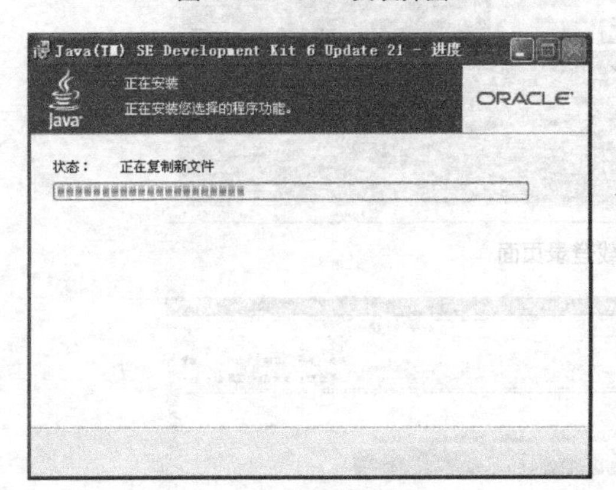

图 1-14　安装过程页面

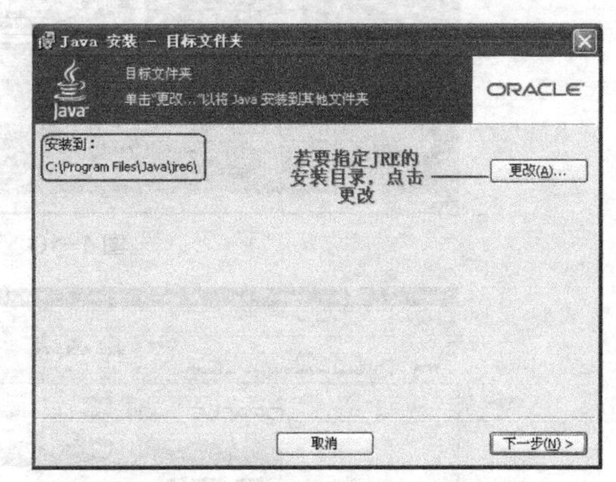

图 1-15　JRE 安装目录设置界面

（5）单击【完成】按钮，完成 JDK 的安装，界面如图 1-16 所示。

3．配置环境变量

JDK 安装成功后，还需要在 Windows 系统中对其进行配置，方可正常使用。主要需要配置两个环境变量：Path 和 CLASSPATH。PATH 用于指定 JDK 命令的所在路径；CLASSPATH 用于指定 JDK 类库的所在路径。

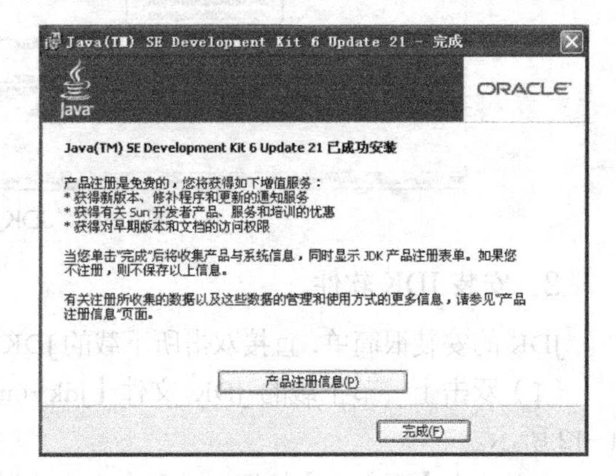

图 1-16　安装完成界面

（1）鼠标右键单击桌面上【我的电脑】图标，选择【属性】命令，可打开系统属性配置界面，如图 1-17 所示，选择【高级】选项卡。

（2）单击【环境变量】按钮，进入环境变量配置界面如图 1-18 所示。本界面已经列出一些定义好的环境变量。在配置界面中有用户变量和系统变量两种，其中用户变量所配置的环境变量适用于某个用户，例如本例中的 Administrator；系统变量所配置的环境变量适用于本机上的所有用户。

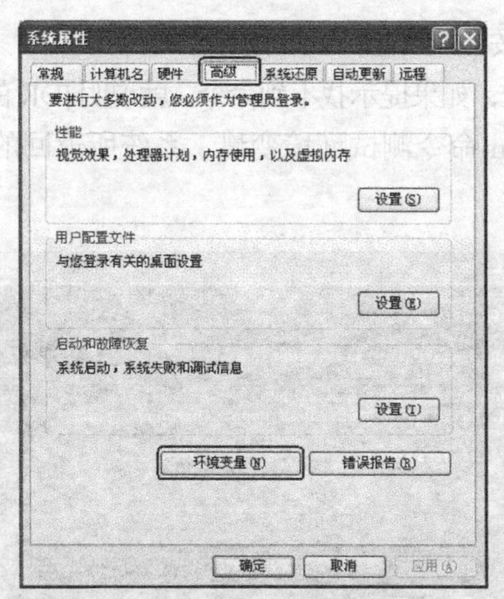

图 1-17　Windows 系统属性界面

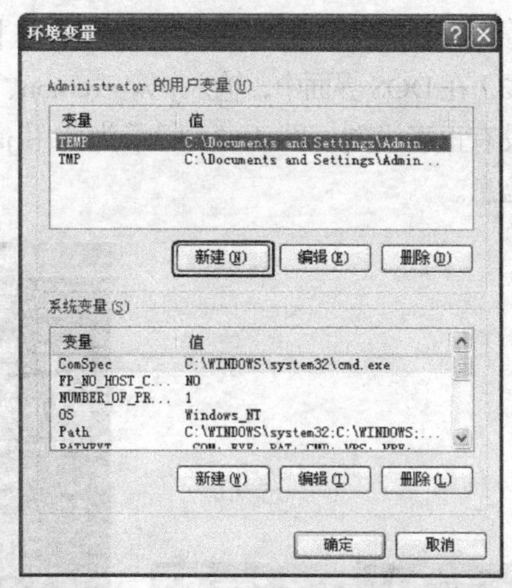

图 1-18　环境变量配置界面

（3）单击【Administrator 的用户变量部分】列表下的【新建】按钮，进入新建用户变量界面，如图 1-19 所示。

（4）输入变量名为 PATH，变量值为 C:\Program Files\Java\jdk1.6.0_21\bin 的用户变量，如图 1-20 所示。

说明

PATH 环境变量的取值和 JDK 的安装目录有关，需要先查找到安装的 JDK 的所在目录，然后再获得 JDK 的 bin 目录的完整路径。

（5）单击【确定】按钮，采用第（3）步的操作，输入变量名为 CLASSPATH，变量值为；C:\Program Files\Java\jdk1.6.0_ 21\lib 的用户变量，如图 1-21 所示。

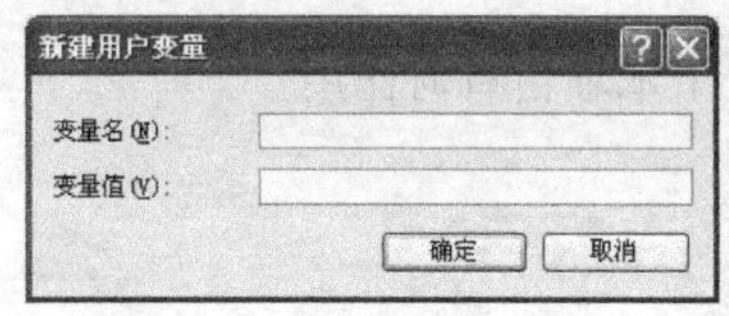

图 1-19　新建用户变量界面

图 1-20　编辑用户变量

图 1-21　新建用户变量

说明

CLASSPATH 变量值中的“.”表示当前目录，用“;”和 JDK 的库目录隔开，和第（4）步类似，注意变量值的位置同样和 JDK 的安装目录有关。

（6）新建用户变量完成后，可看到在【Administrator 的用户变量】列表中增加了两个新的变量，分别为 CLASSPATH 和 PATH，如图 1-22 所示。单击【确定】按钮，完成 JDK 的配置。

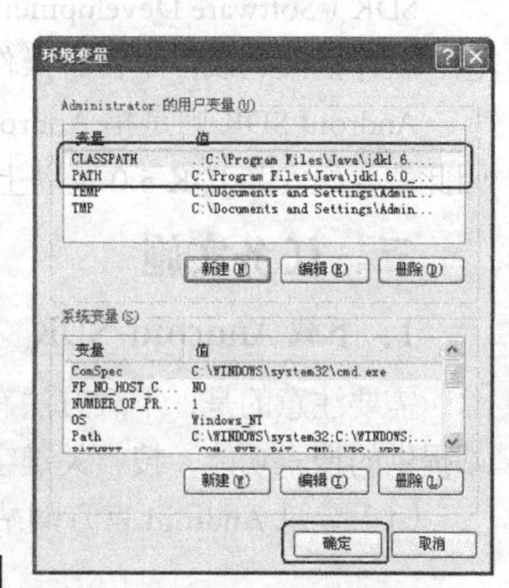

图 1-22　新建用户变量完成

4．检验安装配置

查看 JDK 是否安装配置成功，可在 DOS 环境下进行测试。

（1）进入 DOS 环境，选择 Windows 的【开始】→【运行】

命令，在打开的对话框中输入cmd，单击【确定】按钮，如图1-23所示。

（2）在DOS界面中，输入java -version、javac命令，如果提示找不到命令，就说明JDK的环境变量没有正确设置。图1-24所示为使用java-version命令测试环境变量，系统所返回的JDK版本信息。

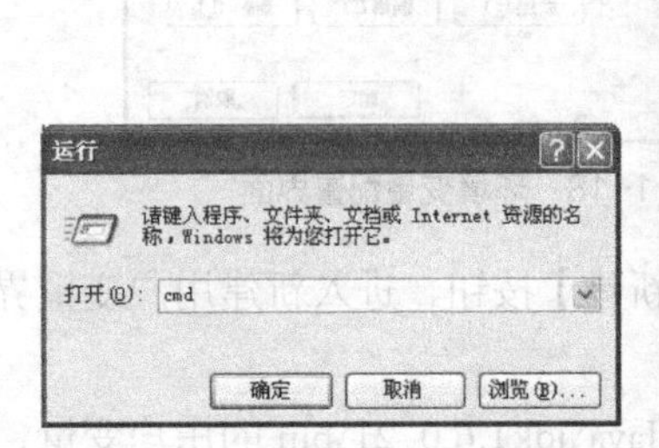

图1-23　进入DOS环境的办法

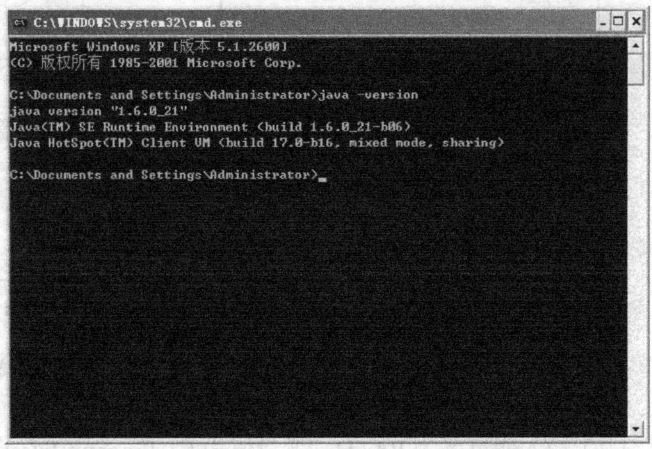

图1-24　使用Java命令测试环境变量的配置

1.3　安装Android SDK

一、任务分析

由于Android SDK自身并没有附带Java的运行环境JDK，在安装过程中，程序会自动检测当前系统已有的Java虚拟机，所以在Android SDK安装之前需要先安装JDK。JDK的安装已经在1.2节完成。本次任务要求完成Android SDK的下载、安装和配置。要完成本次任务，需要思考如下几个问题。

（1）Android SDK是什么软件，与JDK软件有何关系，对于本项目有何作用？

（2）从何处获得正确的Android SDK软件？

（3）Android SDK对计算机的硬件和操作系统有何要求？

（4）如何安装Android SDK？

二、相关知识

SDK（Software Development Kit）软件开发工具包，是软件开发工程师用于为特定的软件包、软件框架、硬件平台、操作系统等建立应用软件开发工具的集合。

Android SDK则是指Android专属的软件开发工具包。Android SDK采用的是Java语言，因此需要先安装JDK 5.0及以上版本。

三、任务实施

1．下载Android SDK

需要注意的是，下载方法有可能因为Android网站的变动而不一致。这时，可以利用网站提供的搜索功能，搜索关键字sdk download。

（1）登录Android官方网站http://developer.android.com。页面效果如图1-25所示。

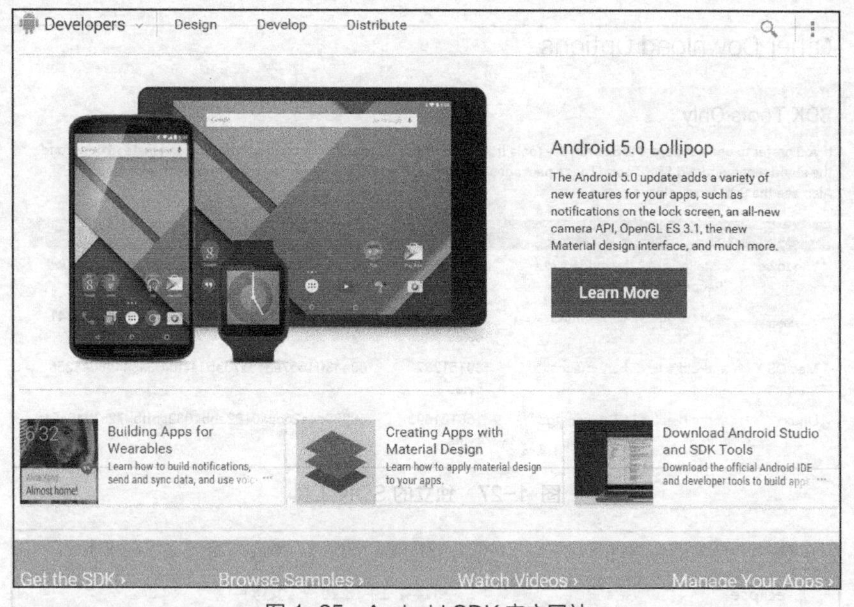

图 1-25 Android SDK 官方网站

（2）单击【Get the SDK】链接，出现 Android SDK 下载页面，如图 1-26 所示。该页面提供下载开发 Android 应用程序的各种软件和工具，其中 Android Studio 为 Google 官方推荐使用的集成开发环境。

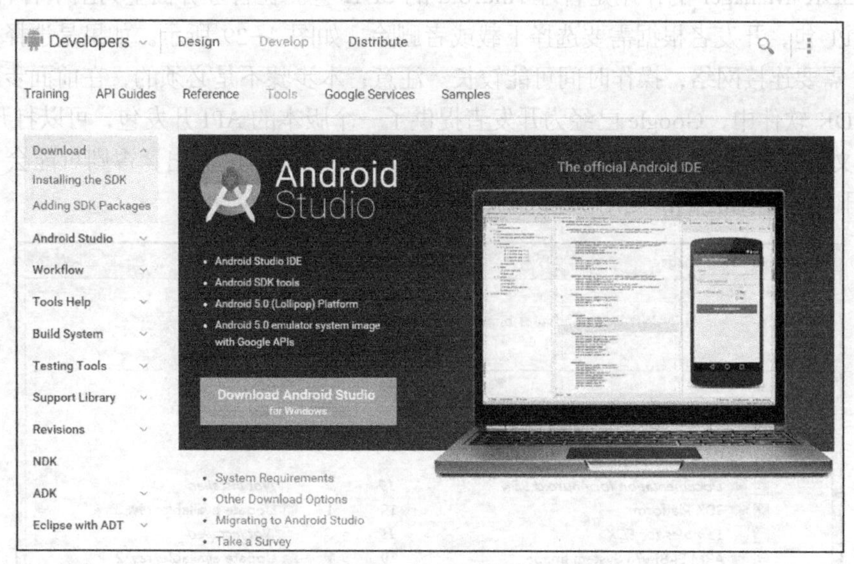

图 1-26 Android SDK 软件下载页面

2．安装 Android SDK

Google 为开发者提供了捆绑好的 Android SDK 的安装软件，安装较为简单，但有一点需要注意，由于 Android SDK 自身并没有附带 Java 的运行环境 JDK，所以在安装 Android SDK 之前需要先安装相应的 JDK 软件。由于 Eclipse+ADT 插件使用方式为广大 Java 程序员所接受，本教材将以该开发环境来进行讲解。

（1）对教材附带的 Android SDK 安装压缩文件进行解压，打开文件夹，可以看到有 eclipse 和 sdk 两个文件夹和 1 个可执行文件 SDK Manager，如图 1-28 所示。

Other Download Options

SDK Tools Only

If you prefer to use a different IDE or run the tools from the command line or with build scripts, you can instead download the stand-alone Android SDK Tools. These packages provide the basic SDK tools for app development, without an IDE. Also see the SDK tools release notes.

Platform	Package	Size	SHA-1 Checksum
Windows	installer_r24.1.2-windows.exe (Recommended)	111364285 bytes	e0ec864efa0e7449db2d7ed069c03b1f4d36f0cd
	android-sdk_r24.1.2-windows.zip	159778618 bytes	704f6c874373b98e061fe2e7eb34f9fcb907a341
Mac OS X	android-sdk_r24.1.2-macosx.zip	89151287 bytes	00e43ff1557e8cba7da53e4f64f3a34498048256
Linux	android-sdk_r24.1.2-linux.tgz	168121693 bytes	68980e4a26cca0182abb1032abffbb72a1240c51

图 1-27　独立的 SDK 工具

名称	修改日期	类型	大小
eclipse	2014-2-19 18:29	文件夹	
sdk	2014-2-19 20:05	文件夹	
SDK Manager	2013-10-31 5:42	应用程序	350 KB

图 1-28　Android SDK 软件包

（2）SDK Manager 的作用是管理 Android 的 SDK 包。在管理界面上列出了各种 Android 版本的 API 包，开发者根据需要选择下载或者删除，如图 1-29 所示。如果是选择安装新的 SDK 包，需要连接网络，操作时间可能较长。注意：本步骤不是必须的，在前面步骤下载的 Android SDK 软件中，Google 已经为开发者提供了一个版本的 API 开发包，可以打开图 1-28 中的 sdk 文件夹来查看。注意：在 SDK 更新过程中，请勿中途退出，否则可能会导致整个 Android 开发环境无法正常使用。

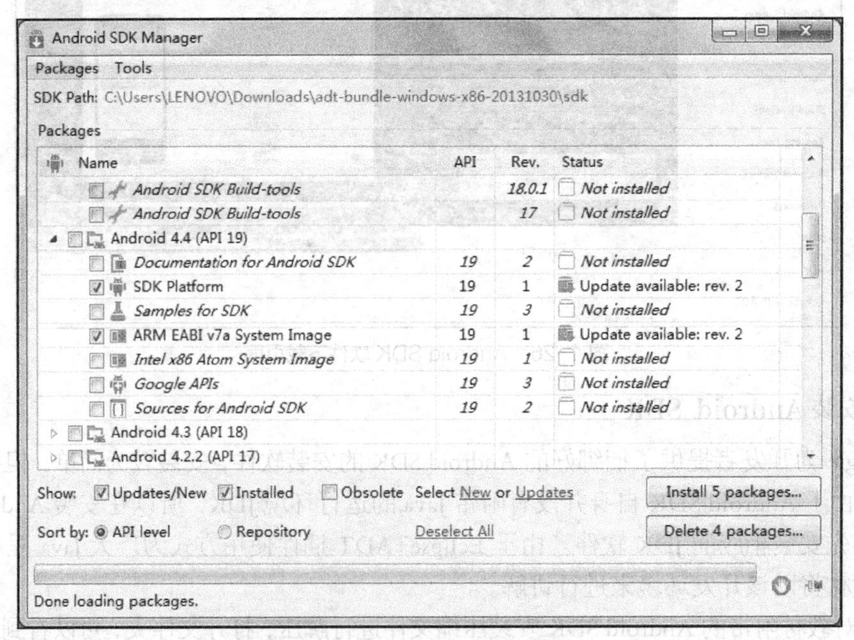

图 1-29　Android SDK 管理界面

（3）打开 eclipse 文件夹，单击 eclipse.exe 可执行文件，可打开 Eclipse 开发工具（见图 1-30）。第一次打开 Eclipse 程序若出现升级提示（见图 1-31），则打开 Android SDK 管理界面，单击 Install 按钮进行升级。

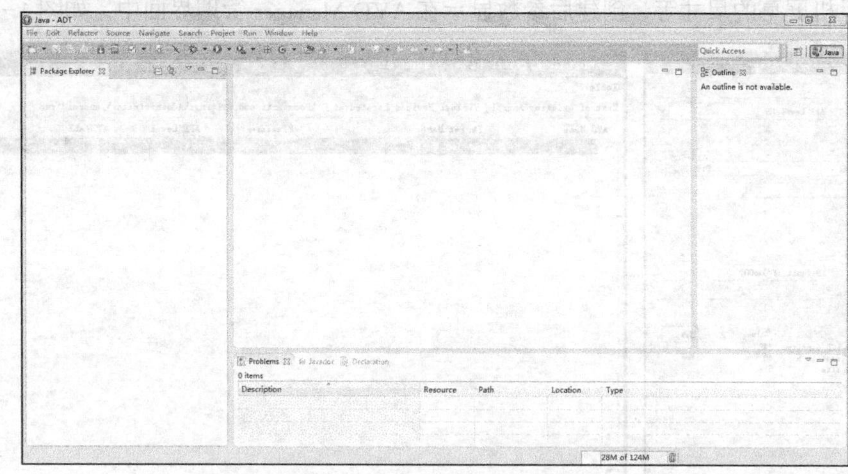

图 1-30 Eclipse ADT 主界面

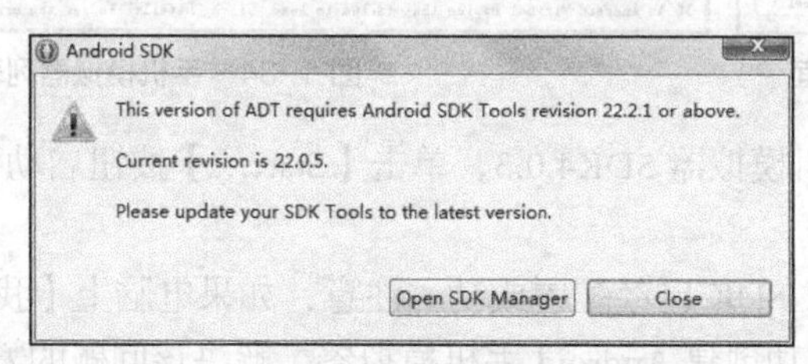

图 1-31 Android SDK 工具升级界面

（4）配置 Android 模拟器。选择 Eclipse 菜单【Window】→【Android Virtual Device Manager】进入模拟器管理界面，如图 1-32 所示。其中 AVD（Android Virtual Device）表示 Android 的虚拟设备，例如 Android 手机模拟器，可用在电脑上运行测试 Android 程序。AVD Manager 是专门用于管理 Android 手机模拟器的软件。

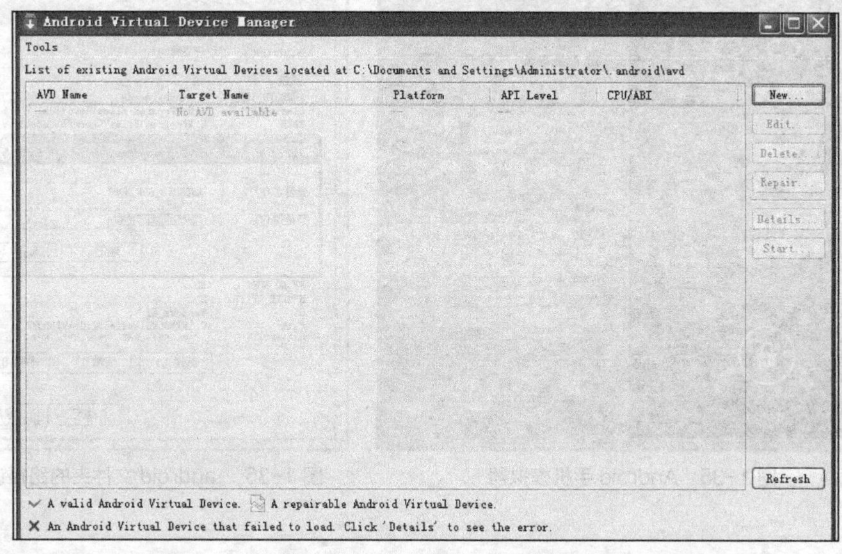

图 1-32 AVD Manager 界面

（5）单击【New...】按钮，进入 Android 手机模拟器配置界面，如图 1-33 所示，选择 Android 4.0.3 SDK 配置虚拟机。注意：该步骤根据自己开发需要选择 SDK 版本。

（6）单击【Create AVD】按钮。创建配置的虚拟机参数，例如，SDK API 的版本，SD 卡的大小，手机屏幕的尺寸等，创建后参数显示在 AVD Manager 管理界面中，如图 1-34 所示。

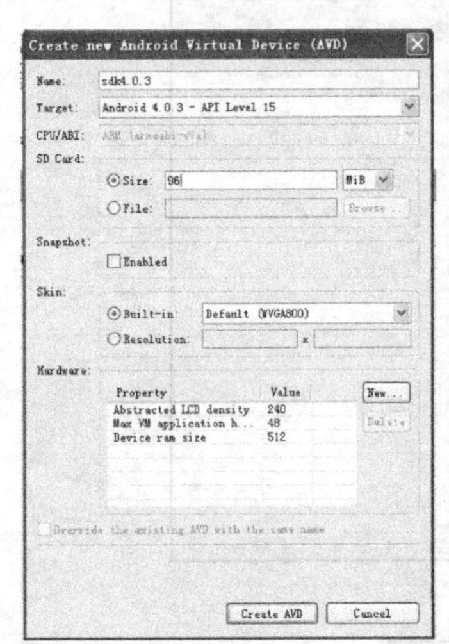

图 1-33　模拟器配置界面

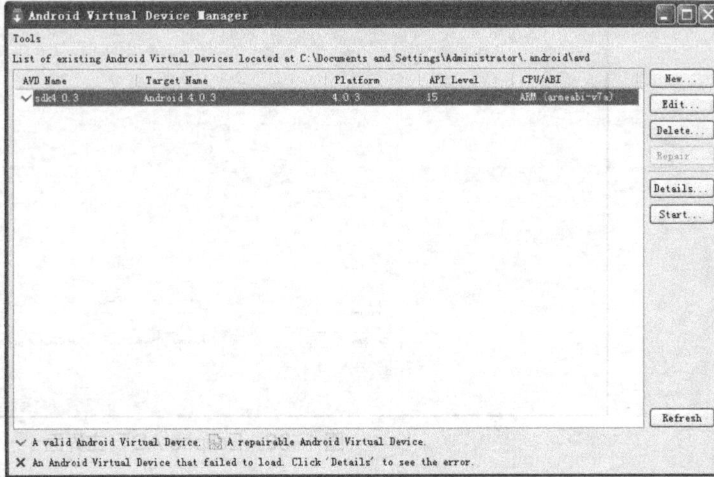

图 1-34　手机模拟器列表

（7）选中创建的手机模拟器 SDK4.0.3，单击【Start...】按钮启动手机模拟器，如图 1-35 所示。

至此，说明 Android SDK 已经配置成功。注意：如果电脑上【我的文档】使用的是中文名，会导致 AVD 无法成功创建 Android 手机模拟器。解决该问题的方法是设置【我的电脑】上的环境变量，即：打开【系统属性】→【环境变量】，在【系统变量】中选择【新建】→变量名为 ANDROID_SDK_HOME（变量名不能改变，只能是这个名字），然后把变量值指定为存放 AVD 所在的.android 文件夹的路径，比如 d:\androidcode，如图 1-36 所示。

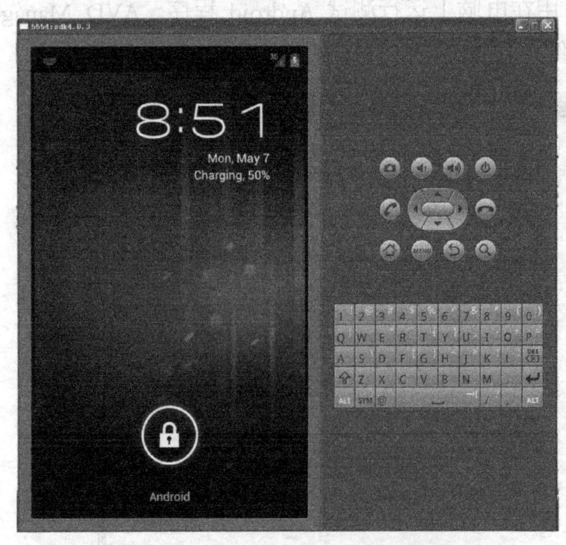

图 1-35　Android 手机模拟器

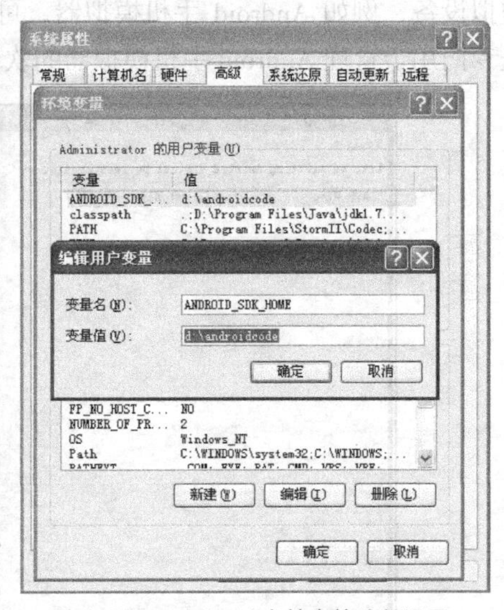

图 1-36　.android 文件夹的路径设置

1.4 安装 Eclipse ADT

一、任务分析

本次任务要求完成 Eclipse ADT 的下载、安装和配置。要完成本次任务，需要思考如下几个问题。

（1）Eclipse ADT 是什么软件，与 JDK、SDK 软件有何关系，对 Android 项目开发有何作用？

（2）从何处获得正确的 Eclipse ADT 软件？

（3）如何安装 Eclipse ADT？

（4）如何配置 Eclipse ADT？

二、相关知识

1．认识 IDE 开发环境

软件开发涉及的环节较多，早期的程序设计将代码编写与编译分开进行，不便于调试，影响了开发的效率。随着程序开发的规模越来越大，对不同功能的程序代码进行有效管理的需求也就越来越大。

集成开发环境（IDE，Integrated Development Environment）旨在提供一个综合的图形用户开发环境，方便程序员进行软件开发。它一般集成了程序生成器、代码编辑器、编译器、调试器和发布器等，具有代码编写、管理、分析、编译、调试和发布等功能。

2．认识 Eclipse

Eclipse 是一个开源社区，所提供的项目致力于建立开放的开发平台，具有可扩展性的框架、工具和运行环境的建立、发布和软件生命周期的管理。Eclipse 社区提供了适合 Android、J2EE、Java、J2ME、C/C++、JavaScript 等语言的 IDE 开发工具。

Eclipse 最初由 IBM 公司所开发，于 2001 年捐献给开源社区，现由 Eclipse 基金会管理。Eclipse 一个很重要的特色是具有通过插件来扩展开发平台的功能。Eclipse 本身只是一个框架平台，运行在 Eclipse 平台上的各种插件提供了开发程序的各种功能。Eclipse 的发行版本带有最基本的插件，软件开发人员可以在此基础上通过开发插件建立自己的 IDE。

Eclipse 提供很多快捷键来帮助开发者更好地编写代码，下面列出几个常用的快捷键。

（1）【Ctrl】+【Shift】+【O】组合键：自动导入代码中用到类所在的包。

（2）【Alt】+【/】组合键：利用代码助手完成一些代码的提示插入，例如类名补全、方法提示等。

（3）【Ctrl】+【Shift】+【F】组合键：格式化当前代码，使代码更加整齐。

（4）【Ctrl】+【/】组合键：注释当前行，若再按则取消注释；选择多行，则可以批量注释。

（5）【Ctrl】+【D】组合键：删除当前行。

（6）【Ctrl】+【T】组合键：快速显示当前类的继承结构。

Eclipse 的使用很直观，本节将对 Eclipse 的主界面（见图 1-37）进行介绍。Eclipse 的主界面主要由菜单、工具栏、视图（View）、透视图（Perspective）切换器、编辑器组成。

菜单包含了 Eclipse 提供的大部分功能，下面介绍几个重要的菜单项操作。

（1）【Window】→【Open Perspective】：打开切换特定的透视图。

（2）【Window】→【Reset Perspective…】：恢复默认的透视图设置。当不小心关闭 Eclipse 的某些默认窗口，可使用该操作进行恢复。

```
/** Called when the activity is first created. */
@Override
public void onCreate(Bundle savedInstanceState) {
    super.onCreate(savedInstanceState);
    setContentView(R.layout.main);
}

@Override
protected void onStart() {
    // TODO Auto-generated method stub
    super.onStart();
}

@Override
protected void onPause() {
    // TODO Auto-generated method stub
    super.onPause();
}

@Override
protected void onResume() {
```

Problems @ Javadoc Declaration Console
Android
Writable Smart Insert 31 : 1 Android SDK Con

图 1-37　Eclipse 主界面

（3）【Window】→【Show View】：打开特定的视图。当视图不小心关闭后，可以通过该菜单再次打开。

（4）【Project】→【Clean...】：将工程中旧的.class 文件删除，同时重新编译工程。当遇到不明原因的报错时，可以执行该操作来尝试是否能够解决该问题。

（5）【Search】→【File】：在项目中遍历出包含特定关键字的代码，方便开发者查找代码。

一个透视图相当于一个自定义的界面，保存了当前的菜单栏、工具栏按钮以及视图的大小、位置、显示与否的所有状态。Eclipse 打开时呈现的界面布局就是一个透视图。Eclipse 提供了多种透视图，以方便用户在不同的开发环境下工作，例如调试透视图。单击 按钮可以在多个透视图间切换。

视图是显示在主界面中的一个小窗口，主要用于信息展示，可以单独最大化、最小化显示，调整显示大小和位置，也可以将其关闭。下面简要列举几个主要的视图。

（1）包浏览视图（Package Explore）以包层次显示程序的各个类。

（2）大纲形式视图（Outline）展示某个类的成员变量和方法。

（3）控制台视图（Console）展示程序的运行输出。

在菜单【Window】→【Preferences】中，可以对 Eclipse 常用的属性进行配置，例如代码字体大小和颜色、代码行数的显示等，配置界面如图 1-38 所示。

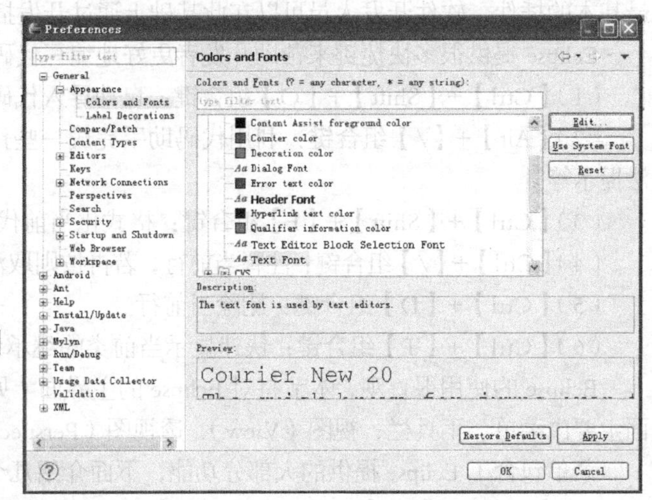

图 1-38　Eclipse 配置界面

3. 认识 ADT

ADT（Android Development Tools）是集成到 Eclipse 上的一个插件，能够提供一个强大的集成开发环境来开发 Android 应用。故可以将 ADT 理解为是 Android SDK 与 Eclipse 建立起联系的一个软件工具。

三、任务实施

1. Eclipse 的下载和安装

Eclipse 的下载比较简单，如果在任务 1.3 的 Android SDK 下载中损坏了 Eclipse 软件，可以通过以下步骤重新获取。

（1）登录 Eclipse 软件下载主页 http://www.eclipse.org/downloads/，页面如图 1-39 所示。

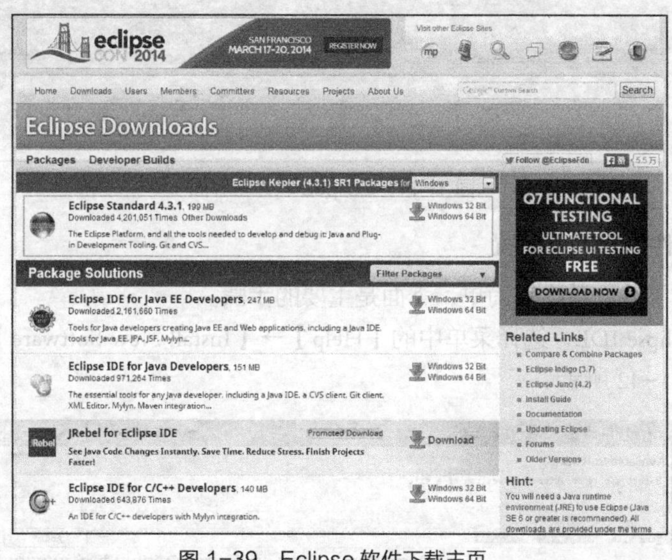

图 1-39　Eclipse 软件下载主页

（2）选择【Eclipse IDE for Java Developers】项，单击相应平台的版本软件，即可进入具体下载位置，如图 1-40 所示。

图 1-40　Eclipse 的下载位置

（3）下载完成后，解压即可使用 Eclipse IDE 开发软件，软件主界面如图 1-41 所示。

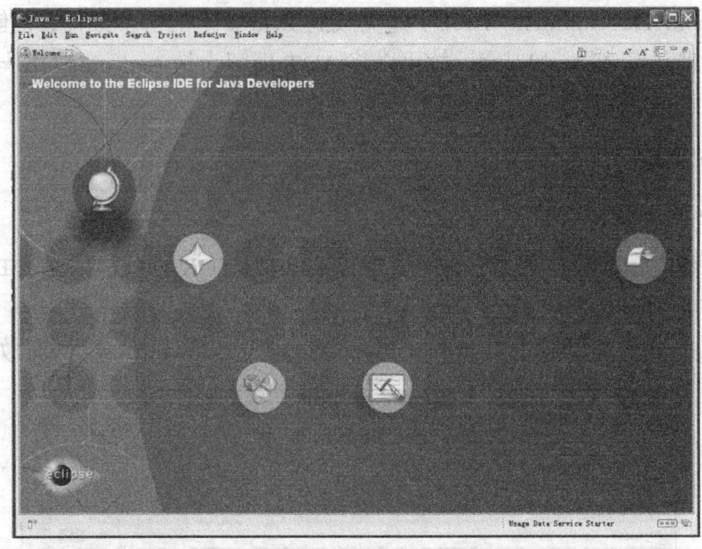

图 1-41　Eclipse 软件主界面

2. ADT 插件的安装及配置

ADT 插件的下载安装比较简单，下面是主要的步骤。

（1）打开 Eclipse IDE，选择菜单中的【Help】→【Install New Software】命令，打开插件安装界面，如图 1-42 所示。

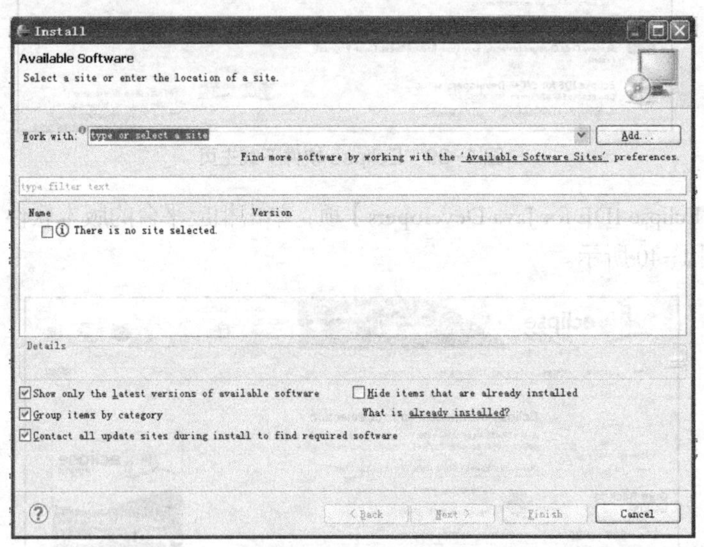

图 1-42　插件安装界面

（2）单击【Add...】按钮，弹出对话框要求输入【Name】和【Location】。【Name】可随意命名,【Location】文本框中输入 ADT 的在线安装地址 http://dl-ssl.google.com/android/eclipse，如图 1-43 所示。

（3）确定返回后，在【Work with】后的下拉列表中选择刚才添加的 ADT，会看到下面出现【Developer Tools】，将其展开，级联选项中包括【Android DDMS】、【Android Development

Tools】、【Android Hierarchy Viewer】和【Android Traceview】选项，将它们选中，如图 1-44 所示。

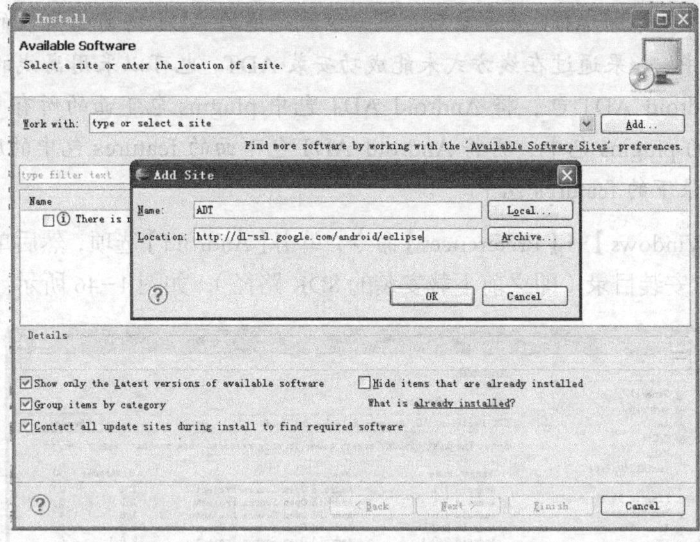

图 1-43　ADT 设置界面

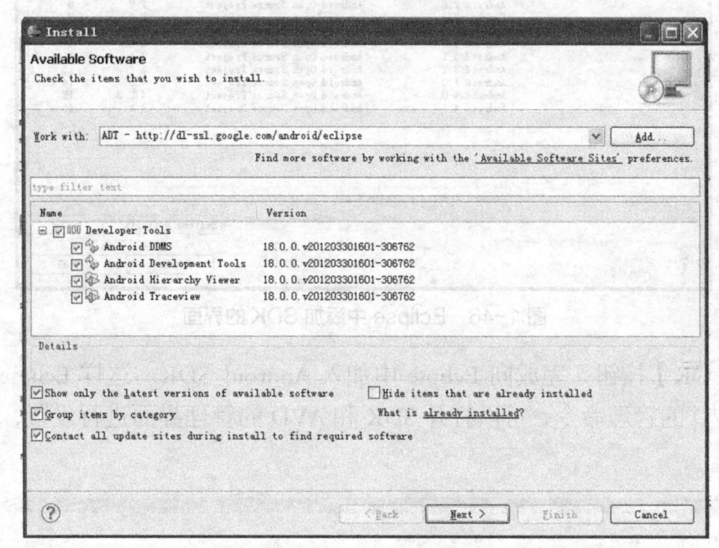

图 1-44　ADT 安装界面

（4）按提示一步一步地单击【Next】按钮，完成后按提示单击【Restart Now】按钮，重启 Eclipse，如图 1-45 所示。

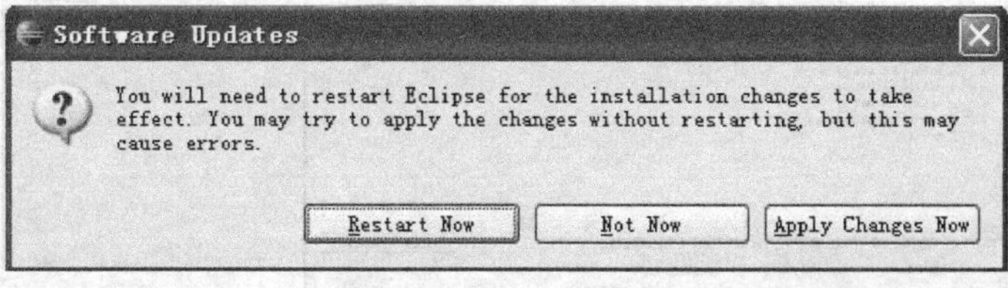

图 1-45　重启提示界面

说明

在 Eclipse 中安装好 ADT 并重启后，下一步就是通过 ADT 将前面安装的 Android SDK 加入到 Eclipse 中。如果通过在线方式未能成功安装 ADT，也可以采用离线的安装方法。先到网上下载 Android ADT 包，将 Android ADT 包中 plugins 包下面的所有 jar 文件拷贝到 Eclipse 目录下的 plugins 包中，再将 Android ADT 包下面的 features 包中的所有 jar 文件拷贝到 Eclipse 目录下的 features 包中。

（5）选择【Windows】→【Preferences】命令，单击【Android】选项，然后单击【Browse...】按钮，找到 SDK 安装目录（即之前下载安装的 SDK 路径），如图 1-46 所示。

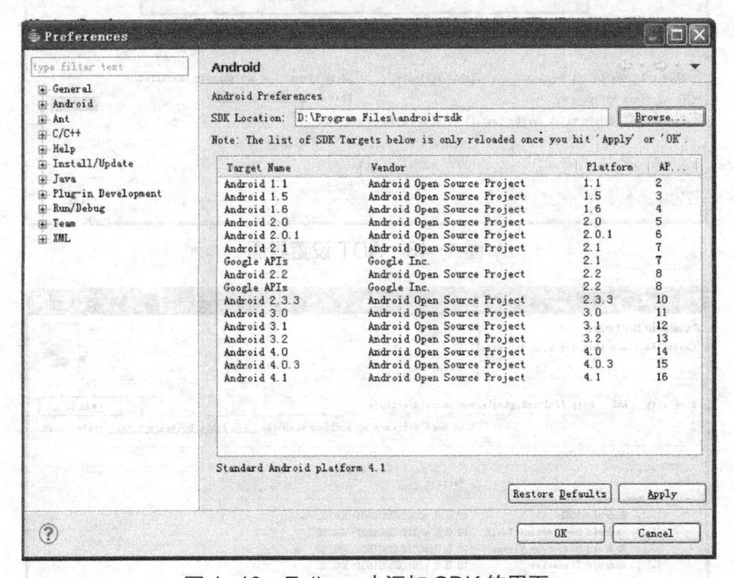

图 1-46　Eclipse 中添加 SDK 的界面

（6）单击【OK】按钮，完成向 Eclipse 中加入 Android SDK。这样 Eclipse 的菜单中就包含了 SDK 和 AVD 的管理命令，可以打开 SDK 和 AVD 的管理界面进行设置，分别如图 1-47 和图 1-48 所示。

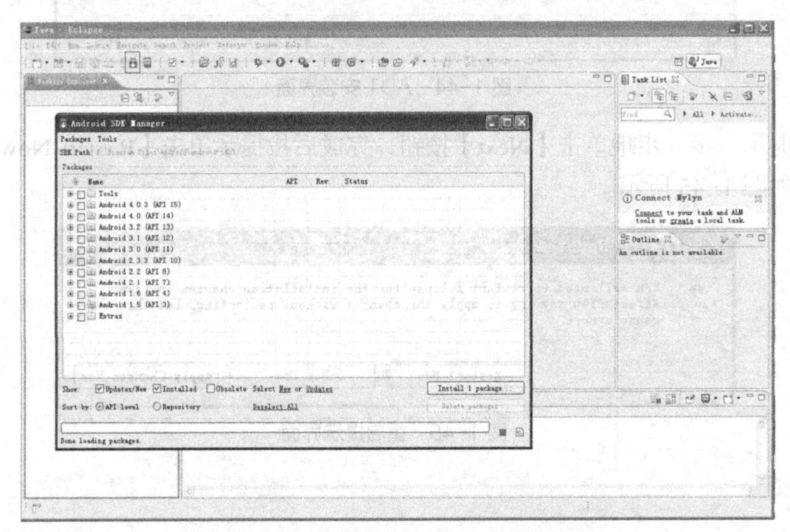

图 1-47　Eclipse 中的 SDK 管理界面

至此，Eclipse 的 Android 开发环境搭建完毕。

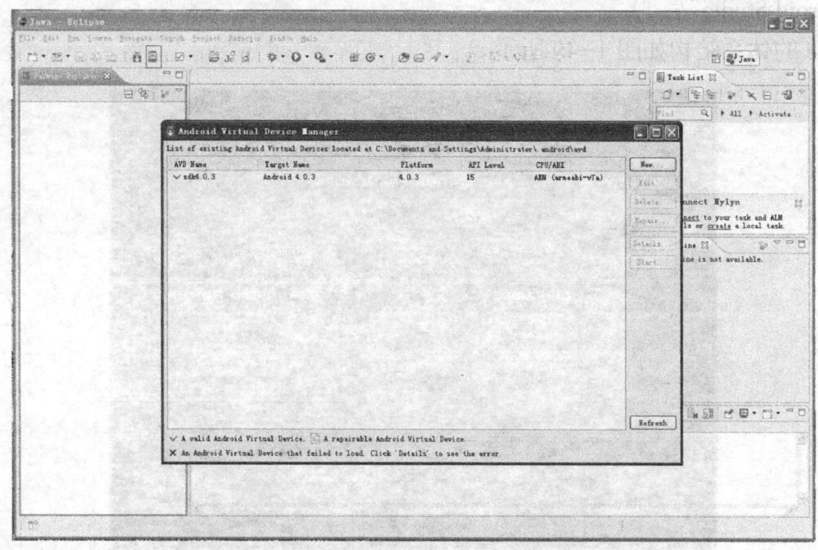

图 1-48　Eclipse 中的 AVD 管理界面

1.5　测试开发环境

一、任务分析

检查 Android 开发环境是否已经完全搭建成功，就是要看前面 1.2、1.3、1.4 节的软件安装配置是否能够正常工作。比较简单直接的方法是在开发环境中尝试开发一个 Android 程序，并测试其是否能够正常运行。要完成本次任务，需要思考如下几个问题。

（1）如何在 Eclipse 中开发一个 Android 程序？

（2）Android 程序具有什么特点？

二、相关知识

1．Android 概述

Android 是一个用于移动设备的软件集，包括操作系统、中间件和关键的应用程序。Android SDK 提供了必需的工具和 APIs，用于在 Android 平台上使用 Java 编程语言开发应用程序。Android 具有如下特征。

- 允许重用和替换组件的应用程序框架。
- 具有专门为移动设备优化的 Dalvik 虚拟机。
- 集成了基于开源引擎 WebKit 的浏览器。
- 通过自定义的 2D 图形库优化显示图形，提供基于 OpenGL ES 规范的 3D 图形支持。
- 用于结构数据存储的 SQLite。
- 为常见的声音、视频和图形格式，如 MPEG-4、H.264、MP3、AAC、AMR、JPG、PNG、GIF，提供多媒体支持。
- 支持 GSM、CDMA、TD SCDMA 电话（依赖于硬件）。
- 支持蓝牙、EDGE、3G 和 WiFi（依赖于硬件）。
- 支持相机、GPS、指南针和加速度传感器（依赖于硬件）。

● 丰富的开发环境，包括设备模拟器、调试工具、内存和性能分析、Eclipse IDE 插件、Android Studio 等。

Android 的体系结构如图 1-49 所示。

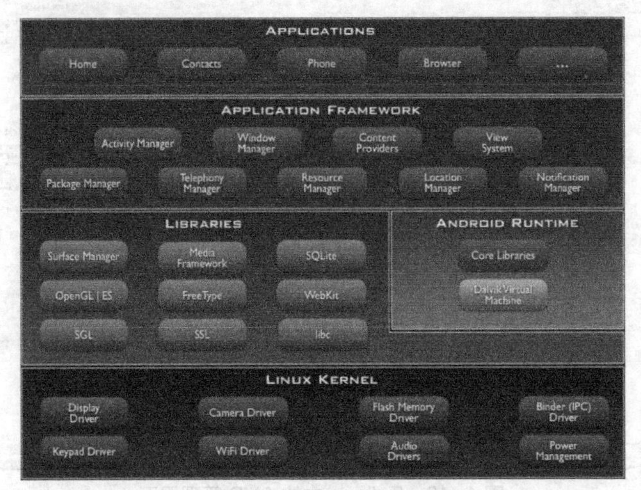

图 1-49　Android 体系结构

下面对框架的各部分做简要的介绍。

（1）应用程序（Applications）：是 Android 推出的一组使用 Java 语言编写的核心应用程序，例如 E-mail 客户端、短信程序、日历、地图、通讯录等。

（2）应用程序框架（Application Framework）：是 Android 为开发者提供的一个开放平台，位于应用程序的下一层。开发者可以通过它们自由地利用设备硬件、访问位置信息、运行后台服务、设置告警、在状态栏上增加通知等。应用程序框架的设计是为了简化组件的重用，允许用户替换组件。

应用程序框架提供的是一组服务和系统，这些会在程序开发过程中直接使用到，具体如下。

● 丰富且具有可扩展性的视图系统（Views System）：可用于构建应用程序的界面，包括列表、文本框、按钮、甚至嵌入的 Web 浏览器。

● 内容提供者（Content Providers）：允许应用程序访问其他应用程序的数据（如通讯录），或者共享数据。

● 资源管理器（Resource Manager）：提供对非代码资源（如本地化字符串、图像和布局文件）的管理。

● 通知管理器（Notification Manager）：允许应用程序在状态栏上显示定制的提示信息。

● 活动管理器（Activity Manager）：管理应用程序的生命周期，提供一个通用的导航回退功能。

（3）库（Libraries）：Android 包括一组 C/C++库，用于 Android 系统中不同的组件。这些功能通过 Android 应用程序框架对开发者开放。下面列出了一些核心库。

● C 语言系统库（System C library）：派生于标准 C 语言系统库，并根据嵌入式 Linux 设备进行调优。

● 多媒体库（Media Libraries）：基于 OpenCore 多媒体开源框架（该产品由 Google 联合 PacketVideo 公司推出）。该库支持许多流行的音频、视频格式以及静态图像文件，包括 MPEG-4、H.264、MP3、AAC 和 AMR、JPG 和 PNG 格式文件的播放和录制。

- 外观管理器（Surface Manager）：管理访问子系统的显示，将 2D 绘图与 3D 绘图进行显示上的合成。
- LibWebCore：是一个现代的 Web 浏览器，为 Android 浏览器和嵌入式 Web 视图提供支持。
- SGL：底层的 2D 图形引擎。
- 3D 库（3D Libraries）：是基于 OpenGL ES API 的实现。该库使用了硬件 3D 加速（如果可用）或高度优化的 3D 软件光栅。
- FreeType：用于位图和矢量字体的渲染。
- SQLite：一个强大的轻量级关系数据库引擎，可以用于各种应用程序。

（4）Android Runtime：Android 的一组核心库，提供大部分 Java 编程语言核心库的功能。Android 应用程序是在 Dalvik 虚拟机的实例下以进程形式运行。Dalvik 可以允许一个设备有效地运行多个虚拟机。Dalvik 虚拟机执行的文件是 Dalvik 可执行格式，文件后缀为.dex，并被优化成最小内存存放。与 Java 虚拟机基于堆栈的原理不同，Dalvik 虚拟机是基于寄存器的，通过转换工具 DX 将 Java 字节码转换成 DEX 格式，它依赖于 Linux 内核的底层功能，例如线程和低级别的内存管理。

（5）Linux 内核（Linux Kernel）：Android 依赖于 Linux 相应版本的核心系统服务，例如安全、内存管理、进程管理、网络堆栈、驱动程序模型。Google 对 Linux 内核做了改动，提供了支持 Android 平台的设备驱动。

Android 应用程序是使用 Java 语言编写的。Android SDK 工具将代码、数据和资源文件编译成一个 Android 包（后缀为.apk 的归档文件）。.apk 归档文件可以理解为一个文件和目录的集合，而这个集合被储存在一个文件中，用于在移动设备上安装应用程序。一旦安装成功，每个 Android 应用即位于自己的安全沙箱中（安全沙箱是分离运行程序的一种安全机制，一般用于对程序访问的资源进行控制）。

- Android 操作系统是一个多用户的 Linux 系统，每一个应用都是一个不同的用户。
- 默认情况下，系统赋予每个应用唯一的 Linux 用户 ID，该 ID 仅用于系统，应用程序不需要知道。系统为应用的所有文件赋予权限，使只有该 ID 的应用才能够对其进行访问。
- 每一个进程有自己的虚拟机，应用程序代码的运行与其他应用隔离。
- 默认情况下，每一个应用运行在自己的进程中。当应用组件中的任意一个需要执行时，Android 开启进程；当不需执行或者系统必须要为其他应用回收内存时，该进程将被关闭。

这样一来，Android 系统实现了最小特权原则。也就是说，每个应用程序在默认情况下只能访问其运行时所需要的部分。这样便创建了一个非常安全的环境，使得应用程序不能访问系统中其他不批准的部分。

当不同应用程序之间需要共享数据时，可以为其安排共享相同的 Linux 用户 ID，在这种情况下，程序能够互相访问对方的文件。为了节省系统资源，具有相同用户 ID 的应用程序可以被安排在相同的 Linux 进程中运行，并共享相同的虚拟机（应用程序也必须使用同一证书签名）。

应用程序是可以访问系统服务的，如应用程序可以请求访问用户的通讯录、短信、SD 存储卡、摄像头、蓝牙以及其他设备，但要求用户在安装应用程序时必须被授予相应的权限。

2．Android 的组件

应用程序组件是 Android 应用程序的重要基石，在编程时需根据需要编写相应的组件。

每个组件具有一个不同的指向，系统可以由此进入应用程序。并非所有的组件都是用户的实际入口点，一些组件是互相依赖的，但每一个都以自己的实体形式存在，并扮演特定的角色。

应用程序组件有四种不同的类型，每种类型服务于不同的目的，并具有独特的定义。下面分别对这四种不同类型的应用程序组件进行介绍。

（1）Activities：一个Activity（活动）表示一个用户界面，可以和用户进行交互。例如，E-mail应用程序可以包括一个显示新邮件的Activity、一个撰写电子邮件的Activity和一个阅读邮件的Activity。虽然这些Activities一起为E-mail应用程序提供用户体验，但每个都彼此独立。正因为如此，如果E-mail应用程序允许，不同的应用程序可以启动这些活动中的任何一个。例如，摄像头应用程序为了分享图片，可以启动E-mail应用程序中撰写新邮件的Activity。

要实现一个Activity，需要定义一个Activity类的子类。如果新建一个活动，必须要在AndroidManifest.xml中进行声明。

（2）Services：Service（服务）是运行在后台的组件，一般用于执行需要长时间运行的操作或远程进程的工作。Service并不提供用户界面。例如，Service在后台播放音乐的同时，用户在使用另一个不同的应用程序，或者Service可能正在获取网络上的数据，但并不阻塞用户交互的活动。作为一个组件，某个Activity可以启动该服务，使之运行或与之绑定以便于交互。

实现一个Service，需要定义一个Service类的子类。

（3）Content Providers：Content Provider（内容提供者）管理应用程序共享的数据集。程序员可以将数据存储在文件系统、SQLite数据库、网络或者任何其他应用程序可以访问到的持久性存储位置。通过Content Provider，其他应用程序可以查询甚至修改Content Provider上的数据（前提是Content Provider允许）。例如，Android系统提供了一个Content Provider管理用户的联系人信息，于是任何具有权限的应用程序可以查询Content Provider（如ContactsContract.Data）的某部分，读取和写入关于特定人的信息。

要实现一个Content Provider，需要定义一个ContentProvider类的子类，并实现一套标准的API，以便使其能够与其他的应用程序进行交易。

（4）Broadcast Receivers：Broadcast Receiver（广播接收器）是一个负责全系统广播通知的组件。许多广播信息都来源于系统，如某个Broadcast宣布屏幕已关闭、电池电量低或图片已被捕捉到。应用程序也可以启动广播，例如，让其他应用程序知道某些数据已被下载到设备。虽然广播接收器不显示用户界面，但可以创建状态通知栏用于在广播事件发生时提醒用户。更普遍的做法是，将广播接收器设计为对其他组件的“网关”，且只需做非常少量的工作，例如，只启动一个服务来执行基于事件的一些工作。

要实现一个Broadcast Receiver，需要定义一个BroadcastReceiver类的子类。每一个Broadcast对象均以Intent对象的形式进行调用。

Android系统设计的一个独特之处是，应用程序可以通过Intent来启动另一个应用程序的组件。例如，程序员希望使用设备上的相机捕捉照片时，并不需要自己额外去开发一个拍摄照片的Activity，只需简单地启动摄像头应用程序中用于捕捉照片的Activity即可。完成后，照片甚至能够返回到应用程序中，并可以被进一步处理。对于用户而言，这一过程好像相机实际上是其应用程序中的一部分。

在系统启动一个组件时，需启动该应用程序的进程（如果进程尚未运行）和实例化组件所需的类。例如，如果应用程序开启摄像头应用程序中捕捉照片的Activity，该Activity运行在摄像头应用程序的进程中，而不是在刚开启的应用程序进程中。因此，不同于其他大多数系统上的应用程序，Android 应用程序不具有单一的入口点（没有 main()函数）。

因为系统在单独的进程中运行每一个应用程序，通过文件权限限制应用程序访问其他应用程序，而用户的应用程序不能直接激活另一个应用程序组件。这就要求程序员必须传递信息到系统，指定启动一个特定组件的意图（使用 Intent），然后系统将会激活相应的组件。

3．认识手机模拟器

手机模拟器（Mobile Emulator）的作用是在计算机上模拟手机环境，从而可以在计算机上进行手机程序开发、调试和发布。不同平台、不同型号的手机有不同的手机模拟器，例如有 Android 手机模拟器、Windows Phone 手机模拟器、Samsung 手机模拟器、Nokia 手机模拟器、BlueStacks 模拟器、Genymotion 模拟器。Android SDK 自带 Android 手机模拟器可以根据配置信息配置不同的手机模拟器，如图 1–50 所示。使用快捷键【Ctrl】+【F11】可以对手机模拟器的横屏和竖屏进行切换。

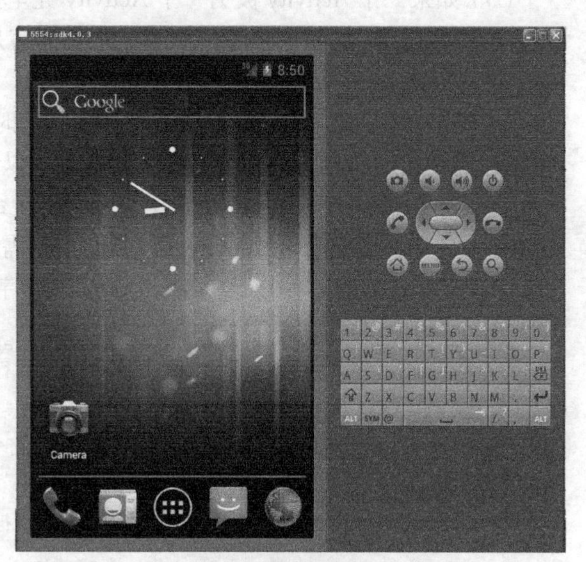

图 1–50　Android SDK 自带的手机模拟器界面

4．Activity 应用程序的生命周期

Activity 是 Android 程序图形用户界面的基本组成部件。Android 程序由一个或者多个 Activity 类组成，而程序是从 Activity 类开始执行的，系统规定了 Activity 的生命周期有创建、开始、唤醒、暂停和销毁五种状态。Activity 的生命周期交给系统统一管理。Android 中所有的 Activity 都是平等的，通过堆栈来管理。当一个新的 Activity 开始时，它被放到堆栈的顶部，并成为运行的 Activity，显示在用户界面上。而前一个 Activity 则移至其下方。只有当新的 Activity 退出后，原 Activity 才会再次被置于堆栈顶部，从而可以再次得以运行。例如，一个程序正在运行，突然手机来电，Android 接收到来电广播后会打开一个接听电话的 Activity，并放入堆栈的栈顶，这样，程序就会被接电话的 Activity 所覆盖，手机会显示接听电话的 Activity。

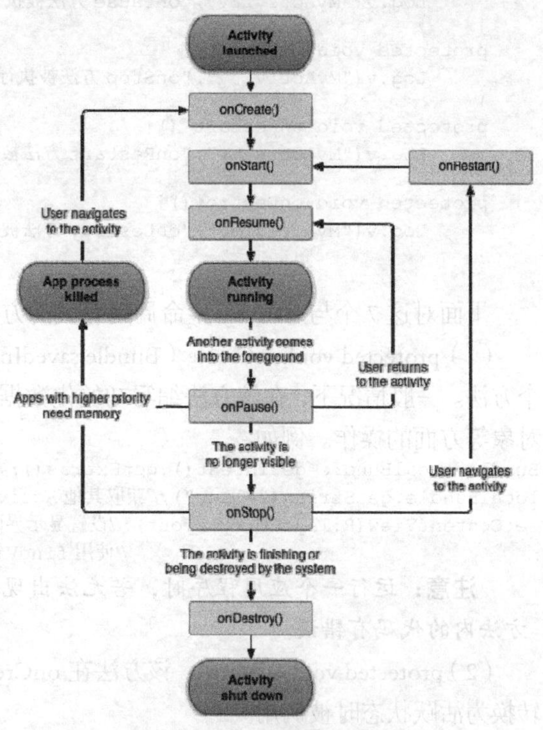

图 1–51　Activity 的生命周期

图 1–51 所示为 Activity 的生命周期图，图中的 onCreate()、onStart()等方法与 Activity 状态变化的处理紧密相关。

Activity 本质上有 4 种状态。

（1）如果一个 Activity 是在屏幕的前台，即在堆栈的上面，那么其处于活动（Active）或者运行的（Running）状态。

（2）如果一个 Activity 已经失去焦点，但依然可见，那么其处于暂停（Paused）状态，依然保留着所有的状态信息，但在内存非常低时，可被系统关闭。

（3）如果一个 Activity 被另一个 Activity 完全掩盖，那么其处于停止状态。这时，Activity 也同样保留了所有的状态信息，但对于用户而言是不可见的，当系统的其他地方有内存需要的时候，会被系统关闭。

（4）当一个 Activity 处于暂停或者停止状态时，系统可以从内存中销毁该 Activity。当该 Activity 需要再次向用户显示时，须重新开启，以恢复其之前的状态。

在 Activity 类中，Android 定义了一系列与生命周期相关的方法。开发一个 Android 程序，首先需要创建一个继承于 Activity 的子类，根据需要重写 Activity 的方法，来实现数据初始化、回收、菜单点击等操作。以下为编写 Activity 程序并实现其生命周期相应操作的示例。

```
public class MyActivity extends Activity {
    protected void onCreate(Bundle savedInstanceState){
        Log.v("MyActivity","onCreate 方法被执行");
    }
    protected void onStart() {
        Log.v("MyActivity","onStart 方法被执行");
    }
    protected void onResume(){
        Log.v("MyActivity","onResume 方法被执行");
    }
    protected void onPause(){
        Log.v("MyActivity","onPause 方法被执行");
    }
    protected void onStop(){
        Log.v("MyActivity","onStop 方法被执行");
    }
    protected void onRestart (){
        Log.v("MyActivity","onRestart 方法被执行");
    }
    protected void onDestroy(){
        Log.v("MyActivity","onDestroy 方法被执行");
    }
}
```

下面对这 7 个与 Activity 生命周期有关的方法进行说明。

（1）protected void onCreate（Bundle savedInstanceState）：Activity 实例启动时调用的第一个方法。一般情况下，在该方法编写初始化数据、从 XML 布局文件中加载设计好的用户控件对象等方面的操作。例如，

```
Bundle localBundle=getIntent().getExtras();
localBundle.getString("vlaue");//获取其他 Activity 传递过来的参数
setContentView(R.layout.mylayout);//设置显示界面
…                                  //使用 findViewById 获得用户界面的组件
```

注意：运行一个应用程序时，若无法出现界面，并直接报错，一般情况下是 onCreate 方法内的代码有错误。

（2）protected void onStart()：该方法在 onCreate()方法执行之后或者在 Activity 从停止状态转换为活跃状态时被调用。

（3）protected void onResume()：该方法在 Activity 从暂停状态转换到活跃状态时或者在 onStart()方法执行之后被调用。onResume()方法也较为适合放入初始化数据的操作。

（4）protected void onPause()：该方法在 Activity 从活跃状态转换到暂停状态时被调用。

（5）protected void onStop()：该方法在 Activity 从活跃状态转换到停止状态时被调用，一般在这里保存 Activity 的状态信息。

（6）protected void onRestart ()：该方法在 Activity 从停止状态重新向用户显示时调用。

（7）protected void onDestroy()：该方法在 Activity 结束时被调用，一般用于释放资源、清理内存等工作。

5．XML 知识

标记语言采用一套标记标签来表示文本信息，标记是由尖括号包围的关键词，如<html>，其作用是描述文本信息。

XML（eXtensible Markup Language）称为可扩展标记语言。XML 文档的后缀是.xml，既可以使用专门的编辑工具，也可以使用文本编辑器对其进行编写。XML 提供统一的方法来描述和交换独立于应用程序或供应商的结构化数据。XML 的标记没有被预定义，用户可以自行定义标记来描述数据。与 HTML（HyperText Markup Language）的差异在于，XML 主要用来存储规范的数据信息。

每个 XML 元素都以一个起始标记开始，以一个结束标记收尾。起始标记以"<"符号开始，以">"符号结束。结束标记以"</"符号开始，以">"符号结束。XML 的标记是可以自定义的，主要用来描述数据。XML 元素可以带有多个属性，属性值需要添加引号。

6．AndroidManifest.xml 文件描述

AndroidManifest.xml 文件可以理解为关于 Android 应用程序的清单文件，使用 XML 向 Android 系统描述关于应用程序的主要功能、执行的动作和处理的信息、所需的权限等重要信息。例如，一个 Android 应用程序中一般包含多个 Activity 类，需要具体区分程序有哪些 Activity 类可由系统识别；程序首先执行的是哪一个 Activity 类；系统有哪些权限可以交由应用程序使用；应用程序的图标及程序名称管理等。任何一个 Android 项目都需要使用该文件，并要求其存放在项目的根目录下。

在 Android 系统开始启动一个应用程序组件时，系统必须读取 AndroidManifest.xml 文件以便知道组件是否存在。因此应用程序必须在该文件中声明其所有组件。

AndroidManifest.xml 包含了很多功能，具体如下。

- 描述应用程序包含的组件，对实现每一个组件的类进行命名，让 Android 系统知道有哪些组件以及在什么条件下可以对其进行加载。
- 给予应用程序请求的使用权限，如连接互联网或以只读方式访问用户的联系人。
- 根据应用程序实际用到的 APIs，声明应用程序要求的最小 API 等级。
- 声明应用程序使用或者请求的硬件和软件特征，例如，相机、蓝牙服务或多点触摸屏。
- 声明应用程序需要链接的外部 API 库，例如，Google 地图类库。

下面介绍如何在 AndroidManifest.xml 文件中进行描述。

（1）声明组件：AndroidManifest.xml 文件的主要任务是通知 Android 系统应用程序包含的组件。可以这样定义一个 Activity：

```
<?xml version="1.0" encoding="utf-8"?>
<manifest ... >
   <application android:icon="@drawable/app_icon.png"
android:label="@string/app_name">
      <activity android:name="com.example.project.ExampleActivity"
```

```
                android:label="@string/example_label" ... >
        </activity>
        ...
    </application>
</manifest>
```


说明

- 在<application>元素中，android:icon 属性指明应用程序图标用到的资源，android:label 属性表示应用程序的标签，可以理解为整个 Android 应用程序的名字；
- 在< activity >元素中，android:name 属性指明 Activity 子类的完整名称，android:label 属性作为 Activity 在手机界面上用户可见的标签，可以理解为用字符串来表示的 Activity 的名字。

定义应用程序不同组件的元素分别如下。

- < activity >元素：用于定义 Activities。
- < service>元素：用于定义 Services。
- <receiver>元素：用于定义 Broadcast Receivers。
- <provider> 元素：用于定义 Content Providers。

若程序使用了 Activities、Services 或者 Content Providers，而这些在 AndroidManifest.xml 文件中又没有定义，则意味着它们对于 Android 系统是不可见的，因此不能够运行。然而 Broadcast Receivers 组件既可以在 AndroidManifest.xml 文件中定义，也可以动态地在代码中定义。注意：作为 BroadcastReceiver 对象，通过调用 RegisterReceiver()向系统注册。

（2）声明组件的能力：在激活组件时可以使用一个 Intent 开始活动、服务和广播接收器。程序员可以在 Intent 中使用组件类名称的方式明确命名目标组件。然而，Intent 的真实能力体现在 Intent 的动作上，因此需要在 AndroidManifest.xml 文件中描述要执行动作的类型和动作需要的数据。如果有多个组件可以执行 Intent 所描述的动作，则系统会让用户选择使用哪一个。

Android 系统识别组件能响应 Intent 组件的方式是通过对接收的 Intent 与在设备其他应用程序上的 AndroidManifest.xml 文件中定义的 Intent Filters 进行比较得出的。

（3）声明应用程序的要求：在各种搭载 Android 系统的设备中，并非所有的设备都能提供同样的功能和能力。为了防止应用程序被安装在缺乏所需功能的设备上，应在 AndroidManifest.xml 文件中通过声明设备和软件要求的方式清楚地定义应用程序支持的设备类型。这些声明大部分是信息性的，Android 系统并不会对其进行读取，但诸如 Google Play 外部服务将会读取它们，用于为用户搜索、过滤满足其设备需求的应用程序，这样应用程序将限定于满足其所有要求的设备中。

例如，如果应用程序需要一个摄像头，并要求使用在 Android 2.1 API（API 等级 7）系统中，则应在 AndroidManifest.xml 文件中声明这些要求。这样，没有摄像头以及系统低于 Android 2.1 版本的设备将无法从 Google Play 中安装该应用程序。

下面是在设计和开发应用程序时需要考虑的一些重要的设备特征。

① 屏幕尺寸和屏幕密度。以屏幕类型对设备进行分类，Android 为每个设备定义了两个特征：屏幕尺寸（屏幕的物理尺寸）和屏幕分辨率（物理屏幕上的像素密度或 dpi（点数/英寸））。为了简化各种不同类型的屏幕配置，Android 系统将其归纳到不同的组，使其更容易对准。

- 屏幕的大小分为小（small）、正常（normal）、大（large）和特别大（extra large）。
- 屏幕的密度分为低密度（low density）、中等密度（medium density）、高密度（high density）

和特别高密度（extra high density）。

默认情况下，应用程序兼容所有的屏幕尺寸和密度，因为 Android 系统会对用户界面布局和图像资源进行适当的调整。开发者可以使用可替代的布局资源为特定的屏幕尺寸提供专门的用户界面布局，为特定的屏幕密度提供专门的图像，并在 AndroidManifest.xml 文件中使用<supports-screens>元素声明应用程序支持的屏幕尺寸。

② 输入配置。许多设备提供不同类型的用户输入机制，例如硬件键盘、轨迹球或 5 方向导航台。如果应用程序还需要特定类型的输入硬件，那么应该在 AndroidManifest.xml 中使用<uses-configuration>元素进行声明。一般情况下，应用程序很少需要声明特定的输入配置。

③ 设备功能。一些硬件和软件功能可能装配在 Android 设备上，例如，摄像头、光传感器、蓝牙、某版本的 OpenGL 或触摸屏的保真度。除了标准的 Android 库是可用的之外，开发者不能够假设 Android 设备应具有哪些功能，所以应使用<uses-feature>元素声明应用程序所使用到的功能。

④ 平台版本。不同的 Android 设备通常运行在不同的 Android 版本上，例如 Android 1.6 或者 Android 2.3。每个后续的版本通常新增了前面版本不能够使用的 API。为了指明哪个集合的 API 是可用的，每个平台版本均需指出 API 的等级，例如，Android 1.0 的 API 等级是 1，Android 2.3 的 API 等级是 9。如果所用 API 是在版本 1.0 之后增加到平台上的，那么应该使用<uses-sdk>元素声明最小的 API 等级。

为了帮助读者更深入地理解 AndroidManifest.xml 文件的定义，下面列出 AndroidManifest.xml 文件的一般结构。当中的元素名称是规定的，不能够添加自己的元素。其中只有<manifest> 和<application>是 AndroidManifest.xml 文件必须的元素，并只能够出现一次。其他大部分元素根据实际项目需要可以出现多次或者一次都不出现。同一个等级的元素在文件中无需按照顺序摆放，例如，<activity>, <receiver>, <provider>和<service>元素可以以任何顺序放置。只有一个例外：<activity-alias>元素必须要跟着<activity>元素。需要注意的是：<uses-permission>元素用于指定程序的权限，和< application >元素处于同一层。一些开发者易误将<uses-permission>元素放在<activity>一层，这会导致 XML 文件报错。

除了根元素<manifest>中的一些属性，其他所有属性都是以 android：prefix 形式命名的，如 android:alwaysRetainTaskState。

```
<?xml version="1.0" encoding="utf-8"?>
<manifest>

    <uses-permission />
    <permission />
    <permission-tree />
    <permission-group />
    <instrumentation />
    <uses-sdk />
    <uses-configuration />
    <uses-feature />
    <supports-screens />
    <compatible-screens />
    <supports-gl-texture />

    <application>

        <activity>
            <intent-filter>
                <action />
                <category />
```

```
            <data />
        </intent-filter>
        <meta-data />
    </activity>

    <activity-alias>
        <intent-filter> . . . </intent-filter>
        <meta-data />
    </activity-alias>

    <service>
        <intent-filter> . . . </intent-filter>
        <meta-data/>
    </service>

    <receiver>
        <intent-filter> . . . </intent-filter>
        <meta-data />
    </receiver>

    <provider>
        <grant-uri-permission />
        <meta-data />
    </provider>

    <uses-library />
  </application>
</manifest>
```

很多元素与 Java 类是对应的，例如<application>、<activity>、<provider>、<receiver>和<service>元素，在程序代码中如果有组件类（Activity、Service、BroadcastReceiver、ContentProvider）的子类，那么需要在元素的属性上对名字（包括包名）进行定义。下面是定义 Service 子类的例子，其中 com.example.project.SecretService 中的 SecretService 是类名，com.example.project 是类所在的包。

```
<manifest . . . >
  <application . . . >
    <service android:name="com.example.project.SecretService" . . . >
        . . .
    </service>
        . . .
  </application>
</manifest>
```

简单起见，可以将包名作为应用程序的属性，下面的做法和上面的效果是相同的。

```
<manifest package="com.example.project" . . . >
  <application . . . >
    <service android:name=".SecretService" . . . >
        . . .
    </service>
        . . .
  </application>
</manifest>
```

如果属性值超过一个，需要重复定义该元素，而不是在一个元素中列出多个值。下面是在 Intent Filter 中列出几个动作的例子。

```
<intent-filter . . . >
  <action android:name="android.intent.action.EDIT" />
  <action android:name="android.intent.action.INSERT" />
  <action android:name="android.intent.action.DELETE" />
  . . .
</intent-filter>
```

配置文件中有一些属性值可以显示在屏幕上，如 Activity 的标记和图标。这些属性的取值应该被本地化，所以应该设置为某个资源或者主题，资源值表达的格式如下：

```
@[package:]type:name
```

- package 表示包名，如果资源和应用程序在同一个包中，那么 package 可以被忽略；
- type 表示资源的类型，例如取值为“string”或者“drawable”；系统会自动到项目的 res 目录进行匹配，字符串的定义放在项目的 res/values 目录下的 strings.xml 文件中，drawable 文件则放在项目的 res/drawable 目录下。
- name 表示用于标识特定资源的名字。

例如，

```
<activity android:icon="@drawable/smallPic" . . . >
```

主题表达与资源值表达类似，只是将“@”改为了“？”，其格式为？[package:]type:name。

AndroidManifest.xml 文件描述示例如下：

```
<?xml version="1.0" encoding="utf-8"?>
<manifest xmlns:android="http://schemas.android.com/apk/res/android"
     package="com.demo"              //包名
     android:versionCode="1"         //程序版本号
     android:versionName="1.0">      //程序版本名
    <uses-sdk android:minSdkVersion="8" /> //使用的 SDK 开发版本
     //应用程序描述
    <application android:icon="@drawable/icon"   //应用程序图标
      android:label= "@string/app_name">          //应用程序名称
      //程序启动时执行的 Activity 的描述
<activity android:name=".MyActivity"
               android:label="@string/app_name">
         <intent-filter>
              <action
                    android:name="android.intent.action.MAIN" />
              <category
               android:name="android.intent.category.LAUNCHER" />
          </intent-filter>
       </activity>
        // 注册应用程序其他可识别的 Activity,否则程序将无法访问该 Activity
       <activity android:name="程序可访问的 Activity 类"
                 android:screenOrientation="portrait">
</activity>
           …
</application>
//程序可用的权限
<uses-permission android:name="程序可用的权限名称">
</uses-permission>
…
</manifest>
```

图 1-52 所示为某应用程序的 AndroidManifest.xml 文件的完整描述。

7．应用程序的资源

一个 Android 应用程序的组成不仅仅是代码，它还要求图片、声音文件以及其他可视部分与源程序分离，因此，应该在 XML 文件中定义动画、菜单、样式、颜色和用户界面的布局。使用资源可以更容易升级应用程序的各种特征，而不需要改变代码，即通过一组可选的资源，能够使开发者根据不同的设备配置（如不同的语言和屏幕尺寸）优化开发的应用程序。

对于 Android 项目中的每一个资源，系统为其定义了唯一的整数 ID，可以在应用程序代码中通过该 ID 引用资源。例如，如果应用程序包含一个名为【logo.png】的图像文件（保存在 res/drawable/目录下），自动生成的资源 ID 名为【R.drawable.logo】，该 ID 可以被用来引用图像，在程序代码中进行处理。

将资源从源代码中分开的最重要的作用是可以为用户提供针对不同设备配置的可替代资源的能力。例如，对于在 XML 中定义用户界面的字符串，可以将该字符串翻译成其他语言，并保存在单独的文件中；然后，根据资源目录名称的语言修饰符（如 res/values-fr/表示法文字符串值）以及用户的语言设置，Android 系统自动将合适的语言字符串引用到用户界面上。

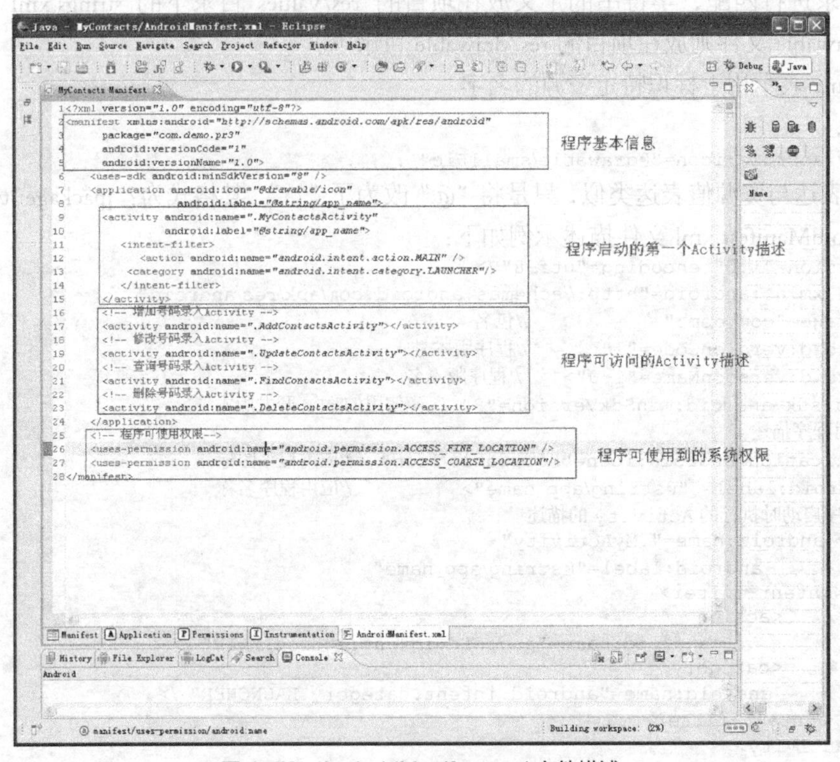

图 1-52　AndroidManifest.xml 文件描述

Android 支持许多不同的用于可替代资源的修饰符。修饰符是一个短的字符串，在资源目录中定义哪些资源在设备配置上使用。如程序员可以根据设备的屏幕方向和大小，为 Activities 创建不同的布局。当设备屏幕为纵向（高）时，可能需要按钮垂直布局，当屏幕为横向（宽）时，需要按钮水平对齐。为了能够根据方向改变布局，可以定义两种不同的形式，并为每个布局目录的名字使用合适的限定符。Android 系统会根据当前设备的方向自动采用相应的布局。

三、任务实施

开发一个 Android 程序主要分为两步：第一步是创建一个 Android 项目；第二步是根据项目的需要在 Activity 子类中编写代码，或者新建其他类，在 Activity 子类中进行引用。

（1）创建一个 Android 项目，在 Eclipse 的 Package 视图中单击鼠标右键，选择【New】→【Android】→【Android Application Project】，如图 1-53 所示。

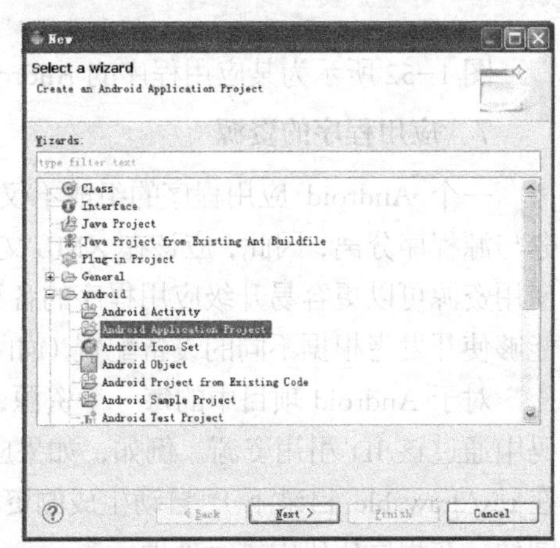

图 1-53　创建 Android 项目

（2）单击【Next】按钮，进行应用程序名称设置，如图 1-54 所示。在此须设置 Application Name（软件名称）、Project Name（项目名称）、Pakage Name

（包名称，是区分 Android 软件的唯一标记，必须填写）、Build SDK（表示目前开发项目所应用的 Android 平台）、Minimum Required SDK（支持的最小 Android 版本，软件无法安装到低于该版本的手机上）。在应用程序参数的命名上，可以根据个人需要起能够反映项目特征的名字。除了 Application Name 可用中文之外，建议为其他属性使用有意义的英文名。这里将应用程序命名为 Hello。

（3）单击【Next】按钮，设置应用程序的图标，如图 1-55 所示。

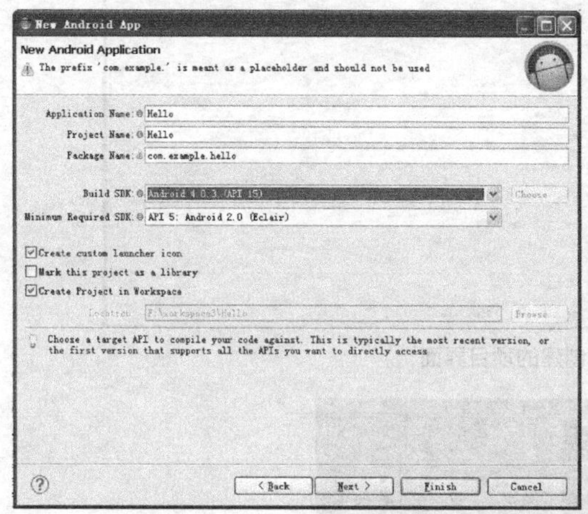

图 1-54 Android 项目参数设置

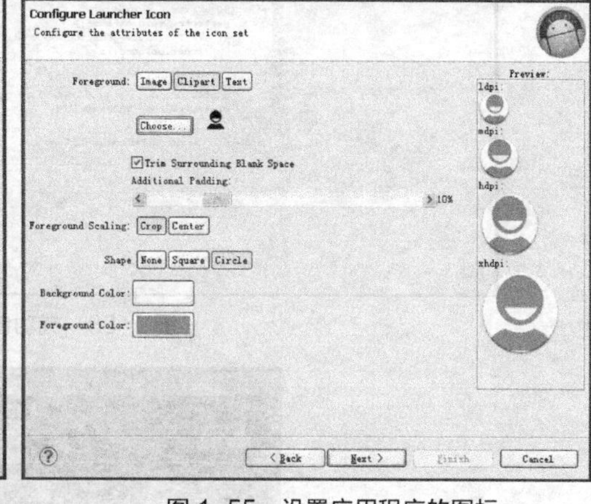

图 1-55 设置应用程序的图标

（4）单击【Next】按钮，创建一个空白的 Activity，如图 1-56 所示。

（5）单击【Next】按钮，配置 Activity 信息，输入 Activity 的名字和对应的布局文件的名字，如图 1-56 所示。

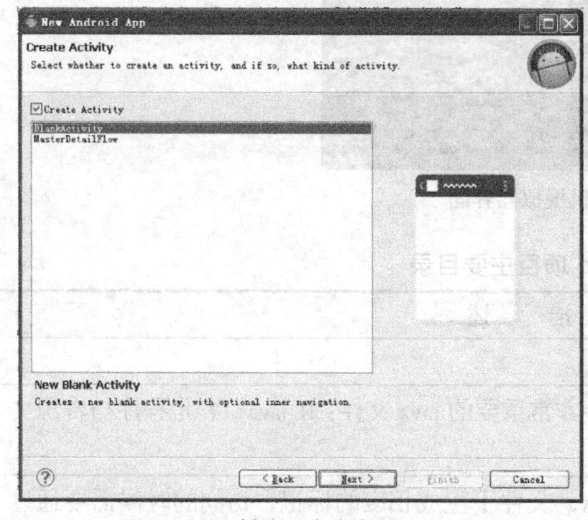

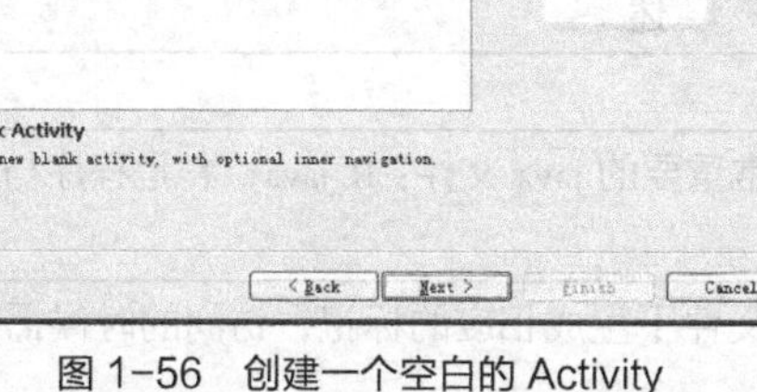

图 1-56 创建一个空白的 Activity

New Android App
New Blank Activity
Creates a new blank activity, with optional inner navigation.
Activity Name: HelloActivity
Layout Name: main
Navigation Type: None
Hierarchical Parent:
Title: HelloActivity
The name of the layout to create for the activity
< Back
Next >
Finish
Cancel

图 1-57 配置 Activity 信息

（6）单击【Finish】按钮，成功建立第一个 Activity 项目后的界面如图 1-58 所示。

（7）单击菜单栏 按钮，运行 Hello 程序或者用鼠标右键单击 Hello 项目，选择【Run As】→【Android Application】命令来运行项目。如果成功，将会弹出如图 1-59 所示的手机模拟器界面，代表运行 Android 程序的手机。本例没有编写任何代码就可以生成一个手机应用程序。

除了生成 AndroidManifest.xml 文件之外，Eclipse 默认为 Android 项目建立如表 1-2 所示的目录。

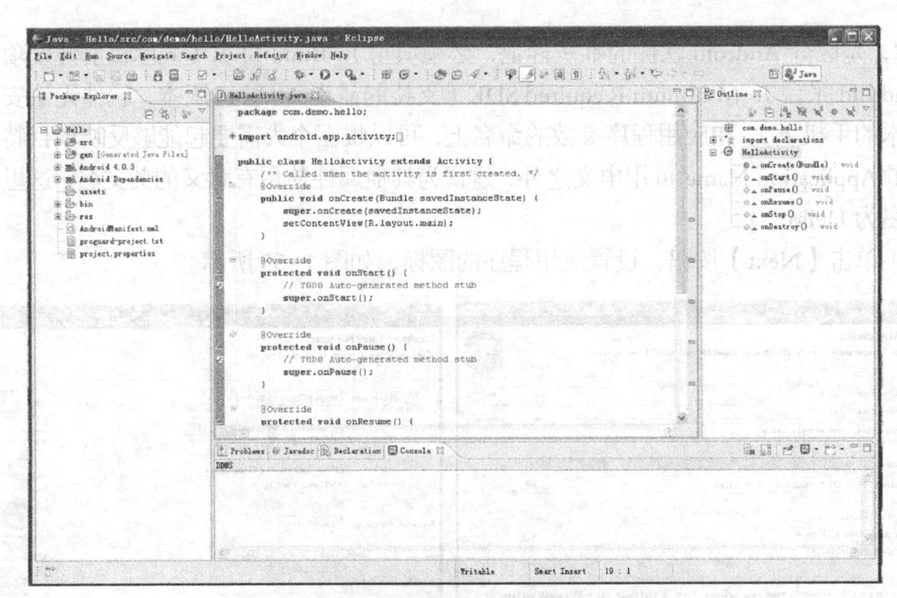

图 1–58　成功创建的项目界面

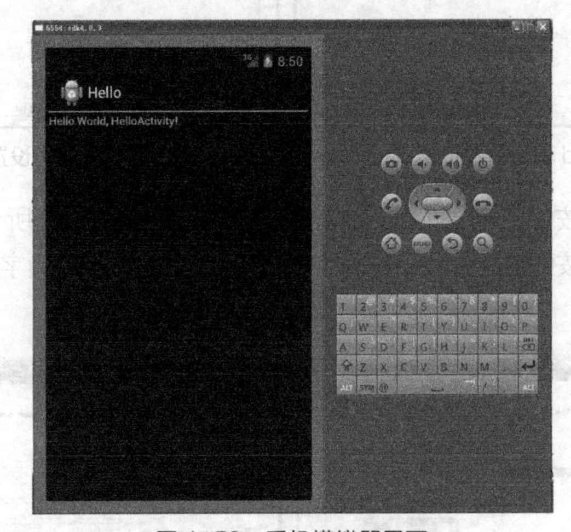

图 1–59　手机模拟器界面

表 1-2　Android 项目主要目录

目录名	描　　述
src	用于存放整个项目源代码
gen	由 Android 自动生成和维护，包含一个非常重要的.java 文件：R.java，它是程序与资源文件建立联系的桥梁，不可改动或删除
assets	存放资源，但这里的资源不会在 R.java 文件下生成相应的标记，访问的时候需要通过 AssetManager 类
bin	在项目运行以后生成，里面包含生成的可执行文件，如可直接在手机上运行的 APK 文件
res	存放资源，这里的资源会被映射到 R.java 文件中，访问时直接使用资源 ID，即 R.id.★ ★的形式。注意：res 目录下的文件名应为小写
drawable	是 res 的子目录，用于存放应用程序使用的图像资源或者图像描述文件。该目录还分为 drawable- hdpi、drawable-ldpi、drawable-mdpi、drawable-xhdpi，分别用于放置不同分辨率的图片资源。在代码中使用 getResources().getDrawable（resourceId）可获取相应 ID 的资源

目录名	描　述
layout	是 res 的子目录，放置 Android 应用程序的布局 XML 文件
values	是 res 的子目录，放置应用程序常量资源，包括字符串、样式和颜色定义。可以用于文本内容的国际化，以及避免出现硬编码（即在代码中直接编写文本内容）。对于 strings.xml 文件，通过使用 getResources().getString（resourceId）来取得字符串内容
menu	是 res 的子目录，用于放置菜单的布局文件

下面是生成的代码。

```
package com.example.hello;
public class HelloActivity extends Activity {
    /** Called when the activity is first created. */
    @Override
    public void onCreate(Bundle savedInstanceState){
        super.onCreate(savedInstanceState);
        setContentView(R.layout.main);
    }
    @Override
    protected void onStart() {
        // TODO Auto-generated method stub
        super.onStart();
    }

    @Override
    protected void onPause() {
        // TODO Auto-generated method stub
        super.onPause();
    }
    @Override
    protected void onResume() {
        // TODO Auto-generated method stub
        super.onResume();
    }
    @Override
    protected void onStop() {
        // TODO Auto-generated method stub
        super.onStop();
    }

    @Override
    protected void onDestroy() {
        // TODO Auto-generated method stub
        super.onDestroy();
    }
}
```

代码分析如下。

（1）HelloActivity 类能够在手机上直接运行，是因为该类继承了 Activity 抽象类。

（2）实现 Activity 生命周期的 6 个方法：onCreate、onStart、onResume、onPause、onStop 和 onDestroy。为了更好地理解上述代码的作用，可以对其进行修改。除了可以利用 Java 的 System.out.println()方法输出信息之外，还可以利用 Android 提供的 Log. v（String tag, String msg）方法以日志方式输出相应的信息。其中，参数 tag 表示标签，参数 msg 表示内容，取值都可以在编程中指定。这些信息可以通过 Eclipse 的 LogCat 视图进行查看。

（3）@Override 表示下面的方法是重写父类 Activity 的方法。可以不要，但该标识符有利于避免方法名的书写错误。

1.6 实训项目

一、建立 Android 开发环境

1．实训目的与要求

学会下载、安装和配置 Android 开发环境所需的软件，建立 Android 程序的开发环境。

2．实训内容

根据 1.2、1.3、1.4 节中的任务，按照规划任务内容，实施实训。

二、开发运行一个简单的 Android 程序

1．实训目的与要求

学会利用 Eclipse 自动生成一个 Android 程序，并使用不同的模拟器运行该程序，通过 LogCat 日志视图查看输出结果，以便对 Android 程序开发的步骤有初步认识。

2．实训内容

根据 1.5 节中的任务实施实训。

3．思考

（1）运行不同的手机模拟器对 Android 的程序开发有什么作用？

（2）Android 开发为何大量使用 XML 文件，这样做有什么作用？

PART 2 项目二 开发标准身高计算器

本项目工作情景的目标是让学生掌握 Android 的界面开发技术。主要的工作任务划分为 ① 开发输入界面；② 进行事件处理；③ 显示计算结果；④ 发布到手机。本项目主要涉及的关键技术包括：Activity 类的使用、布局文件的设计、事件的处理、创建菜单等。

本章将正式开始学习开发 Android 项目，示例中将会多次出现 XML 配置文件，里面有大量的 XML 元素，学习者只需要掌握重要的 XML 元素及其属性的含义，而不必花费过多时间记忆每个元素和属性的拼写，因为在实际项目开发中，可以通过 Android 可视化的布局编辑器来设置布局界面属性，并自动生成 XML 布局文件。

本书会对项目的相关类进行介绍。在一个类的应用中，首先应知道该类的功能作用；然后去掌握如何定义类的对象，实际上就是了解类的构造方法；最后应熟悉调用类的主要方法，及这些方法要求的参数各有哪些。

Android 程序需要用到大量的包，在实际编写代码时，可以在 Eclipse 中使用【Ctrl】+【Shift】+【O】组合键自动导入相应的包，无需人工输入导入包的语句。

2.1　背景知识

一、常见的手机硬件参数知识

手机软件与手机的硬件密切相关，如果了解手机常见的硬件参数，将便于程序员更好地开发适合于手机硬件的软件产品，也有利于提高软件产品的可移植性。

1．分辨率

手机屏幕尺寸分为物理尺寸和显示分辨率两个概念。物理尺寸是指屏幕的实际大小，以屏幕的对角线长度作为依据（比如 3.5 英寸、4.0 英寸）。在屏幕上看到的画面其实都是由一个个小点组成，这些小点又称为像素。每个像素点可以近似看作屏幕上的一个发光点，点的密度越大，显示效果越清晰，单位面积下显示的内容也就越多。屏幕分辨率反映的是在物理尺寸下可以显示的像素数量，以乘法形式表现，比如手机常见的分辨率 320 像素 × 480 像素，其中“320”表示屏幕上水平方向显示的像数个数，“480”表示垂直方向的像数个数。分辨率越大表示像素的数量越多，图像就越清晰，因此这个指标是决定画面质量的最主要因素。两台手机的物理尺寸相同并不表示其分辨率相同，不同物理尺寸的屏幕，也可以具有相同的分辨率，例如，4.3英寸的Samsung i9100和4.0英寸的HTC G11 都具有480 像素× 800像素（WVGA）。

分辨率比值是分辨率中横向像素与纵向像素的比值，例如320像素×480像素的分辨率比值为2∶3。

目前，手机主流分辨率很多，也有很多的名词术语，下面对此进行解释。

- GA（Video Graphics Array）：支持640像素×480像素，是IBM计算机的一种显示标准，也是现在绝大多数分辨率的基准。
- QVGA（Quarter VGA）：VGA分辨率的四分之一，支持240像素×320像素。传统的手机一般都采用这种分辨率，例如，Sony Ericsson S500c、Samsung S3650C、Motorola A1800、Nokia E66V等。
- HVGA（Half-size VGA）：VGA分辨率的一半，即480像素×320像素，iPhone、第一款Google手机T-Mobile G1均采用这种分辨率。
- WVGA（Wide VGA）：扩大了的VGA分辨率，支持480像素×800像素，例如，Samsung的I9000、HTC Desire HD等。
- FWVGA（Full Wide VGA）：扩大了的WVGA分辨率，支持480像素×854像素，例如，Motorola Milestone 2、Nokia N9、小米M1等。

分辨率的高低直接影响了产品外型的大小及画质的好坏。由于手机型号和操作系统的多样性，要求同一款游戏需要考虑在不同手机上的运行效果。目前，大部分的软件开发是以兼容分辨率480像素×800像素和480像素×854像素的手机为标准。对于美术设计人员而言，需要在设计之初就考虑屏幕的自适应问题。

2．色彩数量

屏幕颜色是由色阶来决定的。色阶是表示手机液晶显示屏亮度强弱的指标，也就是通常所说的色彩指数，表示色彩的丰满程度。

目前手机的色阶指数从低到高可分为最低单色、256色、4096色、65536色、26万色、1600万色。$256=2^8$，即8位彩色；依此类推，65536色$=2^{16}$，即通常所说的16位真彩色；26万色$=2^{18}$，也就是18位真彩色；1600万色$=2^{24}$，也就是24位真彩色。

目前手机能达到的色彩数量也是限制美术人员发挥的一个重要瓶颈。将色阶高的图片放到色阶低的手机上，会导致图片色彩失真，有的颜色无法区分，且色偏严重。所以，设计人员应根据手机的实际情况进行图片绘制。

此外，液晶屏幕由于其独特的发光原理，颜色的明亮度不高。在强光下色彩丰富的图像不能显示出原有的效果，户外显示时这一缺点尤其明显。因此设计人员在设计手机游戏图片时需要考虑这一点，避免将色彩对比度设置得过于接近。

3．CPU

手机同计算机一样具有CPU和内存，特别是随着智能手机越来越普及，更高的CPU硬件配置将成为手机发展的一个趋势，四核、八核的手机CPU已出现。CPU具有运算器和控制器功能，是手机的心脏，构成了系统的控制中心，可对各部件进行统一协调和控制。主频是衡量手机CPU性能的一个重要技术参数。频率越高，表明指令的执行速度越快，指令的执行时间也就越短，对信息的处理能力与效率就越高。

从技术发展趋势来看，手机和计算机正逐渐走向融合，手机CPU的处理性能在近几年得到了较大提高。下面对业界较有名气的手机CPU厂商进行介绍，此外samsung、Apple等公司也有自己的CPU。

① 德州仪器公司：提供 OMAP 系列处理器，能够兼容 Linux、Symbian、Windows Mobile、Android 等主流操 作系统，是手机 CPU 的主要供应商。

② Marvell 公司：2006 年购买了 Intel 公司的通信及应用处理器业务，得到了 Intel 著名的针对嵌入式设备的 XScale 处理器技术。

③ 高通公司：包括 Mobile Station Modems（MSM 芯片组）、单芯片（QSC）以及 Snapdragon 平台，根据手机定位的不同，推出经济型、多媒体型、增强型和融合型四种不同的芯片。

在手机游戏特别是 3D 游戏中，动画帧数不流畅的问题多数是由于运算速度慢造成的，从而对游戏动画效果产生很大的影响。对此程序员应该采取优化算法来改进画面质量，例如局部刷帧、缓存技术等。

4．内存

手机上的内存分为 RAM 和 ROM。其中，RAM 是动态内存，相当于计算机的内存，是影响手机程序运行性能的重要指标，它决定了手机可以同时运行多少应用程序。RAM 中的数据在手机关闭后丢失。目前来说，应选用 512 MB 以上的 RAM 以确保手机使用的流畅性。

ROM 则相当于计算机的硬盘，用于存储手机操作系统、应用程序和用户的文件，ROM 中的信息即使在手机断电后也不会丢失。随着手机上安装程序的增多，以及数据信息的累加，ROM 的可用空间会不断减少。如果 ROM 的空间太少，就会影响到手机的操作速度。这一点和计算机引导盘空间（如 Windows 系统 C 盘空间）的减少会影响计算机的使用类似。目前，对 Android 手机而言，应选择 1 GB 以上的 ROM 空间。

手机大小的内存可以通过插入 SD 卡来进一步扩充。SD 卡一般用于存放用户的文件，如视频、音频文件，可以将 SD 卡理解为计算机上的外接移动硬盘。

二、Android 的像素单位

Android 的像素单位有 dip、px、pt、sp 等，下面分别介绍它们各自的主要用途。

（1）dip（Device Independent Pixels）：设备独立像素。Android 设备的屏幕尺寸有很多种，为了使显示能尽量与设备无关，提出了 dip 的概念，这样可以使程序能够较好地在分辨率不同的手机上运行，并保持类似的外观。即在分辨率不同的屏幕上，相同 dip 的控件占据屏幕的百分比相等。在程序中一般建议采用该像素单位进行设置。

（2）px：表示像素，采用的是绝对大小，不同设备的显示效果相同，即大小不随设备分辨率的改变而改变，是 Android 默认的像素单位。

（3）pt：表示磅，是一个标准的长度单位，1 pt=1/72 英寸，用于印刷业。

（4）sp：放大像素，与 dip 概念类似，推荐用于设置字体，TextView 控件的字体最好以 sp 为单位。

（5）in：表示英寸。

（6）mm：表示毫米。

2.2 开发输入界面

一、任务分析

本任务需要实现的效果如图 2-1 所示。

应用程序主要由界面和功能组成。标准身高计算器的开发可分为信息的录入界面、数据处理和结果反馈 3 个模块。本次任务是为用户提供数据录入界面，需要思考如下几个问题。

（1）如何使用 Android 常用的界面类，它们有哪些重要的方法？

（2）如何使用 XML 描述布局？

（3）如何在手机上显示所开发的界面？

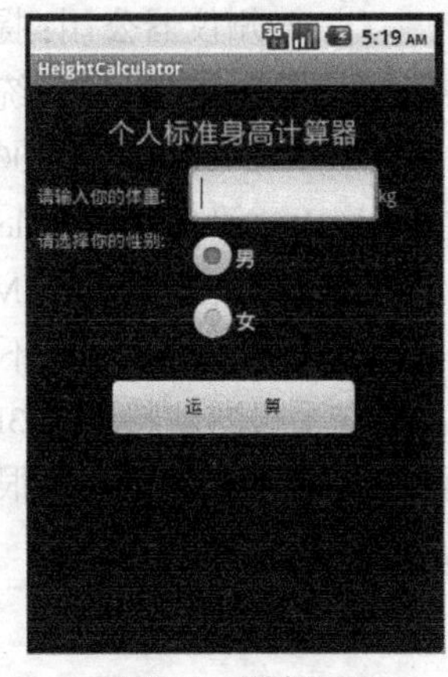

图 2-1 用户界面

二、相关知识

1. 用户界面设计

用户界面（UI，User Interface）是用户使用程序的桥梁，良好的界面能够使用户更乐意去接受和使用程序。由于不同程序的功能要求差异较大，复杂程度不一，所以评价用户界面并没有绝对的标准。设计一个良好的用户界面不应一味地追求漂亮的外观。下面给出用户界面设计的 4 项原则。

（1）满足系统功能的需求。这是最基本的原则，用户界面反映了程序对外提供的功能，如果不能符合系统功能的需求，将会直接影响到程序的使用效果。

（2）能够为用户提供准确的信息。不会对用户使用程序起到误导。

（3）布局合理，易于使用。根据信息显示的载体特点来进行界面布局，例如手机和计算机的屏幕大小差异较大，需要在布局上做更精心的设计，易于用户快速找到所需要的信息；具有良好的交互性，使用户不需要太多的培训就可以直接使用程序。

图 2-2 导航软件界面

（4）界面风格要一致，符合用户的使用习惯。在进行软件开发时，如果感觉到没有头绪，也可以通过网络查找业界同类型的软件，进行参考。图 2-2 所示为使用 Android 开发的一款导航软件的界面。

一个 Android 应用是类 android. app. Application 的一个实例。Application 中可以包含多个 Activity 实例。系统给每个 Activity 分配一个默认的窗口，而窗口中的内容则需要调用 setContent View() 方法将其放在一个显式的视图 ContentView 中，该视图描述了界面上具体的 UI 控件，如标签、文本框、单选框、多选框等。上述关系如图 2-3 所示，理解该图有助于读者在后面编写代码时理解各个类之间的关系。

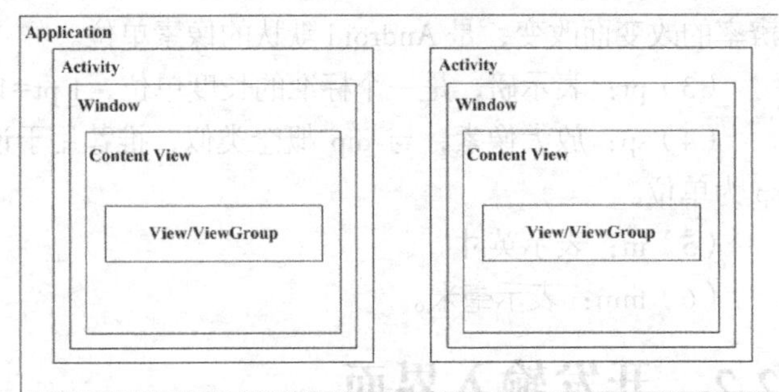

图 2-3 应用程序界面的主要组成部分

2. View 类

View 类是 Android 所有 UI 控件的父类。View 控件和 Swing 编程中的 JPanel 类似，代表一个空白的矩形区域，并负责绘图和事件处理。ViewGroup 是 View 类的一个重要子类，通常作为其他控件的容器使用，可定义其子 View 对象的布局。图 2-4 所示为 Android UI 的控件关系，其中 View 对象是指在界面上可直接看到的控件，例如，文本框、单选框、多选框、图片、按钮等。

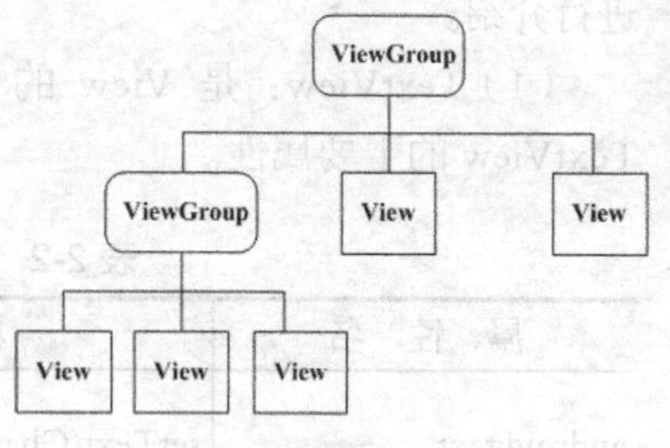

图 2-4 Android UI 的控件关系

Android 的所有 UI 控件都建立在 View、ViewGroup 基础之上。所有继承于 View 类的控件都可以使用其属性和方法。View 类典型的 XML 属性和方法如表 2-1 所示。

表 2-1 View 类典型的 XML 属性和方法

属 性 名	相关方法	说 明
android:alpha	setAlpha(float)	设置组件的透明度 例如，取值为 android:alpha="0.3"
android:background	setBackgroundResource（int）	设置该组件的背景颜色。可将一个 Drawable 作为背景，Drawable 可以看作某种可被绘制的资源，例如图像、XML 文件
android:clickable	setClickable（boolean）	定义视图是否响应单击事件
android:focusable	setFocusable（boolean）	控制视图是否有焦点
android:id	setId（int）	为视图提供一个标识名字。例如，android: id="@ + id/my_id "，其中 my_id 表示元素 ID 的唯一标识，用户可以自行修改。可通过 View.findViewById()或者 Activity.findViewById()查找
android:minHeight	setMinHeight（int）	设置视图的最小高度
android:minWidth	setMinWidth（int）	设置视图最小的宽度
android:padding	setPadding Relatives(int, int, int, int)	设置上下左右的边距，以像素为单位填充空白。指该控件内部内容，如文本距离该控件的边距
android:tag		为标签设置一个字符串标签，可以通过 View.getTag()或者 View.findViewWithTag() 搜索
android:visibility	setVisibility (int)	控制视图是否可见

属性说明如下。

（1）在用户界面设计中一方面可以在 XML 中指定属性的值，另一方面也可以在编程过程中通过相关方法动态地修改该属性。对于开发者来说，建议只需掌握常用的属性，并不需要花费时间去记忆所有的属性。可以根据属性名字和方法的英文含义去理解其作用。

（2）属性 android:id 的命名和程序变量的命名类似，需要赋予一个有含义的标识，以便于之后在程序代码中调用。

（3）Android 包下面的 android.R.styleable 类包含系统所有控件相关的 XML 属性，在编程时可以直接调用。

View 的子类也有很多自己的 XML 属性，下面以 TextView 和 EditText 两个视图控件为例进行介绍。

（1）TextView：是 View 的子类，其作用是显示文本，相当于一个标签。表 2-2 给出 TextView 的主要属性。

表 2-2 TextView 类典型的 XML 属性和方法

属 性 名	相 关 方 法	说 明
android:text	setText(CharSequence)	控件显示的文本。尽量将该信息放置在 Strings.xml 中，这样更具可扩展性
android:textColor	setTextColor(int)	设置文本的颜色
android:textSize	setTextSize	设置文字的大小
android:textStyle	setTypeface(Typeface)	设置文字的样式，取值有三种：normal、bold 和 italic
android:typeface	setTypeface(Typeface)	设置文字的字体类型
android:height	setHeight(int)	设置文本区域的高度
android:width	setWidth(int)	设置文本区域的宽度
android:gravity	setGravity(int)	设置文本的对齐方式
android:lineSpacingExtra	setLineSpacing(float,float)	设置文本的行间距，如“5 dip”表示每行之间间隔 5 个像素
android:lineSpacing Multiplier	setLineSpacing(float,float)	设置行间距的倍数，如：“1.5”表示 1.5 倍行距
android:lines	setLines(int)	设置文本显示的行数
android:ems	setEms(int)	设置 TextView 的宽度为 N 个字符的宽度
android:drawableBottom android:drawableLeft android:drawableRight android:drawableTop	setCompoundDrawablesWithIntrinsic Bounds(int,int,int,int)	分别在文本的下方、左侧、右侧、上方绘制一个指定的图片
android:drawablePadding	setCompoundDrawablePadding(int)	设置绘制图片与文本之间的间隔

android:gravity 属性的取值类型较多，下面是它们取值的含义。

- android：gravity="top"：不改变大小，位于容器的顶部。
- android：gravity="bottom"：不改变大小，位于容器的底部。
- android：gravity="left"：不改变大小，位于容器的左边。
- android：gravity="right"：不改变大小，位于容器的右边。
- android：gravity="center_vertical"：不改变大小，位于容器的纵向中央部分。
- android：gravity="center_horizontal"：不改变大小，位于容器的横向中央部分。
- android：gravity="center"：不改变大小，位于容器横向和纵向的中央部分。
- android：gravity="fill_vertical"：可能的话，纵向延伸以填满容器。
- android：gravity="fill_horizontal"：可能的话，横向延伸以填满容器。
- android：gravity="fill"：可能的话，纵向和横向延伸以填满容器。

（2）EditText：TextView 的子类，其功能是向用户提供输入框。表 2-3 给出 EditText 的主要属性。

表 2-3　EditText 类典型的 XML 属性和方法

属 性 名	相 关 方 法	说　明
android:enabled	setEnabled(boolean)	设置文本框是否可以编辑，例如，android:enabled="false"表示文本框不可编辑
android:hint	setHint(CharSequence)	设置用户输入的提示信息
android:maxLength	setMaxLines(int)	设置最大输入字符个数，例如，android:maxLength="5"表示最多能输入 5 个字符
android:inputType	setInputType(int)	设置虚拟键盘的类型。inputType 的取值类型很多，有"text", "textE-mailAddress", "textUri", "number", "phone"等。如果需要取多个值，则用"|"间隔开 ndroid:inputType="textPostalAddress| textCapWords| textNoSuggestions"
android:imeOptions	setImeOptions(int)	设置用户输入信息后执行的动作。可以通过 OnEditorActionListener 接口来进行监听

在 Android 源程序中 EditText 类的常用方法主要有两种。

① selectAll()：文本全选。

② getText().toString()：以字符串形式返回文本框的文本。其中 getText()返回的是 CharSequence 类型的接口，用 toString()方法可将其转换成 String 类型。

对于涉及颜色赋值的 XML 属性，例如，android:textColor、android:textColorHighlight、android: textColorHint、android:background，使用六位十六进制数表示 RGB 颜色，其中前两位表示红色（R），中间两位表示绿色（G），最后两位表示蓝色（B），每两位的取值范围都是 00～FF（相当于十进制的 0～255）。红、绿、蓝三种颜色的十六进制表示方法是：本位取最大值 FF，其他位取最小值 00，即：

#FF0000 红色，#00FF00 绿色，#0000FF 蓝色

另外两个比较特殊的颜色表示为#000000 黑色，#FFFFFF 白色。

在程序代码中，若要对颜色进行赋值，可以使用 Color 类。

- 使用 Color 类的常量，例如，int color=Color.RED;　//创建一个红色
- 使用 Color 类的 argb 方法，例如，int color=Color.argb(145, 123, 34, 68);// argb 方法的 4 个参数的取值范围都是[0~255]

3．布局方式

设计 Android 用户界面有三种方法。最常用的方法是使用 XML 来描述 UI。一个 XML 元素的名字实际上是对应一个 Java 类，元素的属性对应 Java 类的成员属性，例如，一个<EditText>元素在 UI 中相当于创建一个 EditText 类。当在程序中加载一个布局资源时，Android 系统会初始化这些运行的对象，实例化 UI 布局元素，并操作其属性。这样做的好处是应用 MVC 设计模式将用户界面和程序逻辑分开，使得改变用户界面不会影响到程序逻辑，同样地，程序逻辑的变动也不会影响用户界面。Activity 为 MVC 中的 Controller，Activity 的 ContentView 则是 MVC

中的 View。另外一种用户界面设计方法是使用 Java 代码来创建 UI，这种方法较为复杂，且模块之间的耦合度较高。第三种方法是综合使用前面两种方法，把变化小、行为比较固定的控件放在 XML 布局文件中实现，而变化较多、行为控制复杂的控件则由代码进行管理。

布局是一个程序用户界面的架构，它定义了界面的布局结构且存储所有显示给用户的元素，例如文本框、按钮等控件。布局文件使用 XML 来定义自己的布局和表达层次视图，以.xml 为扩展名，存放在项目的 res/layout 目录下，以便自动被编译工具编译。Eclipse 在创建 Activity 时，会自动为其生成一个布局文件。如果需要新增加布局文件，可以通过图 2-5 和图 2-6 所示的方式来创建。需要注意的是，布局文件的命名不能使用大写字母。

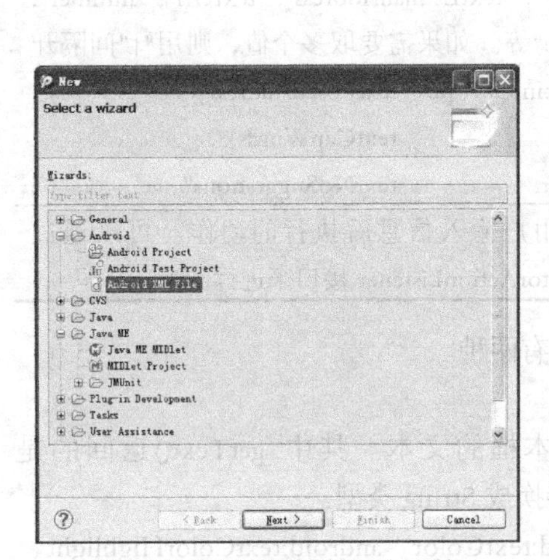

图 2-5　通过向导创建 XML 文件

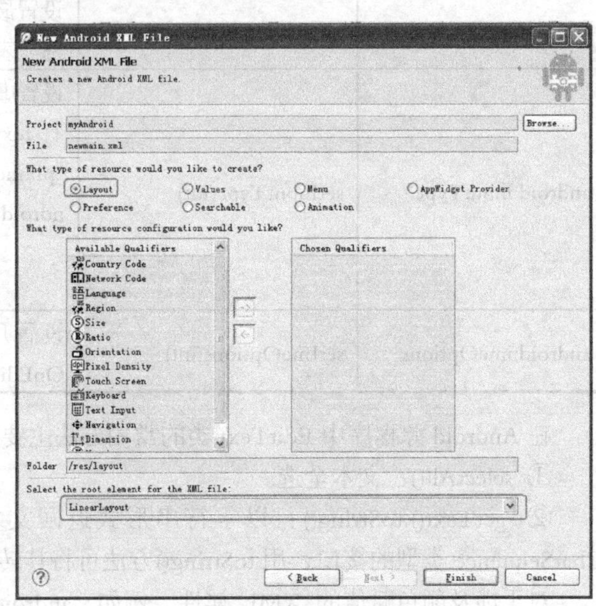

图 2-6　选择 XML 文件的类型

利用 Android 预先定义的 XML 元素，可以快速地设计 UI 界面，这一点和 Web 页面的设计使用 HTML 相似。每个布局文件必须包含一个根元素，根元素必须是一个 View 或 ViewGroup 对象。一旦定义了根元素，就可以添加其他 View 元素，逐步构建一个视图层次的界面布局。下面的 main_layout.xml 布局示例展示了使用纵向的 LinearLayout 方式，添加一个文本框。

```
<LinearLayout xmlns:android="http://schemas.android.com/apk/res/android"
   xmlns:tools="http://schemas.android.com/tools"
   android:id="@+id/LinearLayout1"
   android:layout_width="match_parent"
   android:layout_height="match_parent"
   android:background="#000000"
   android:orientation="vertical" >
   <EditText
      android:id="@+id/editText1"
      android:layout_width="match_parent"
      android:layout_height="wrap_content"
      android:ems="10" >
      <requestFocus />
   </EditText>
</LinearLayout>
```

定义好的布局文件是如何被加载的呢？在编译 Android 应用程序时，每个 XML 布局文件将被编译成一个 View 资源。由 Activity 代码中的 onCreate()函数调用 setContentView()实现布局资源的加载，以“R.layout.布局文件名”的形式作为函数的参数值。如对名为 main_layout.xml 的布局文件，加载如下代码：

```
public void onCreate(Bundle savedInstanceState){
        super.onCreate(savedInstanceState); //调用父类的 onCreate 方法
    setContentView(R.layout.main_layout);
    }
```

设置好用户界面后就可以进一步操作布局文件中的元素。上面已经介绍过，XML 中的每一个元素对应一个 Java 类，那么如何确定元素对应的 UI 类呢？这就涉及 XML 的 ID 属性。XML 中 ID 属性语法如下：

```
android:id="@+id/editText1"
```

其中，“@”表示 XML 解析器应该解析和扩展剩下的 ID 字符串，并将其用作 ID 资源。“+”符号表示这是一个新的资源名字，必须被创建且加入到 R.java 文件中。R.java 文件是一个最终类，保存了应用程序用到的图标、常量等各种资源 ID，被放置在“项目名称/gen/项目包名/R.java”目录下。R.java 是编译器根据用户的 XML 文件或资源文件自动创建的，应避免手工修改 R.java 文件。通过 R.java，应用程序能方便地查找到对应的资源，调用的格式是 R.×××.×××。第一个“×××”在实际使用时应填写具体的值，如 R.drawable. ×××、R.string. ×××、R.layout. ×××、R.id. ×××，分别对应于“项目目录/res/drawable/”下的资源、“/res/string/”下的字符串、“/res/layout/”下的布局文件及 XML 文件中控件元素的 ID 等。第二个“×××”表示要找的资源 ID，如 R.layout.main_layout 表示名为 main_layout 的布局文件。另外编译器也会检查 R.java 列表中的资源是否被调用，没有被调用的资源将不会被编译进软件中，从而减少了应用程序占用的空间。下面的代码通过 findViewById 方法从布局文件中找出文本输入框，并向其中填写内容，editText1 为文本框的 ID。图 2-7 所示为整个程序的效果图。

图 2-7　程序效果

```
public void onCreate(Bundle savedInstanceState){
        super.onCreate(savedInstanceState);
         setContentView(R.layout.main_layout);
    //对 findViewById 方法返回的结果需要根据 View 的类型进行强制类型转换，以下代码是转换为文本框
    EditText tEdit=(EditText)findViewById(R.id.editText1);
    tEdit.setText("您好！");
}
```

关于手机屏幕上的控件摆放设置，Android 提供了 5 种布局方式，分别为线性布局（LinearLayout）、表格布局（TableLayout）、相对布局（RelativeLayout）、单帧布局（FrameLayout）、绝对布局（AbsoluteLayout）。可以将它们理解为上一节讲到的 ViewGroup。这些布局也可以嵌套在一起使用，以便于灵活地生成各种布局样式。本节主要讨论线性布局。需要注意的是，对需要设置成某种布局的控件，应将其 XML 配置代码作为布局的子元素。布局属性的设置会影响内部控件的排列，所以在修改控件的属性时，如果无法出现预期效果，应进一步检查其上一级布局属性的设置。

1．线性布局

线性布局是 Android 默认的布局方式，分为垂直线性布局和水平线性布局，分别表示为 android:orientation=“vertical”和 android:orientation=“horizontal”。前者表示控件以垂直方式排列，即每一行放一个控件；后者表示控件按照水平方式排列，即所有控件都

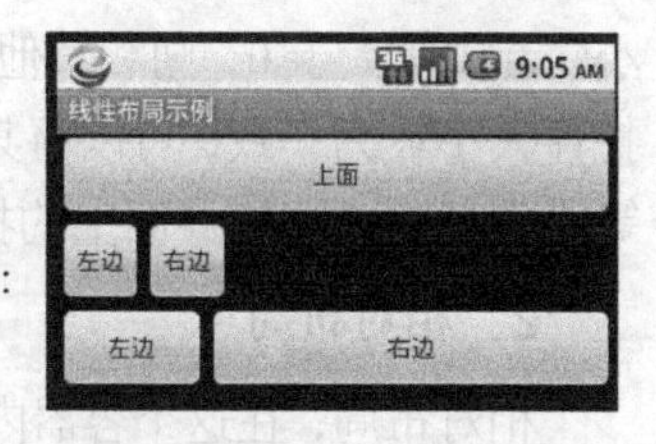

图 2-8　复杂的布局界面

放在同一行，超出部分会被遮盖。

图 2-8 所示为综合利用多个 LinearLayout 实现较为复杂的布局界面。

```
<?xml version="1.0" encoding="utf-8"?>
<LinearLayout xmlns:android="http://schemas.android.com/apk/res/android"
    android:orientation="vertical" android:layout_height="fill_parent"
    android:layout_width="fill_parent">
    <Button android:id="@+id/Button01" android:layout_height="wrap_content"
        android:text="上面" android:layout_width="fill_parent"></Button>
    <LinearLayout android:id="@+id/LinearLayout01"
        android:layout_height="wrap_content" android:orientation="horizontal"
        android:layout_width="wrap_content">
        <Button android:id="@+id/Button03" android:layout_height="wrap_content"
            android:text="左边" android:layout_width="wrap_content"></Button>
          <Button android:id="@+id/Button02" android:layout_height="wrap_content"
            android:text="右边" android:layout_width="wrap_content"></Button>
    </LinearLayout>
    <LinearLayout android:id="@+id/LinearLayout02"
        android:layout_width="fill_parent" android:layout_height="fill_parent">
        <Button android:id="@+id/Button05" android:layout_height="wrap_content"
            android:layout_weight="1" android:text="左边"
             android:layout_width="wrap_content"></Button>
        <Button android:id="@+id/Button04" android:layout_height="wrap_content"
            android:text="右边"
             android:layout_weight="5"
android:layout_width="wrap_content"></Button>
    </LinearLayout>
</LinearLayout>
```

放到布局中的 View 控件，需要设置其主要的布局属性。

① android:layout_height 属性：设置控件的高度。

② android:layout_width 属性：设置控件的宽度。

上面两个属性，都有 3 种取值其中，fill_parent 表示强制性使控件扩展，以填充布局单元内尽可能多的空间，在空间允许情况下，可以占满屏幕；match_parent 与 fill_parent 的意思一样，在 Android 2.2 之后版本中建议采用 match_parent；wrap_content 表示只需显示控件的全部内容即可，例如对于 TextView，将完整显示其内部的文本；对于 ImageView，将完整显示其图像。

③ android:layout_gravity 属性：设置控件显示的位置，默认取值为 top，表示顶部对齐。如果希望居中对齐，可取值为 center_vertical 表示垂直居中，或 center_horizontal 表示水平居中。

④ android:layout_margin 属性：设置控件上、下、左、右边框的边距。

⑤ android:layout_marginBottom 属性：设置控件下边框的边距，如取值为"5.0 dip"，表示下边距为 5 个像素。

（6）android:layout_marginLeft 属性：设置控件左边框的边距。

（7）android:layout_marginRight 属性：设置控件右边框的边距。

（8）android:layout_marginTop 属性：设置控件上边框的边距。

（9）android:layout_weight 属性：默认值是 0，意味着只在屏幕上占据需要显示的空间大小。若赋值大于 0，则和其他值大于 0 的视图按照取值的比例来分割可用的空间。如果屏幕上控件大小较为一致，可以将其值都设置为 1，控件的大小将会被等比例划分，这样界面上的控件布局看起来会较为整齐。

图 2-9 相对布局示例

2．相对布局

相对布局，在这个容器内部的子元素们可以使用彼此之间的相对位置或者与容器间的相对位置来进行定位。使用相对布局的好处是位置控制比较灵活。如图 2-9 所示，其布局代码如下。

```
<?xml version="1.0" encoding="utf-8"?>
<RelativeLayout xmlns:android="http://schemas.android.com/apk/res/android"
    android:layout_width="fill_parent"
    android:layout_height="wrap_content"
    android:background="#0000FF"
    android:padding="10px">

    <TextView
        android:id="@+id/tv01"
        android:layout_width="wrap_content"
        android:layout_height="wrap_content"
        android:layout_marginBottom="30dp"
        android:text="请输入:" />

    <EditText
         android:id="@+id/txt01"
         android:layout_width="match_parent"
         android:layout_height="wrap_content"
         android:layout_below="@id/tv01"/>

    <Button
        android:id="@+id/btn01"
        android:layout_width="wrap_content"
        android:layout_height="wrap_content"
        android:layout_alignParentRight="true"
        android:layout_below="@id/txt01"
        android:text="确认" />

    <Button
        android:id="@+id/btn02"
        android:layout_width="wrap_content"
        android:layout_height="wrap_content"
        android:layout_below="@id/txt01"
        android:layout_marginRight="30dp"
        android:layout_toLeftOf="@id/btn01"
        android:text="取消" />

</RelativeLayout>
```

相对布局的主要属性包括 4 类：① 设置控件与控件之间的位置关系（见表 2-4）；② 设置控件与控件之间对齐的方式，是顶部、底部还是左、右对齐（见表 2-5）；③ 设置控件与父控件之间对齐的方式，是顶部、底部还是左、右对齐（见表 2-6）；④ 设置控件方向的属性（见表 2-7）。可以通过组合这些属性来实现各种各样的布局效果。

表 2-4 控件之间位置的关系属性及描述

属性名称	描　　述
Android:layout_above	将该控件置于给定 ID 控件的上面
Android:layout_below	将该控件置于给定 ID 控件的下面
Android:layout_toLeft0f	将该控件置于给定 ID 控件的左边
Android:layout_toRight0f	将该控件置于给定 ID 控件的右边

说明：属性的取值为某个控件的 ID，如：Android:layout_above="@id/exit"，"exit"为某个 Button 按钮的 ID。

表 2-5　控件之间对齐的属性及描述

属性名称	描　述
android:layout_alignBaseline	将该控件的 Baseline 与给定 ID 控件的 Baseline 对齐
android:layout_alignTop	将该控件的顶部与给定 ID 控件的顶部对齐
android:layout_alignBottom	将该控件的底部与给定 ID 控件的底部对齐
android:layout_alignLeft	将该控件的左边缘与给定 ID 控件的左边缘对齐
android:layout_alignRight	将该控件的左边缘与给定 ID 控件的右边缘对齐

说明：属性的取值为某个控件的 ID，如：Android: layout_alignLeft="@id/exit"，其中"exit"为某个 Button 按钮的 ID。

表 2-6　与父控件间对齐的属性及描述

属性名称	描　述
android:layout_alignParentTop	将该控件的顶部与父控件的顶部对齐
android:layout_alignParentBottom	将该控件底部与父控件的底部对齐
android:layout_alignParentLeft	将该控件的左边边缘与父控件的左边边缘对齐
android:layout_alignParentRight	将该控件的右边边缘与父控件的右边边缘对齐

说明：属性的取值为 true 和 false，如：Android: layout_alignParentBottom ="true"，表示将该控件底部与父控件的底部对齐。

表 2-7　控件方向位置的属性及描述

属性名称	描　述
android:layout_centerHorizontal	将该控件置于水平方向的中心
android:layout_centerVertical	将该控件置于垂直方向的中心
android:layout_centerInParent	将该控件置于父控件水平方向和垂直方向的中心

说明：属性的取值为 true 和 false，如：Android: layout_centerInParent ="true"，表示将该控件置于父控件水平方向和垂直方向的中心。

3．表格布局

表格布局的样式与一个表格类似。通过 TableRow 来定义行，有几行就有几个 TableRow。图 2-10 中所示的表格布局的代码如下。

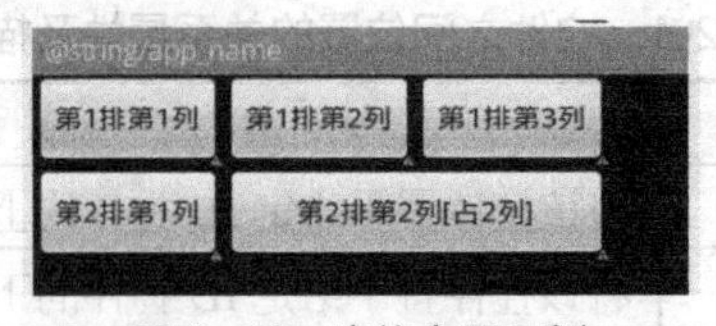

图 2-10　表格布局示例

```
<?xml version="1.0" encoding="utf-8"?>
<TableLayout xmlns:android="http://schemas.android.com/apk/res/android"
    android:layout_width="fill_parent"
    android:layout_height="fill_parent" >
     <TableRow>
         <Button
             android:text="第 1 排第 1 列"
             />
         <Button
             android:text="第 1 排第 2 列"
```

```
            />
        <Button
            android:text="第 1 排第 3 列"
            />
    </TableRow>
    <TableRow>
        <Button
            android:text="第 2 排第 1 列"
            />
        <Button
            android:layout_span="2"
            android:text="第 2 排第 2 列[占 2 列]"
            />
    </TableRow>
</TableLayout>
```

表格布局中常用的属性如下，它们的取值从序号 0 开始。

① android:collapseColumns：隐藏指定的列。

② android:shrinkColumns：收缩指定的列以适合屏幕，以避免被挤出屏幕。

③ android:stretchColumns：用指定的列尽量填充屏幕的空白部分。

④ android:layout_column：控件放在指定的列。

⑤ android:layout_span：该控件所占的列数，取值越大，控件占的空间越大。

此外，Android 的 UI 布局还有单帧布局（FrameLayout）、绝对布局（AbsoluteLayout）两种方式。单帧布局是将所有添加到这个布局中的视图都以层叠的方式显示，第一个添加的组件放到最底层，最后添加到框架中的视图显示在最上面，上一层的会覆盖下一层的控件。绝对布局是指定子控件的 x-y 精确坐标的布局方式，但这种方式不便于程序适应不同分辨率的手机，因此不推荐使用。

4. RadioButton 单选框

要实现单选功能，需要使用 RadioGroup 和 RadioButton 控件。RadioGroup 是一个容器，用于管理一组 RadioButton，用户只能选择其中的一个 RadioButton。在没有 RadioGroup 的情况下，RadioButton 可以被全部选中。下面举例说明如何在布局文件中创建单选框，效果如图 2-11 所示。

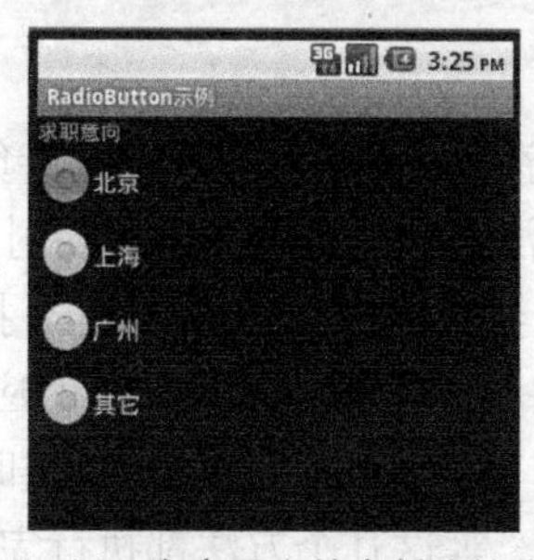

图 2-11　在布局文件中创建单选框

```
<?xml version="1.0" encoding="utf-8"?>
<LinearLayout xmlns:android="http://schemas.android.com/apk/res/android"
    android:layout_width="wrap_content" android:layout_height="wrap_content"
    android:orientation="vertical">
    <TextView android:id="@+id/TextView01" android:layout_width="wrap_content"
        android:layout_height="wrap_content" android:text="求职意向"></TextView>
    <RadioGroup android:id="@+id/RadioGroup01"
        android:layout_width="wrap_content" android:layout_height="wrap_content"
        android:orientation="vertical">
        <RadioButton android:id="@+id/RadioButton01"
            android:layout_width="wrap_content" android:layout_height="wrap_content"
```

```
            android:text="北京"></RadioButton>
        <RadioButton android:id="@+id/RadioButton02"
            android:layout_width="wrap_content" android:layout_height="wrap_content"
            android:text="上海"></RadioButton>
        <RadioButton android:id="@+id/RadioButton03"
            android:layout_width="wrap_content" android:layout_height="wrap_content"
            android:text="广州"></RadioButton>
        <RadioButton android:id="@+id/RadioButton04"
            android:layout_width="wrap_content" android:layout_height="wrap_content"
            android:text="其它"></RadioButton>
    </RadioGroup>
</LinearLayout>
```

下面对 XML 代码进行解释。

① RadioGroup 也是一个布局管理器，即 ViewGroup，可以通过改变其 orientation 属性的值来设置在 RadioButton 的布局。

② RadioButton 作为 RadioGroup 的子元素，主要用来设置 text 属性，用于展示按钮选项的内容。

如果希望在程序代码中获得 ID 为 groupSex 的 RadioGroup 对象，那么可采取如下方法：

```
RadioGroup  group =(RadioGroup)findViewById(R.id.groupSex);
```

RadioGroup 类中的方法 getChildCount()用于获得按钮组内的按钮数量，getChildAt(index) 方法用于获得索引值为 index 的按钮。

类似地，在程序代码中获得 ID 为 male 的 RadioButton 对象的方法如下：

```
RadioButton  maleButton =(RadioButton)findViewById(R.id.male);
```

这样就可以在程序中对 RadioButton 对象做进一步的处理，例如使用 isChecked()方法判断按钮是否被选中，使用 getId()方法获得按钮的 ID 等。

5．标准身高计算公式

目前有很多种标准身高计算公式，本书选择其中一种常用计算公式作为示例。具体计算方法如下：

```
男性标准身高=170-(62-体重)/0.6
女性标准身高=158-(52-体重)/0.5
```

6．Activity 类

Activity 类是 Android 应用程序中非常重要的类，直接负责与用户进行交互。项目一已经对 Activity 的生命周期和有关的方法进行了介绍。本节将对 Activity 做进一步的补充。

Activity 方法的主要功能分为与事件处理有关的操作、按键操作、菜单操作、系统较为底层的操作等。掌握并利用这些方法，可以编写较为复杂的应用。

（1）Activity 的事件处理方法：当启动一个 Activity 类时，onCreate 方法首先启动，接着是 onStart 和 onResume。一般来说，等这几个方法都执行完成之后 Activity 就被显示出来。在 Acitivity 销毁的时候会调用 onPause、onStop、onDestory 3 种方法。在执行完 onDestory 之后，Acitivity 完成销毁。

除了显示或者被销毁之外，Activity 类还可能处于其他状态。例如，当 Activity A 启动了另一个 Activity B 时，这个 Activiy A 会调用 onPause 方法，进入停滞状态，被放到系统的堆栈中，等被启动的另一个 Activity B 返回时，Activiy A 才会调用 onResume 方法重新被显示出来。

onSaveInstanceState 与 onRestoreInstanceState 的用途是：当设备的配置（如屏幕方向、语言等）发生变化而使 Activity 自动重启或者 Activity 从前台转到后台（如按下 Home 键）时，Activity 会调用 onSaveInstanceState(Bundle)方法将当前 Activity 的状态保存到 Bundle 变量中。

这个 Bundle 变量与 onCreate（Bundle）方法、onRestoreInstanceState（Bundle）中的 Bundle 参数是同一个，也就是说，后两个方法通过传入的 Bundle 参数能够还原 Activity 之前的状态。

Activity 的状态被保存到 Bundle 中。要想手动设置 Activity 重启后需要的状态数据，可以直接调用该 Bundle 实例的 putXXX 方法存入额外的数据信息。

下面给出的示例建议放在真实手机上进行测试，通过切换屏幕方向和切换不同的应用程序，来观察界面上的两个文本在不同操作下的提示差异（见图 2-12、图 2-13、图 2-14），从而更深入地理解此类事件的含义。

图 2-12 程序刚打开时的界面

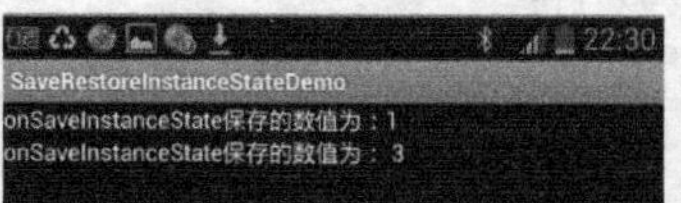

图 2-13 切换屏幕方向后数值的差异

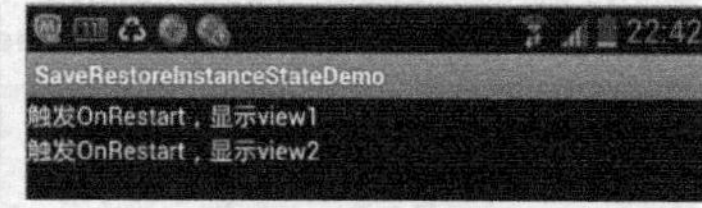

图 2-14 切换其他应用程序之后的显示

布局文件 main.xml：

```
<LinearLayout xmlns:android="http://schemas.android.com/apk/res/android"
    xmlns:tools="http://schemas.android.com/tools"
    android:id="@+id/LinearLayout1"
    android:layout_width="match_parent"
    android:layout_height="match_parent"
    android:background="#000000"
    android:orientation="vertical" >
    <TextView
        android:id="@+id/textView1"
        android:layout_width="match_parent"
        android:layout_height="wrap_content"
        android:text="TextView" />
    <TextView
        android:id="@+id/textView2"
        android:layout_width="match_parent"
        android:layout_height="wrap_content"
        android:text="TextView" />
</LinearLayout>
```

程序代码：

```
public class SaveRestoreInstanceStateActivity extends Activity {
private TextView tv1;
private TextView tv2;
private int iNum1,iNum2;
@Override
public void onCreate(Bundle savedInstanceState){
  super.onCreate(savedInstanceState);
  setContentView(R.layout.main);
  tv1 =(TextView)findViewById(R.id.textView1);
  tv2 =(TextView)findViewById(R.id.textView2);
  tv1.setText("在 onCreate 方法中赋的值");
  tv2.setText("在 onCreate 方法中赋的值");
  iNum1=0;
  iNum2=0;
  Toast.makeText(this,"onCreate",Toast.LENGTH_LONG);
  }

@Override
protected void onRestart() {
  super.onRestart();
  tv1.setText("触发 OnRestart,显示 view1");
  tv2.setText("触发 OnRestart,显示 view2");
}

@Override
protected void onSaveInstanceState(Bundle outState){
```

```
  iNum1++;
  iNum2++;
  outState.putString("tv1","onSaveInstanceState 保存的数值为"+iNum1);
  outState.putString("tv2",String.valueOf(iNum2));
  super.onSaveInstanceState(outState);
}

@Override
protected void onRestoreInstanceState(Bundle savedInstanceState){
  super.onRestoreInstanceState(savedInstanceState);
   tv1.setText(savedInstanceState.getString("tv1"));
   tv2.setText("onSaveInstanceState 保存的数值为
"+savedInstanceState.getString("tv2"));
   iNum2=Integer.parseInt(savedInstanceState.getString("tv2"));
}

}
```

（2）按键事件的处理方法：Activity 提供了对按键的捕捉。按键的捕捉分为两个层面：一是对输入法、输入值的捕捉；二是功能键操作，例如回退、主菜单、声音调节等。对于某些输入法而言，按键事件未必能够捕捉到，建议在程序中使用 Android keyboard 输入法进行测试。

① Boolean onKeyDown（int keyCode，KeyEvent event）：当按键被按下时调用。参数 keyCode 表示按键的编码；参数 event 表示按键事件对象。KeyEvent 类不仅包含了触发按键的详细信息，例如事件的状态、事件的类型、事件发生的时间，还定义了一些 keyCode 的常量，表示标准 Android 设备中的功能按键。

② Boolean onKeyUp（int keyCode，KeyEvent event）：当按键被释放时调用。

③ Boolean onTouchEvent（MotionEvent event）：当触屏事件发生时调用。参数 event 表示移动事件对象。MotionEvent 类包含笔、手指、轨迹球等对象的移动信息，例如坐标。

④ onKeyLongPress（int keyCode，KeyEvent event）：当长按某个键时被调用。

⑤ Void onBackPressed()：按下【Back】键时被调用，一般用来屏蔽【Back】键或者添加后退功能，例如，弹出确认退出的对话框，如果用户选择退出，则执行 finish()方法关闭 Activity。注意：与 System.exit(0)方法的退出整个应用程序不同，finish()方法只是将 Activity 推向后台，并没有立即将活动的资源释放。

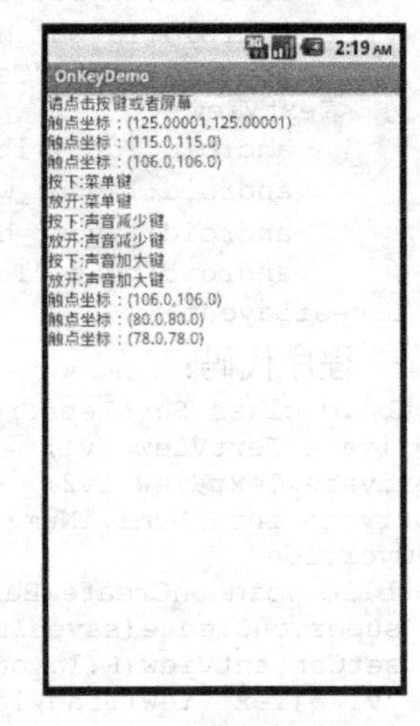

图 2-15　按键效果示例

按键捕捉示例如下，捕捉用户按键以及屏幕的触点坐标，并将这些信息进行显示，实现如图 2-15 所示的效果。

```
public class OnKeyDemo extends Activity {
private TextView tView;
   @Override
   public void onCreate(Bundle savedInstanceState){
       super.onCreate(savedInstanceState);
       //在程序中创建一个 EditText 控件
       tView=new TextView(this);
       tView.setText("请单击按键或者屏幕");
       //不需要布局文件,将创建的 EditText 控件作为显示视图
       setContentView(tView);
   }

   /**按键按下时触发的事件*/
```

```
public boolean onKeyDown(int keyCode,KeyEvent event)
{
 switch(keyCode)
 {

 case KeyEvent.KEYCODE_HOME:
      SetMessage("按下:Home 键");
      break;
 case KeyEvent.KEYCODE_MENU:
      SetMessage("按下:菜单键");
      return true;
 case KeyEvent.KEYCODE_BACK:
      SetMessage("按下:回退键");
      break;
 case KeyEvent.KEYCODE_VOLUME_UP:
      SetMessage("按下:声音加大键");
      break;
 case KeyEvent.KEYCODE_VOLUME_DOWN:
      SetMessage("按下:声音减小键");
      event.startTracking();
    return true;
 default:
      SetMessage("按下的键码是"+keyCode);
          break;
 }
 return super.onKeyDown(keyCode,event);
}

/**按键弹起时所触发的事件*/
public boolean onKeyUp(int keyCode,KeyEvent event)
{
 switch(keyCode)
 {

 case KeyEvent.KEYCODE_HOME:
      SetMessage("放开:Home 键");
      break;
 case KeyEvent.KEYCODE_MENU:
      SetMessage("放开:菜单键");
      return true;
 case KeyEvent.KEYCODE_BACK:
      SetMessage("放开:回退键");
      break;
 case KeyEvent.KEYCODE_VOLUME_UP:
      SetMessage("放开:声音加大键");
      break;
 case KeyEvent.KEYCODE_VOLUME_DOWN:
      SetMessage("放开:声音减少键");
      break;
 default:
      SetMessage("放开的键码是"+keyCode);
          break;
 }
 return super.onKeyUp(keyCode,event);
}
/**长按键事件*/
public boolean onKeyLongPress(int keyCode,KeyEvent event){
      SetMessage("长时间按键");
      return super.onKeyLongPress(keyCode,event);
 }

/**触屏事件*/
public boolean onTouchEvent(MotionEvent event)
{
```

```
    int action=event.getAction();
    if(action == MotionEvent.ACTION_CANCEL||action ==
MotionEvent.ACTION_DOWN||action == MotionEvent.ACTION_MOVE)
    {
        return false;
    }
    //得到触点的位置
    String x=String.valueOf(event.getX());
    String y=String.valueOf(event.getY());
    SetMessage("触点坐标: ("+x+","+y +")");
    return super.onTouchEvent(event);
   }

  @Override
public void onBackPressed() {
    super.onBackPressed();
    SetMessage("按下返回键了");
}

/**显示触发事件的信息*/
 public void SetMessage(String str)
{
    String oldStr=tView.getText().toString();
    String newStr=oldStr+"\n"+str;

    tView.setText(newStr);
  }
}
```

代码分析如下。

- 在 onKeyDown、onKeyUp 方法中，分别根据 keyCode 判断触发的是哪个功能键，如菜单 Menu 键、Back 键、声音键等，并且通过更改文本控件的内容来进行按键提示。
- 如果想避免用户执行功能键的实际功能，则只需在onKeyDown()的回调方法中返回true值，让系统知道当前捕捉的是 onkeyDown 事件即可。若返回 false，则表明还会执行用户功能键的实际功能。例如，在对 KeyEvent.KEYCODE_VOLUME_DOWN 的捕捉事件中，返回的值不是 true，而是 false，则表明由系统直接处理该功能事件，当用户按下声音键时，实际上已经通过系统调用降低了音量。
- 要让系统调用 onKeyLongPress()方法，必须要在 onKeyDown()方法中对需要监控的按键调用 event.startTracking()方法并返回 true 值。

（3）系统主要的管理方法包括以下几种。

① void addContentView（View view，ViewGroup.LayoutParams params）：将 View 添加到 Activity 的内容视图。参数 view 表示想要显示的视图；参数 params 表示想要显示的视图的布局参数。

② View findViewById（int id）：根据 XML 文件中的 ID 属性，找到相应的视图对象。需要在方法的前面指定视图对象的类型，进行相应的强制类型转换。

③ void finish ()：关闭 Activity。

④ void finishActivity（int requestCode）：强制关闭之前通过 startActivityForResult（Intent，int）方法打开的 Activity。

⑤ View getCurrentFocus()：取得当前焦点的 View。

⑥ Intent getIntent()：获取启动 Activity 的 Intent。

⑦ void setIntent（Intent newIntent）：为 Activity 设置新的 Intent。

⑧ LayoutInflater getLayoutInflater()：获得一个 LayoutInflater 对象，该对象可以解释 XML

布局文件，将其实例化为一个视图对象。

⑨ MenuInflater getMenuInflater()：获得一个 MenuInflater 对象，该对象可以解释菜单 XML 文件，将其实例化为菜单对象。

⑩ setRequestedOrientation（int requestedOrientation）：设置 Activity 显示的方向。参数 requestedOrientation 是常量，在 ActivityInfo.screenOrientation 中定义，如 ActivityInfo.SCREEN_ORIENTATION_LANDSCAPE 表示横屏；ActivityInfo.SCREEN_ORIENTATION_PORTRAIT 表示竖屏。在 AndroidManifest.xml 文件的 activity 中加入 android:screenOrientation="landscape"，表示横屏显示（显示时宽度大于高度）；加入 android:screenOrientation="portrait"表示竖屏显示（显示时高度大于宽度），例如，

```
<activity android:name=".ChangeOrientationDemo"
                  android:label="@string/app_name"
                  android:screenOrientation="portrait">
…
 </activity>
```

⑪ int getRequestedOrientation()：返回 Activity 当前显示的方向。

⑫ Object getSystemService（String name）：根据名字返回系统的服务对象。该方法返回 Object 类型，一般需要根据服务对象进行相应的强制类型转换。

⑬ void setTitle（int titleId）和 setTitle（CharSequence title）：设置 Activity 的标题。

⑭ void setVisible（boolean visible）：控制 Activity 的主窗口是否可见。

⑮ Application getApplication()：获取此 Activity 所属的 Application。

⑯ WindowManager getWindowManager()：获取窗口管理器。

⑰ Window getWindow()：获取 Activity 的当前窗口。

⑱ void onLowMemory()：当整个系统运行在低内存状态时，调用该方法，通过重写该方法来清除缓存或者其他不必要的资源。当方法返回时，系统会进行一次垃圾回收。

⑲ void setDefaultKeyMode（int mode）：用来设置 Activity 默认的按键模式，也就是说当 Activity 中发生了一些按键事件，但是没有任何 Listener 响应的时候就执行 mode 设置的动作。mode 的值可以有 5 种选择。

- DEFAULT_KEYS_DISABLE：关闭按键，相当于丢弃。
- DEFAULT_KEYS_DIALER：将键盘事件传入拨号器进行处理。
- DEFAULT_KEYS_SHORTCUT：将键盘输入作为当前窗体上注册的快捷键，进行快捷键处理。
- DEFAULT_KEYS_SEARCH_LOCAL：将键盘输入作为搜索内容，执行应用程序所定义的搜索。
- DEFAULT_KEYS_SEARCH_GLOBAL：将键盘输入作为搜索内容，进行全局搜索。

⑳ boolean requestWindowFeature（int featureId）：允许扩展窗口的特征，例如，程序全屏显示、自定义标题（使用按钮等控件）。参数 featureId 的取值在 Window 类中定义，常用的有：FEATURE_NO_TITLE、FEATURE_PROGRESS、FEATURE_LEFT_ICON、FEATURE_INDETERMINATE_PROGRESS。

㉑ void setFeatureDrawableResource（int featureId，int resId）：对窗口特征设置图像资源。featureId 在 Window 类中定义，resId 是 res 目录下的图片资源 ID。这个方法应在 requestWindowFeature()方法调用之后才能使用。

㉒ startActivityForResult（Intent intent，int requestCode）：开启一个新的 Activity，并接收

其执行完毕后返回的结果。如果 requestCode 的值小于 0，该方法的作用就和 startActivity(Intent intent) 一样。

㉓ public SharedPreferences getPreferences (int mode)：获得只属于 Activity 的 SharedPerference。

㉔ void setProgressBarVisibility (boolean visible)：设置标题栏中长条形进度条的可见性。ProgressBar 主要有圆形进度条（progressBarStyleLarge）和长条形进度条（progressBarStyle Horizontal）两种形态。圆形进度条通常用于显示未确定何时结束的进度，一直以动画方式沿圆形轨迹旋转。长条形的进度条通常用于显示明确的进度。要想在标题中显示长条形进度条，应先执行 requestWindowFeature (Window.FEATURE_PROGRESS)。

㉕ void setProgressBarIndeterminateVisibility (boolean visible)：设置标题栏中圆形进度条的可见性。要想在标题中显示圆形进度条，应先执行 requestWindowFeature(Window.FEATURE_INDETERMINATE_PROGRESS)。

㉖ void setProgressBarIndeterminate (boolean indeterminate)：设置进度条的进度是否是不确定的。

㉗ void setProgress (int progress)：设置标题栏长条形进度条的当前数值。参数 progress 的取值范围为 0～10000。

㉘ void setSecondaryProgress (int secondaryProgress)：设置第二个进度条的数值。主要进度数值在 setProgress 方法中设置，如在播放网络媒体时，默认进度显示播放进度，第二进度条作为背景显示下载缓冲的进度。

㉙ setVolumeControlStream (int streamType)：设置由手机上音量控制键控制的 Activity 中音频流声音的大小。参数 streamType 的取值在 AudioManager 用静态常量定义了需要调整音量的音频流。

- AudioManager.STREAM_MUSIC：音乐。
- AudioManager.STREAM_RING：铃声。
- AudioManager.STREAM_ALARM：警报。
- AudioManager.STREAM_NOTIFICATION：窗口顶部状态栏通知声。
- AudioManager.STREAM_SYSTEM：系统。
- AudioManager.STREAM_VOICECALL：通话。
- AudioManager.STREAM_DTMF：双音多频。

㉚ void runOnUiThread (Runnable action)：任务 action 将在 UI 线程中运行。

三、任务实施

从标准身高计算公式可看出，计算标准身高需要获得用户的体重和性别，因此在用户界面设计中需要提供用户录入体重和性别的界面。用户的体重要求是数字，所以 EditText 对象需要控制录入范围（必须是数字）。用户的性别要求可以是“男”、“女”或者两者都可以同时计算。

1．创建项目

创建一个 Android Project，项目名称为【HeightCalculator】，包名为【com.demo.pr2】，选择 Android SDK2.2 版本，并创建一个 Activity 名为【HeightCalculatorActivity】。

2．创建用户界面布局

根据项目分析，分 5 个步骤来设置 XML 的布局，如图 2-16 所示。

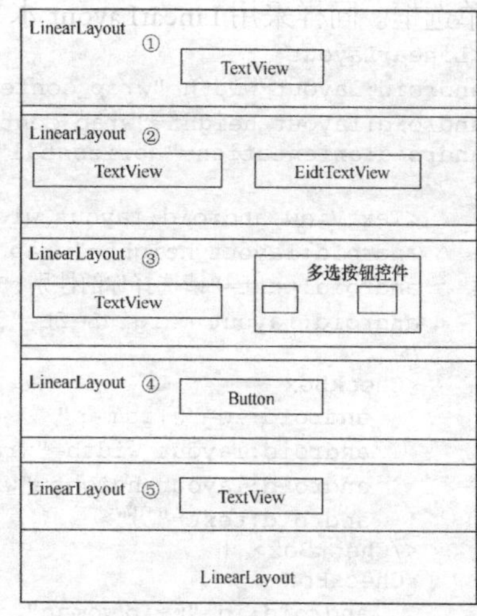

图 2-16 XML 布局图

① 设置项目主题区域。

② 设置体重输入框区域。

③ 设置选择性别（可多选）区域。

④ 设置计算按钮区域。

⑤ 设置结果显示区域。

根据布局图，本任务采用线性布局 LinearLayout。线性布局的好处是提供了控件水平和垂直排列的模型，同时可以通过设置子控件的 weight 布局参数来控制各个控件在布局中的相对大小。下面具体介绍如何实现上面规划的界面布局。

（1）设置项目主题区域。主题“个人标准身高计算器”放在布局的第 1 部分，要求字体居中，大小为 22 像素点，字体样式为 bold 类型，距离顶部 20 像素点。因此选择 LinearLayout 和 TextView 两种 XML 元素进行布局，并将 TextView 作为 LinearLayout 的子元素。

```
<LinearLayout
android:layout_width="fill_parent"              //宽度占满屏幕
android:layout_height="wrap_content"            //高度自适应
android:orientation="horizontal"                //布局内的控件水平分布
android:layout_marginTop="20.0dip"              //距离顶部 20 像素点
android:gravity="center_horizontal">                //水平居中
    <TextView android:layout_width="wrap_content"
    android:layout_height="wrap_content"
    android:text="个人标准身高计算器"
    android:textSize="22.0dip"                  //字体大小为 22.0dip
    android:textStyle="bold"                    //字体样式为 bold
    />
</LinearLayout>
```

（2）设置体重输入框区域。该区域放在布局的第 2 部分，按设计要求，输入框只能输入数字。输入框左右两侧分别显示文字信息“请输入你的体重:”和“kg”，因此采用 LinearLayout 水平分布嵌套两个 Textview 和一个 EditText 子元素。

```
<LinearLayout
  android:layout_width="wrap_content"
  android:layout_height="wrap_content"
  android:orientation="horizontal"
  android:layout_marginTop="10.0dip"                 //距离顶部 10 像素点
  >
      <TextView android:layout_width="120.0dip"           //控件宽度 120 像素点
      android:layout_height="wrap_content"
      android:text="请输入你的体重:"
      android:layout_marginLeft="5.0dip"               //左移动 5 像素点
      />
        <EditText android:layout_width="150.0dip"       //控件宽度 150 像素点
            android:id="@+id/weight"
            android:layout_height="wrap_content"
                android:inputType="number" />            //输入类型为数字类型
    <TextView android:layout_width="wrap_content"     //宽度自适应
              android:layout_height="wrap_content"       //高度自适应
               android:text="kg"
    />
</LinearLayout>
```

（3）设置选择性别区域。该区域放在布局的第 3 部分，按设计要求分别设置男、女性别单选框。同样采用LinearLayout水平布局嵌套RadioGroup和RadioButton元素。

```
<LinearLayout
android:layout_width="wrap_content"
android:layout_height="wrap_content"
android:orientation="horizontal"
>
    <TextView android:layout_width="120.0dip"
    android:layout_height="wrap_content"
    android:text="请选择你的性别:"
    android:layout_marginLeft="5.0dip"
    />
    <CheckBox
        android:id="@+id/man"
        android:layout_width="wrap_content"
        android:layout_height="wrap_content"
        android:text="男">
    </CheckBox>
    <CheckBox
        android:id="@+id/woman"
        android:layout_width="wrap_content"
        android:layout_height="wrap_content"
        android:text="女">
        </check Box>
</LinearLayout>
```

（4）设置计算按钮区域。该区域放在布局的第4部分，代码如下：

```
<LinearLayout
android:layout_width="fill_parent"
android:layout_height="wrap_content"
android:orientation="horizontal"
android:gravity="center_horizontal"
>
   <Button
   android:layout_marginTop="20.0dip"
   android:layout_width="200.0dip"
   android:layout_height="wrap_content"
   android:id="@+id/calculator"
   android:text=" 运算 " />
</LinearLayout>
```

（5）设置结果显示区域。该区域放在布局的第5部分，代码如下：

```
<LinearLayout
android:layout_width="fill_parent"
android:layout_height="wrap_content"
android:orientation="horizontal"
android:gravity="center_horizontal"
>
    <TextView  android:layout_width="wrap_content"
        android:layout_height="wrap_content"
        android:id="@+id/result"
        android:layout_marginTop="10.0dip"/>
</LinearLayout>
```

如何将上面5个水平的LinearLayout布局统一起来呢？同样使用LinearLayout布局。不同的是，这时需要使用垂直布局将5个水平布局组合起来。

```
<LinearLayout
    xmlns:android="http://schemas.android.com/apk/res/android"
android:orientation="vertical"                //垂直布局
android:layout_width="fill_parent"
android:layout_height="fill_parent" >
//将上面5种布局依次内嵌到这里
</LinearLayout>
```

在 Eclipse 中双击上面的布局文件，会显示如图 2-17 所示的项目界面设计效果。Android 提供了可视化的布局编辑器，建议读者在实际的项目开发中通过可视化的布局编辑器来制作应用程序的界面布局，自动生成 XML 布局文件，从而不需要编写复杂的 XML 元素，提高编程的效率。此外，读者也可以到网上下载一个名为 DroidDraw 的软件，它是一个有名的 Android 界面设计器，可以用来方便地生成复杂的 Android Layout XML 文件。

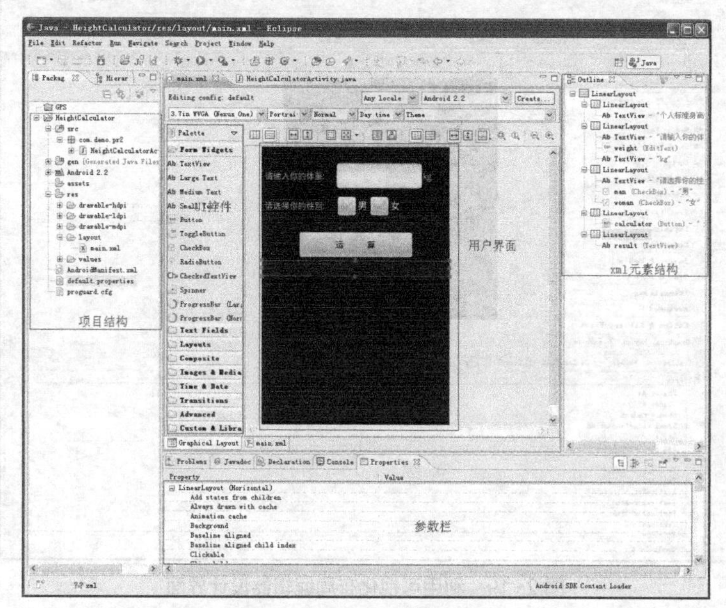

图 2-17　项目界面设计效果

下面简要描述如何利用 Android 的可视化编辑器直接拖曳 UI 控件进行布局。以设置项目主题区域的“个人标准身高计算器”为例，需选择 LinearLayout 和 TextView 两种 UI 控件。

① 从 UI 工具栏 Layouts 下拖曳 LinearLayout UI 控件到布局页面窗口，如图 2-18 所示。

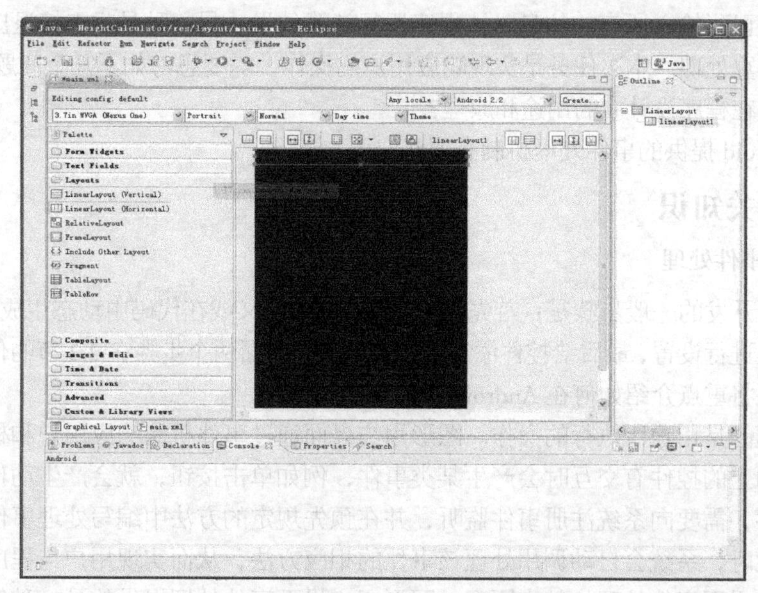

图 2-18 拖曳 LinearLayout 效果图

② 从 UI 工具栏 Form Widgets 下拖曳 TextView UI 控件到布局页面窗口，如图 2-19 所示，作为 LinearLayout 的内嵌控件，并在可视属性参数栏中设置参数。将 Text 属性设置为“个人标准身高计算器”。具体参数设置请查看本章 View 中的相关知识。

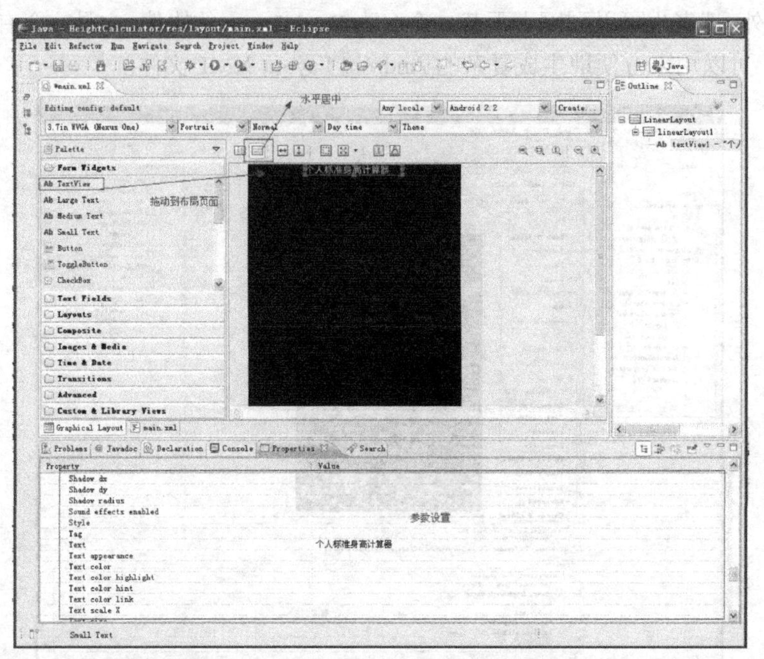

图 2-19　利用可视化布局编辑器设计界面

2.3　进行事件处理

一、任务分析

2.2 节的任务只提供了用户信息的录入界面，在用户录入信息后，有两种可能的操作：一种是告诉程序已经输入所有的信息，让程序运行计算结果；另一种是想直接退出程序，不需要程序进行计算处理。本次任务是实现响应用户的操作，需要思考如下几个问题。

（1）事件在程序设计中的用途和原理？

（2）Android 提供的事件处理机制有哪些？

二、相关知识

1．单击事件处理

用户界面开发的一般步骤是：首先通过 XML 文件定义或在代码中动态生成 UI 控件，然后对控件属性进行设置，最后给控件添加事件监听器。前面两个步骤在 2.2 节的任务中已经做了介绍，下面将重点介绍如何在 Android 中实现事件处理。

事件描述的是用户所执行的操作。图形用户界面通过事件机制响应用户和程序的交互。当用户和界面上的控件有交互时会产生某类事件，例如单击按钮，就会产生动作事件。要处理产生的事件，需要向系统注册事件监听，并在预先规定的方法中编写处理事件的代码。当某种事件发生时，系统会自动调用处理该事件的相应方法，从而实现用户与程序的交互，这就是图形用户界面事件处理的基本原理。图形用户界面事件处理提供的是一种触发响应的交互式机制，增加了程序的灵活性和可扩展性。

Android 的事件处理机制和 J2SE 的事件处理机制很接近，都是 Java 事件（消息）机制实现。Java 有四种事件监听实现方式，分别为自身类作为事件监听器、外部类作为事件监听器、内部类作为事件监听器、匿名内部类作为事件监听器。它们的主要实现步骤是类似的，首先确定需要哪种事件监听方式，选择对应的事件监听接口；然后定义一个类实现该监听接口；最后为需要进行事件监听的控件添加监听器（每个控件都有各自添加监听器的方法）。此监听器为在第二步中实现监听接口的类的对象。表 2-8 为 Android 常见的监听器接口。

表 2-8 常见的监听器接口

接 口 名	接口需要实现的方法	添加监听的方法	用 途
OnClickListener	onClick	setOnClickListener	按钮单击
OnKeyListener	onKey	setOnKeyListener	设备上某个按键被按下或者释放
OnCheckedChangeListener	onCheckedChanged	setOnCheckedChangeListener	单（多）项选择
OnItemSelectedListener	onItemSelected	setOnItemSelectedListener	下拉列表选项选择
TimePicker	onTimeChanged	setOnTimeChangedListener	时间变化

OnClickListener 为处理单击 View 对象事件的接口，是 Android 常用的按钮事件，它需要实现 onClick（View v）方法，其中参数 v 表示被单击的 View 对象。该方法的功能是处理单击 View 对象的操作。SetOnClickListener 方法用于为按钮添加监听器。下面举例说明在 Android 中实现监听的四种方法。

（1）自身类作为事件监听器，实现代码如下。

```
package com.test;
import android.app.Activity;
import android.os.Bundle;
import android.view.View;
import android.view.View.OnClickListener;
import android.widget.Button;
public class ThisClassEvent extends Activity implements OnClickListener
{
    @Override
    protected void onCreate(Bundle savedInstanceState){

        super.onCreate(savedInstanceState);
        //设置页面布局
        setContentView(R.layout.main);
Button button=(Button)findViewById(R.id.button);
        button.setText("单击");
        //注册页面中的 button 按钮事件
        button.setOnClickListener(this);//注册自身类的监听器
    }

    @Override
    public void onClick(View v){
        //在这里写单击触发后的逻辑代码
        Toast.makeText(this, "已单击我...", Toast.LENGTH_SHORT).show();
        System.out.println("已单击我...");
    }
    @Override
    protected void onDestroy() {

        super.onDestroy();
    }
}
```

（2）外部类作为事件监听器，实现代码如下。

```
package com.test;
import android.app.Activity;
import android.os.Bundle;
import android.view.View;
import android.view.View.OnClickListener;
import android.widget.Button;
public class OuterClassEvent extends Activity {

    @Override
    protected void onCreate(Bundle savedInstanceState){

        super.onCreate(savedInstanceState);
        //设置页面布局
        setContentView(R.layout.main);
Button button=(Button)findViewById(R.id.button);
        button.setText("单击");
        //注册页面中的button按钮事件
        button.setOnClickListener(new OuterClass());//注册外部类的监听器
    }
}
//外部类
class  OuterClass implements OnClickListener
{
        @Override
        public void onClick(View v){
            //在这里写单击触发后的逻辑代码
  Toast.makeText(OuterClassEvent.this, "已单击我...", Toast.LENGTH_SHORT).show();
            System.out.println("已单击我...");
        }
}
```

（3）内部类作为事件监听器，实现代码如下。

```
package com.test;
import android.app.Activity;
import android.os.Bundle;
import android.view.View;
import android.view.View.OnClickListener;
import android.widget.Button;
public class InnerClassEvent extends Activity {

    @Override
    protected void onCreate(Bundle savedInstanceState){

        super.onCreate(savedInstanceState);
        //设置页面布局
        setContentView(R.layout.main);
        Button button=(Button)findViewById(R.id.button);
        button.setText("单击");
        //注册页面中的button按钮事件
        button.setOnClickListener(new OuterClass());//注册内部类的监听器
    }
    //内部类
    class  OuterClass implements OnClickListener
    {
        @Override
        public void onClick(View v){
            //在这里写单击触发后的逻辑代码
Toast.makeText(InnerClassEvent.this, "已单击我...", Toast.LENGTH_SHORT).show();
            System.out.println("已单击我...");
        }
    }
}
```

（4）匿名内部类作为事件监听器，实现代码如下。

```
package com.test;
```

```
import android.app.Activity;
import android.os.Bundle;
import android.view.View;
import android.view.View.OnClickListener;
import android.widget.Button;
public class AnonymousEvent extends Activity {

    @Override
    protected void onCreate(Bundle savedInstanceState){

        super.onCreate(savedInstanceState);
        //设置页面布局
        setContentView(R.layout.main);
Button button=(Button)findViewById(R.id.button);
        button.setText("单击");
        //注册页面中的 button 按钮事件
        button.setOnClickListener(new OnClickListener() {//注册匿名内部类的监听器

            @Override
            public void onClick(View v){
                //在这里写单击触发后的逻辑代码
Toast.makeText(AnonymousEvent.this, "已单击我...", Toast.LENGTH_SHORT).show();
                System.out.println("已单击我...");
            }

        });
    }
}
```

除了可以在 Activity 的按钮对象中设置监听事件之外，还可以在按钮对象所在的 XML 布局文件中使用 android:onClick 赋予按钮执行方法，例如，

```
<Button
android:layout_height="wrap_content"
android:layout_width="wrap_content"
android:text="@string/self_destruct"
android:onClick="selfDestruct" />
```

当一个用户单击按钮时，Android 系统将调用 Activity 的 selfDestruct（View）方法。该方法要求必须将 selfDestruct（View）方法定义为 public，并且将 View 对象作为该方法的唯一参数。例如，

```
public void selfDestruct(View view){
    //在此处编写用户单击按钮后,具体的动作执行代码
}
```

2. AlertDialog 提示框

AlertDialog 用于弹出提示框，本身并没有构造函数，不可以直接通过构造函数 new AlertDialog(...)的方式来初始化，首先需要通过 AlertDialog.Builder 来创建一个 Builder，然后有两种方法来完成一个 AlertDialog：一种方法是使用 Builder 对象中的方法定义 AlertDialog 的属性，然后再使用 show()方法来创建并显示 AlertDialog 提示框；另一种方法是先使用 create()方法获得 AlertDialog 对象，然后使用 AlertDialog 对象的方法定义 AlertDialog 属性，最后使用 show()方法显示 AlertDialog 提示框。

AlertDialog.Builder 构造函数有两种：

① AlertDialog.Builder（Context context）;

② AlertDialog.Builder（Context context，int theme）。

其中 Context 类表示应用程序的上下文环境，是一个抽象类。Activity、Service 以及 BroadcastReceiver 类是实现 Context 的子类，因此在 Activity 程序中直接用 this 代表调用的实例为 Activity。如果在内部类中，如在 Button 按钮的 onClick（View view）方法中，直接使用

this 就会报错，这时可以使用 Activity 类名.this 来解决。

AlertDialog 的主要方法及用途介绍如下。

① void dismiss()：将对话框从屏幕上移除。

② void setButton（int whichButton，CharSequence text，DialogInterface.OnClickListener listener）：设置对话框的按钮，一个单选框或复选框列表的对话框可以有 0～3 个按钮。参数 whichButton 表示按钮的类型，在 DialogInterface 接口中定义了 BUTTON_POSITIVE、BUTTON_NEGATIVE 和 BUTTON_NEUTRAL 3 种取值，主要影响按钮的位置；参数 text 表示按钮的显示文字；参数 listener 表示按钮触发的监听器。DialogInterface.OnClickListener 接口需要实现一个方法 onClick（DialogInterface dialog，int which）。

③ void setView（View view）：将视图控件显示在对话框中。参数 view 为某种视图控件。

④ void setMessage（CharSequence message）：设置对话框的显示内容。

⑤ setTitle（CharSequence title）：为对话框设置标题。参数 title 表示标题内容。

⑥ void setIcon（Drawable icon）和 void setIcon（int resId）：为对话框设置图标。参数 icon 表示图像；参数 resId 表示图像资源的 ID，如果取值为 0 表示无需使用图标。

如果希望在对话框中实现更复杂的功能，则可以使用 AlertDialog.Builder 中的方法，具体用法介绍如下。

① AlertDialog.Builder setMultiChoiceItems（CharSequence[] items，boolean[] checkedItems，DialogInterface.OnMultiChoiceClickListener listener）：在对话框中显示多选框。参数 items 表示选项内容数组；参数 checkedItems 用于指出哪些选项默认处于选中状态，取值为 null 则表示没有选项被选中；参数 listener 表示选项被选中的监听器。DialogInterface. OnMultiChoice ClickListener 接口需要实现一个方法 onClick（DialogInterface dialog，int which，boolean isChecked）。

② AlertDialog.Builder setSingleChoiceItems（CharSequence[] items，int checkedItem，DialogInterface.OnClickListener listener）：在对话框中显示单选框。参数 items 表示选项内容数组；参数 checkedItem 用于指出哪个选项默认被选中，取值为-1 表示没有选项被选中；参数 listener 表示选项被选中的监听器。

③ AlertDialog.Builder setItems（CharSequence[] items，DialogInterface. OnClickListener listener）：在对话框中显示列表。参数 items 表示选项内容数组；参数 listener 表示选项被选中的监听器。

下面利用 AlertDialog 创建出 7 种不同风格的对话框，效果如图 2-20 所示。

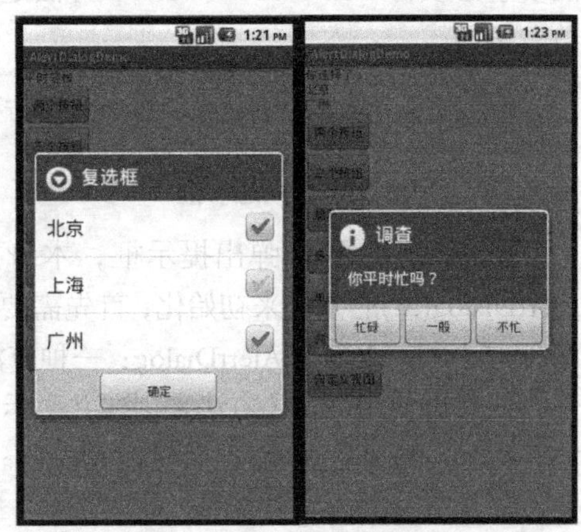

图 2-20 对话框效果图

（1）布局文件 alertdialog.xml。在界面中定义一个文本框用于显示对话框执行的操作。另外定义 7 个按钮，分别用来显示不同类型的对话框。

```
<LinearLayout xmlns:android="http://schemas.android.com/apk/res/android"
    xmlns:tools="http://schemas.android.com/tools"
    android:id="@+id/LinearLayout1"
    android:layout_width="match_parent"
    android:layout_height="match_parent"
    android:orientation="vertical" >
```

```
    <TextView
        android:id="@+id/textView1"
        android:layout_width="wrap_content"
        android:layout_height="wrap_content"
        android:text="TextView" />
    <Button
        android:id="@+id/button1"
        android:layout_width="wrap_content"
        android:layout_height="wrap_content"
        android:text="两个按钮" />

    <Button
        android:id="@+id/button2"
        android:layout_width="wrap_content"
        android:layout_height="wrap_content"
        android:text="三个按钮" />
    <Button
        android:id="@+id/button3"
        android:layout_width="wrap_content"
        android:layout_height="wrap_content"
        android:text="显示视图" />
    <Button
        android:id="@+id/button4"
        android:layout_width="wrap_content"
        android:layout_height="wrap_content"
        android:text="多选列表" />
    <Button
        android:id="@+id/button5"
        android:layout_width="wrap_content"
        android:layout_height="wrap_content"
        android:text="单选列表" />
    <Button
        android:id="@+id/button6"
        android:layout_width="wrap_content"
        android:layout_height="wrap_content"
        android:text="列表数据" />
    <Button
        android:id="@+id/button7"
        android:layout_width="wrap_content"
        android:layout_height="wrap_content"
        android:text="自定义视图" />

</LinearLayout>
```

（2）程序代码如下。

```
public class AlertDialogDemo extends Activity {

    private AlertDialog dialog;
    private AlertDialog.Builder builder;
    private TextView tView;

    @Override
    public void onCreate(Bundle savedInstanceState){
        super.onCreate(savedInstanceState);
        setContentView(R.layout.alertdialog);
        tView =(TextView)this.findViewById(R.id.textView1);
        Button button1 =(Button)this.findViewById(R.id.button1);
        Button button2 =(Button)this.findViewById(R.id.button2);
        Button button3 =(Button)this.findViewById(R.id.button3);
        Button button4 =(Button)this.findViewById(R.id.button4);
        Button button5 =(Button)this.findViewById(R.id.button5);
        Button button6 =(Button)this.findViewById(R.id.button6);
        Button button7 =(Button)this.findViewById(R.id.button7);
        View.OnClickListener listener=new View.OnClickListener() {
            public void onClick(View v){
```

```
                switch(v.getId()){
                case R.id.button1:
                    dialog1();
                    break;
                case R.id.button2:
                    dialog2();
                    break;
                case R.id.button3:
                    dialog3();
                    break;
                case R.id.button4:
                    dialog4();
                    break;
                case R.id.button5:
                    dialog5();
                    break;
                case R.id.button6:
                    dialog6();
                    break;
                case R.id.button7:
                    dialog7();
                    break;
                }
            }
        };
        button1.setOnClickListener(listener);
        button2.setOnClickListener(listener);
        button3.setOnClickListener(listener);
        button4.setOnClickListener(listener);
        button5.setOnClickListener(listener);
        button6.setOnClickListener(listener);
        button7.setOnClickListener(listener);
    }

    public void dialog1() {
        dialog=new AlertDialog.Builder(this).create();
        dialog.setTitle("提示");
        dialog.setMessage("确认退出吗？");
        dialog.setIcon(android.R.drawable.ic_dialog_alert);
        // 创建按键监听器
        DialogInterface.OnClickListener listenter=new DialogInterface.OnClickListener() {
            public void onClick(DialogInterface dialog,int which){
                if(which == DialogInterface.BUTTON_POSITIVE){// 按下确定
                    dialog.dismiss();
                    AlertDialogDemo.this.finish();
                } else if(which == DialogInterface.BUTTON_NEGATIVE){// 按下取消
                    dialog.dismiss();
                }
            }

        };
        dialog.setButton(DialogInterface.BUTTON_POSITIVE,"确定",listenter);
        dialog.setButton(DialogInterface.BUTTON_NEUTRAL,"取消",listenter);
        dialog.show();
    }

    public void dialog2() {
        dialog=new AlertDialog.Builder(this).create();
        dialog.setTitle("调查");
        dialog.setMessage("你平时忙吗？");
        dialog.setIcon(android.R.drawable.ic_dialog_info);
        // 创建按键监听器
        DialogInterface.OnClickListener listenter=new DialogInterface.OnClickListener() {
            public void onClick(DialogInterface dialog,int which){
                String str="";
                switch(which){
                case DialogInterface.BUTTON_POSITIVE:
```

```
                    str="平时很忙";
                    break;
                case DialogInterface.BUTTON_NEUTRAL:
                    str="平时一般";
                    break;
                case DialogInterface.BUTTON_NEGATIVE:
                    str="平时轻松";
                    break;
                }
                tView.setText(str);
            }
        };
        dialog.setButton(DialogInterface.BUTTON_POSITIVE,"忙碌",listenter);
        dialog.setButton(DialogInterface.BUTTON_NEUTRAL,"一般",listenter);
        dialog.setButton(DialogInterface.BUTTON_NEGATIVE,"不忙",listenter);
        dialog.show();
    }

    public void dialog3() {
        dialog=new AlertDialog.Builder(this).create();
        dialog.setTitle("请输入");
        dialog.setMessage("你平时忙吗？");
        dialog.setIcon(android.R.drawable.ic_dialog_info);
        final EditText tEdit=new EditText(this);
        dialog.setView(tEdit);
        // 创建按键监听器
        DialogInterface.OnClickListener listenter=new DialogInterface.OnClickListener() {
            public void onClick(DialogInterface dialog,int which){
                tView.setText("输入的是："+tEdit.getText().toString());
            }

        };
        dialog.setButton(DialogInterface.BUTTON_POSITIVE,"确定",listenter);
        dialog.show();

    }

    public void dialog4() {

        final String item[]=new String[] { "北京","上海","广州" };
        final boolean bSelect[]=new boolean[item.length];
        DialogInterface.OnMultiChoiceClickListener  mListenter=new  DialogInterface.OnMulti
ChoiceClickListener() {
            public void onClick(DialogInterface dialog,int which,
                    boolean isChecked){
                // 用一个数组记录下选择的所有选项
                bSelect[which]=isChecked;
            }
        };
        builder=new AlertDialog.Builder(this);
        builder.setMultiChoiceItems(item,null,mListenter);
        dialog=builder.create();
        dialog.setTitle("复选框");
        DialogInterface.OnClickListener listenter=new DialogInterface.OnClickListener() {
            public void onClick(DialogInterface dialog,int which){
                String str="你选择了：";
                for(int i=0; i < bSelect.length; i++){
                    if(bSelect[i]){
                        str=str+"\n"+item[i];
                    }
                }
                tView.setText(str);
            }
        };
        dialog.setButton(DialogInterface.BUTTON_POSITIVE,"确定",listenter);
        dialog.show();
```

```
}

public void dialog5() {
    final String item[]=new String[] { "北京","上海","广州" };
    final boolean bSelect[]=new boolean[item.length];
    DialogInterface.OnClickListener sListenter=new DialogInterface.OnClickListener() {
        public void onClick(DialogInterface dialog,int which){
            bSelect[which]=true;
        }
    };
    builder=new AlertDialog.Builder(this);
    builder.setSingleChoiceItems(item,-1,sListenter);
    dialog=builder.create();
    dialog.setTitle("单选框");
    DialogInterface.OnClickListener listenter=new DialogInterface.OnClickListener() {
        public void onClick(DialogInterface dialog,int which){
            String str="你选择了：";
            for(int i=0; i < bSelect.length; i++){
                if(bSelect[i]){
                    str=str+"\n"+item[i];
                }
            }
            tView.setText(str);
        }
    };
    dialog.setButton(DialogInterface.BUTTON_POSITIVE,"确定",listenter);
    dialog.show();
}

public void dialog6() {
    final String item[]=new String[] { "北京","上海","广州" };
    final boolean bSelect[]=new boolean[item.length];
    DialogInterface.OnClickListener sListenter=new DialogInterface.OnClickListener() {
        public void onClick(DialogInterface dialog,int which){
            String str="你选择了："+item[which];
            tView.setText(str);
        }
    };
    builder=new AlertDialog.Builder(this);
    builder.setItems(item,sListenter);
    dialog=builder.create();
    dialog.setTitle("列表框");
    DialogInterface.OnClickListener listenter=new DialogInterface.OnClickListener() {
        public void onClick(DialogInterface dialog,int which){
            dialog.dismiss();
        }
    };
    dialog.setButton(DialogInterface.BUTTON_NEGATIVE,"取消",listenter);
    dialog.show();
}

public void dialog7() {
    LayoutInflater inflater=getLayoutInflater();
    View layout=inflater.inflate(R.layout.diydialog,null);
    final EditText tEdit =(EditText)layout.findViewById(R.id.editText1);
    dialog=new AlertDialog.Builder(this).create();
    dialog.setTitle("自定义布局");
    dialog.setView(layout);
    DialogInterface.OnClickListener listenter=new DialogInterface.OnClickListener() {
        public void onClick(DialogInterface dialog,int which){
            tView.setText("输入的是："+tEdit.getText().toString());
        }

    };
    dialog.setButton(DialogInterface.BUTTON_POSITIVE,"确定",listenter);
    dialog.setButton(DialogInterface.BUTTON_NEGATIVE,"取消",listenter);
    dialog.show();
```

```
    }
}
```

三、任务实施

根据 2.2 节给出的标准身高计算公式，应该定义一个方法来实现该公式的业务逻辑。这里将方法的名字命名为 evaluateHeight。该方法有体重和性别两个参数，返回值为 double 类型。

```
private double evaluateHeight(int weight,String sex){
    double height;
    if(sex=="男"){
    height =170-(62- weight)/0.6
    }else{
    height=158-(52- weight)/0.5
    }
    return height;
}
```

当体重输入框没有数据时或者没选择性别时，定义一个 showMessage 方法给予信息提示。

```
private  void  showMessage(String message)
{
    //提示框
    AlertDialog  alert=new AlertDialog.Builder(this).create();
    alert.setTitle("系统信息");
    alert.setMessage(message);
    alert.setButton("确定",new DialogInterface.OnClickListener() {
    public void onClick(DialogInterface dialog,int whichButton){
          可在此方法内编写按下确定按钮的处理代码，但在本项目中不需要编写处理代码
          }
    });
    alert.show();// 显示窗口
}
```

在 Button 按钮的事件处理方法中，首先判断体重输入框的值和性别选择框的值是否为空，若为空则调用 showMessage 以提示信息；否则则传递身高和性别两个参数值，并调用 evaluateHeight 方法。

```
//判断是否已填写体重数据
if(!weightEditText.getText().toString().trim().equals("")){
    //判断是否已选择性别
  if(manCheckBox.isChecked()||womanCheckBox.isChecked())
{
    double weight=0;
    weight=Double.parseDouble(weightEditText.getText().toString());
    StringBuffer  sb=new StringBuffer();//定义一个记录计算结果的变量
    sb.append("------------评估结果----------- \n");//
    if(manCheckBox.isChecked())
    {
        sb.append("男性标准身高：");
        double result=evaluateHeight(weight,"男");//执行运算
        sb.append((int)result+"(厘米)");
    }
    if(womanCheckBox.isChecked())
    {
        sb.append("女性标准身高：");
        double result=evaluateHeight(weight,"女");//执行运算
        sb.append((int)result+"(厘米)");
    }
    //显示结果，在下一节做详细介绍
    String  result=sb.toString();
     .......
}else
{
    showMessage("请选择性别！");
}
}else
```

```
{
    showMessage("请输入体重！");
}
```

2.4 显示计算结果

一、任务分析

本任务需要实现的效果示意图如图 2-21 所示。

图 2-21 身高计算结果

完成了用户信息的录入和处理后，接下来的任务是将程序运行的结果返回给用户，并为用户提供退出操作。要完成本次任务，需要思考如下几个问题。

（1）如何在手机屏幕上显示信息处理结果？

（2）如何退出标准体重计算器程序？

（3）如何制作一个菜单？

二、相关知识

1. 创建菜单

菜单是用户应用程序界面中最常见的元素之一，使用非常频繁。在 Android 系统中，菜单分为三种：选项菜单（OptionsMenu）、上下文菜单（ContextMenu）和子菜单（SubMenu）。

选项菜单通过按下 MENU 键来显示。选项菜单默认在屏幕底部弹出，每个菜单项依次排列在底部，这些菜单项有文字有图标，也被称作图标菜单项（Icon Menus），如图 2-22 所示。选项菜单一般情况下最多显示 2 排，每排 3 个菜单项，如果多于 6 项，从第 6 项开始会被隐藏，并在第 6 项的位置提示 More 信息，单击 More 才出现第 6 项及以后的菜单项，这些菜单项也被称作展开的菜单项。

上下文菜单与 Windows 中的右键弹出菜单类似，即用户长按注册了上下文菜单的视图对象，程序界面将会出现一个提供相关功能的浮动菜单。菜单选项在此不能附带图标，但是可以为其标题指定图标，如图 2-23 所示。

子菜单就是将相同功能的分组进行多级显示的一种菜单，可以添加到选项菜单或上下文菜单的选项中，如图 2-24 所示。

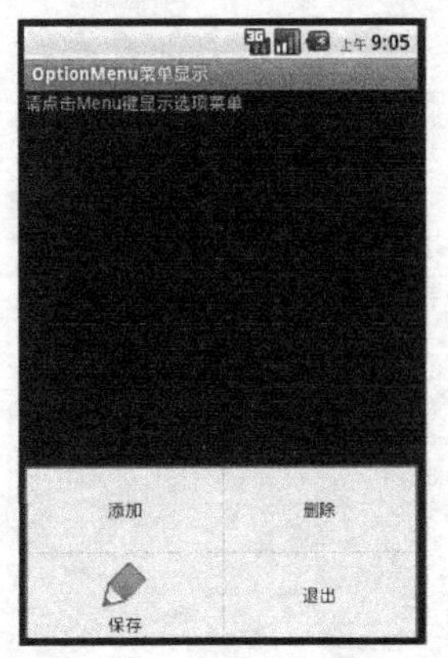

图 2-22 选项菜单示例

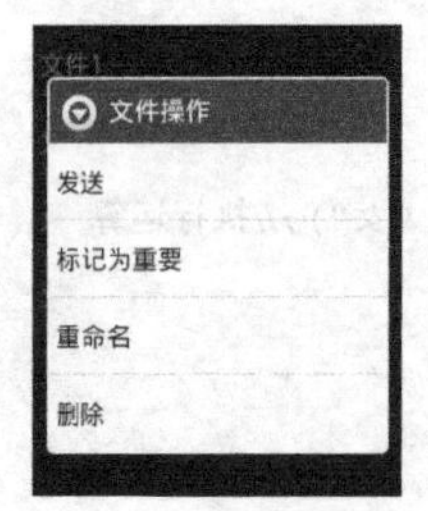

图 2-23 上下文菜单示例

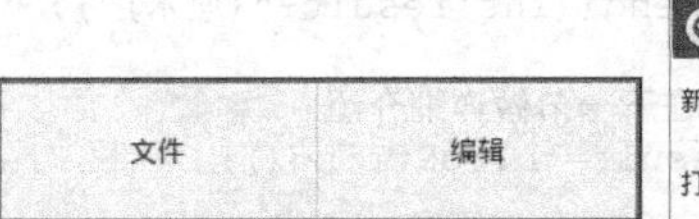

图 2-24 子菜单示例

在程序中最常用的菜单是选项菜单，下面对其进行详细的描述。管理操作 OptionMenu

只需重载 Activity 类中的相应方法。选项菜单的主要方法介绍如下。

① public boolean onCreateOptionsMenu（Menu menu）：用于对 Activity 程序的菜单进行初始化。创建菜单的代码通常放在该方法中，该方法仅在菜单项第一次显示时调用。

② public boolean onOptionsItemSelected（MenuItem item）：用于处理用户选中某个菜单项后执行的动作。

③ public void onOptionsMenuClosed（Menu menu）：用于处理菜单关闭后执行的动作。

④ public boolean onPrepareOptionsMenu（Menu menu）：用于在菜单显示前的相关事件处理，每次菜单显示前都会调用该方法。若想在菜单显示前根据某些条件更新菜单内容，可以重写该方法。例如，将某些菜单项设置为可用或者不可用状态，或者修改其内容。

⑤ public boolean onMenuOpened（int featureId，Menu menu）：用于判断用户打开菜单后发生的动作。

⑥ onOptionsMenuClosed（Menu menu）：用于菜单正在被关闭时发生的动作。

⑦ openOptionsMenu()：通过程序打开选项菜单。选项菜单一般由用户单击键盘上的 MENU 键打开，但如果希望由程序控制来打开菜单，则可以调用该方法。

⑧ closeOptionsMenu()：通过程序关闭选项菜单。

在上面的方法中有 6 个将 Menu 和 MenuItem 用作它们当中的参数。对于 Menu 接口，主要方法有以下几种。

① add 方法：用于添加菜单项，该函数有四个重载方法，返回值都是 MenuItem 类，下面分别介绍。

- add（CharSequence title）：添加菜单项，菜单项上的内容为标题。这种方法添加的菜单项默认以垂直方式排列。

```
public boolean onCreateOptionsMenu(Menu menu){
        menu.add("一个参数的 add 方法创建的菜单项 1");
        menu.add("一个参数的 add 方法创建的菜单项 2");
        return super.onCreateOptionsMenu(menu);
}
```

效果如图 2-25 所示。

- add（int groupId，int itemId，int order，int titleRes）：参数 groudId 是组的标识符，用于表示菜单项属于哪一个组，可批量对菜单项进行状态处理；参数 itemId 是菜单项的标识符，如果不需要表示唯一的标识符，那么可以使用 Menu.NONE；参数 order 用于指定菜单项在选项菜单中的排列顺序，如果不关心其排序，可以使用 Menu.NONE。参数 titlerRes 是菜单项标题字符串的资源标识符，需在 strings.xml 中对其进行定义。如在 strings.xml 文件中添加如下元素：

```
<string name="itemtitle1">菜单项 1</string>
<string name="itemtitle2">菜单项 2</string>
```

在 Activity 程序中添加如下代码：

```
public boolean onCreateOptionsMenu(Menu menu){
    menu.add(1,Menu.NONE,Menu.NONE,R.string.itemtitle1);
    menu.add(1,Menu.NONE,Menu.NONE,R.string.itemtitle2);
    return super.onCreateOptionsMenu(menu);
}
```

效果如图 2-26 所示。

- add（int titleRes）：参数 titleRes 和方法 b 的第 4 个参数相同。
- add（int groupId，int itemId，int order，CharSequence title）：是最常用的添加菜单项方法。前 3 个参数和方法 b 相同；第 4 个参数 title 是 CharSequence 类型，表示菜单项的标题。

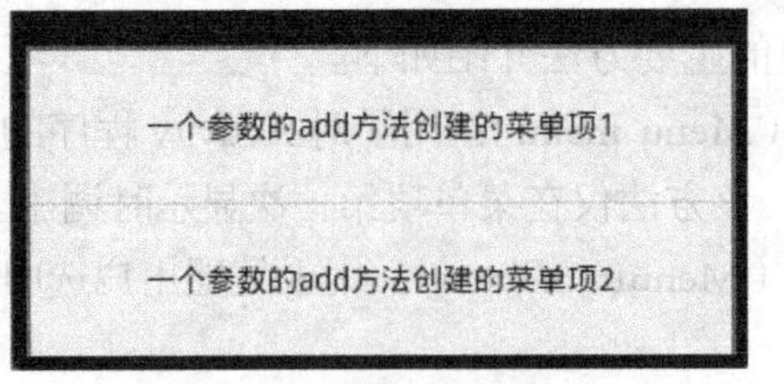

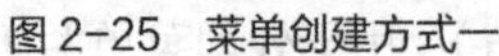

图 2-25　菜单创建方式一

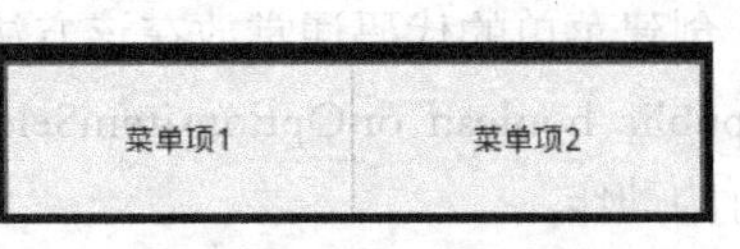

图 2-26　菜单创建方式二

下面的代码可以实现如图 2-22 的效果。

```
public boolean onCreateOptionsMenu(Menu menu){
  menu.add(Menu.NONE,1,Menu.NONE,"添加");
  menu.add(Menu.NONE,2,Menu.NONE,"删除");
  // setIcon()方法为菜单设置图标,这里使用系统自带的图标
  //系统的图标都是以 android.R.开头的,自身项目图标以 R.drawable 开头
  menu.add(Menu.NONE,3,Menu.NONE,"保存").setIcon(
  android.R.drawable.ic_menu_edit);  menu.add(Menu.NONE,4,Menu.NONE,"退出");
return super.onCreateOptionsMenu(menu);
}
```

② findItem（int id）：获得指定 id 的菜单项。

③ getItem（int index）：获得给定索引 index 的菜单项。

④ size()：获得菜单项的数量。

⑤ addSubMenu()：用于创建具有子菜单的菜单项，具有四种重载方法，参数的含义和 add 方法中的参数相同。该方法的返回值是一个 SubMenu 接口，可以继续使用 add 方法创建子菜单项。例如，

```
public boolean onCreateOptionsMenu(Menu menu){
  SubMenu file=menu.addSubMenu("文件");
  SubMenu edit=menu.addSubMenu("编辑");
  file.add(0,1,0,"新建");
  file.add(0,2,0,"打开");
}
```

效果如图 2-27 所示。

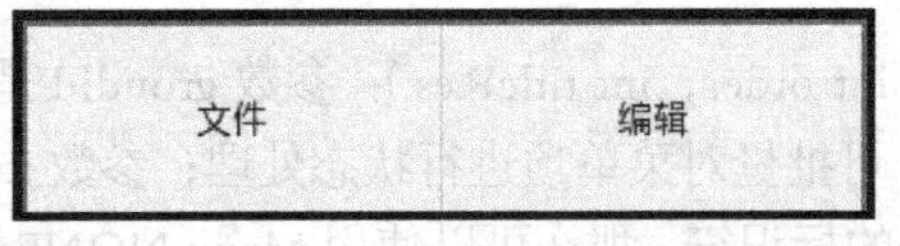

图 2-27　菜单创建方式三

对于 MenuItem 接口实现使用的主要方法如下。

① setEnabled（boolean enabled）：设置菜单项是否可用。参数 enabled 如果取值为 true，则菜单项可用；如果取值为 false，则菜单项不可用。

② setIcon（Drawable icon）：为菜单项设置图标，参数 icon 是类型为 Drawable 的图标。

③ setVisible（boolean visible）：设置菜单项是否可见。参数 visible 如果取值为 true，则菜单项可见；如果取值为 false，则菜单项不可见。

④ isEnabled()：判断菜单项是否可用。

⑤ isVisible()：判断菜单项是否可见。

对于上下文菜单，Activity 也提供了一组处理方法，如下。

① public void onCreateContextMenu（ContextMenu menu，View v，ContextMenuInfo menuInfo）：创建上下文菜单。和选项菜单不同，每次上下文菜单要显示的时候都会调用该方法。

② public void registerForContextMenu（View view）：为某个 View 注册上下文菜单。多个视图可以显示同一个上下文菜单。

③ public void unregisterForContextMenu（View view）：阻止在 View 上显示上下文菜单

④ public void openContextMenu（View view）：通过程序打开上下文菜单。

⑤ public void closeContextMenu()：通过程序关闭上下文菜单。

下面给出一个显示选项菜单和上下文菜单的示例，实现效果如图 2-28 所示。

图 2-28　组合菜单示例

（1）布局文件 menu.xml 代码如下。

```
<LinearLayout xmlns:android="http://schemas.android.com/apk/res/android"
    xmlns:tools="http://schemas.android.com/tools"
    android:id="@+id/LinearLayout1"
    android:layout_width="match_parent"
    android:layout_height="match_parent"
    android:orientation="vertical" >
    <EditText
        android:id="@+id/editText1"
        android:layout_width="match_parent"
        android:layout_height="wrap_content"
        android:layout_marginTop="18dp"
        android:ems="10" >
        <requestFocus />
    </EditText>
    <EditText
        android:id="@+id/editText2"
        android:layout_width="match_parent"
        android:layout_height="wrap_content"
        android:ems="10" />
</LinearLayout>
```

（2）主程序代码如下。

```
public class MenuDemo extends Activity {
private EditText tEdit1;
private EditText tEdit2;
    @Override
    public void onCreate(Bundle savedInstanceState){
        super.onCreate(savedInstanceState);
        setContentView(R.layout.menu);
        tEdit1=(EditText)this.findViewById(R.id.editText1);
        tEdit2=(EditText)this.findViewById(R.id.editText2);
        registerForContextMenu(tEdit1);
        registerForContextMenu(tEdit2);
    }

    @Override
public void onCreateContextMenu(ContextMenu menu,View v,
            ContextMenuInfo menuInfo){
        switch(v.getId()){
            case R.id.editText1:
          menu.add(0,1,0,"菜单项 1");
          menu.add(0,2,0,"菜单项 2");
          menu.add(0,3,0,"菜单项 3");
          break;
      case R.id.editText2:
          menu.add(0,4,0,"菜单项 4");
```

```
        menu.add(0,5,0,"菜单项 5");
        break;
    }
    super.onCreateContextMenu(menu,v,menuInfo);
}

@Override
    public boolean onCreateOptionsMenu(Menu menu){
        menu.add(Menu.NONE,1,Menu.NONE,"添加");
        menu.add(Menu.NONE,2,Menu.NONE,"删除");
        // setIcon()方法为菜单设置图标,这里使用系统自带的图标
        //系统的图标都是以 android.R.开头的,自身项目的图标以 R.drawable 开头
        menu.add(Menu.NONE,3,Menu.NONE,"保存").setIcon(
        android.R.drawable.ic_menu_edit);
        menu.add(Menu.NONE,4,Menu.NONE,"退出");
        return true;
}

}
```

2. 菜单单击事件

在上一节介绍的 onCreateOptionsMenu 方法只能在程序中显示菜单,但单击菜单后执行的操作还没有实现。Activity 提供的 onOptionsItemSelected（MenuItem item）方法则用于处理菜单项单击事件。这个方法将创建一个参数 MenuItem 对象，表示被选中的菜单项，通过调用 getItemId()可以获得相应 ID（该 ID 值用在 add 方法创建菜单项的时候已经指定），这样就可以判断选中的菜单项。下面是单击菜单示例，可添加到上一节的示例代码中。

```
public boolean onOptionsItemSelected(MenuItem item){
        // TODO Auto-generated method stub
        switch(item.getItemId()){
        case 1:
            setTitle("菜单添加事件！");
            break;
        case 2:
            setTitle("菜单删除事件！");
            break;
        case 3:
            setTitle("菜单保存事件！");
            break;
        case 4:
            setTitle("菜单退出事件！");
            break;
        }
        return true;
    }
```

对于上下文菜单，可采用 onContextItemSelected（MenuItem item）方法响应菜单项的单击，在方法的实现上与选项菜单类似。

```
public boolean onContextItemSelected(MenuItem item){
        switch(item.getItemId()){
        case 1:
        case 2:
        case 3:
            setTitle(item.getTitle()+"被按下");
                break;
    case 4:
        case 5:
            setTitle(item.getTitle()+"被按下");
            break;
        }
        return true;
    }
```

三、任务实施

2.3 节介绍了从 UI 控件读取用户输入数据的方法，下面介绍如何改变用户 UI 控件的值。与读取数据类似，UI 控件有相应的赋值方法，如 TextView 类就是使用 setText 方法来设置文本内容的。下面代码是放在单击按钮事件中，向一个 TextView 控件写入结果。

```
String  result=sb.toString();//sb是2.3节中定义的一个记录计算结果变量
TextView  resultTextView=(TextView)findViewById(R.id.result);
resultTextView.setText(result);//使用setText方法向UI控件设置文本内容
```

接着创建一个退出软件功能的菜单。

```
//重载onCreateOptionsMenu(Menu menu)方法,并在此方法中添加菜单项
public boolean onCreateOptionsMenu(Menu menu){
    menu.add(Menu.NONE,1,Menu.NONE,"退出");
        return super.onCreateOptionsMenu(menu);
}
//重载onOptionsItemSelected(MenuItem item)方法,为菜单项注册处理事件
public boolean onOptionsItemSelected(MenuItem menuItem){
      switch(menuItem.getItemId()){
              case 1://退出软件
                int pid=android.os.Process.myPid();//返回当前进程的ID
                android.os.Process.killProcess(pid);//根据给定的进程ID,终止进程
                break;
        }
        super.onOptionsItemSelected(menuItem);
}
```

2.5 发布到手机

一、任务分析

程序员通常在计算机上开发 Android 程序，并使用手机模拟器进行调试。开发完毕后，需要将程序发布到手机上进行测试。最终版测试通过的 Android 程序才能交付给普通用户使用。发布 Android 程序，需先对要发布的程序进行打包，然后通过数据线、WiFi、蓝牙等传输到手机上进行安装。要完成本次任务，需要思考以下两个问题。

（1）如何对 Android 程序进行打包？

（2）在手机上运行 Android 程序，需要哪些文件？

二、任务实施

APK（Android Package Kit）文件是 Android 安装包，其本质是 ZIP 格式的文件。将其后缀改为 zip，然后再对文件进行解压，从中可找到名为 classes 的 DEX（Dalvik VM executes）文件，该文件就是 Android Dalvik 执行程序，使用 Dalvik 字节码表示。此外，APK 里面还包含：存放签名信息的 META-INF 目录、存放资源文件的 res 目录、程序全局配置文件 AndroidManifest.xml，编译后的二进制资源文件 rcsources.arsc。

在 Android 系统中，数字签名用于标识应用程序的作者和在应用程序之间建立信任关系，而非用于决定用户可否安装该应用程序。此项签名由应用程序的作者完成，并不需要权威的数字证书签名机构认证，只是用来让应用程序包自我认证。

在编译完成的 Android 项目的 bin 目录下，Eclipse 会自动生成一个已签名的 APK 文件。当程序代码有改动时文件会重新自动编译。开发者也可以自己创建数字签名文件，方法是右键选中项目，此时弹出 Andoid Tools 菜单，然后单击 Export 子菜单导出签名的或未签名的 APK。

下面演示创建数字签名 APK 文件的步骤。

① 选中项目，然后用鼠标右键单击，选中【Android Tools】→【Export Signed Application Package...】命令，如图 2-29 所示。

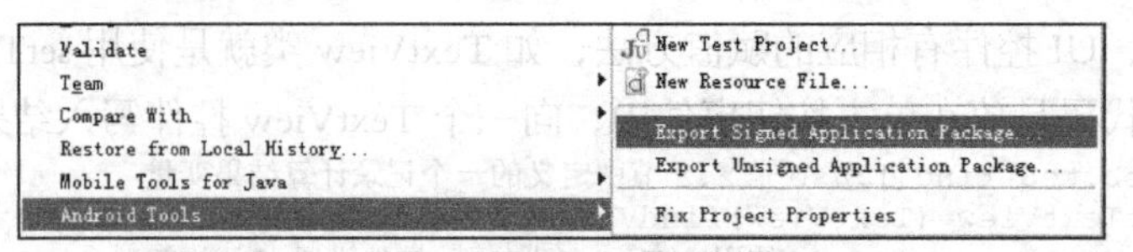

图 2-29　选择导出签名包

② 确定需要导出的项目名称，如图 2-30 所示。

③ 选择密钥库。有两种选项：一种是选择已经存在的密钥库，另一种是重新创建一个密钥库。用户在首次导出签名包时，一般情况下并不存在密钥库。若要新创建密钥库，则需要输入密钥库的存放位置和名字，并设置 6 位密码，如图 2-31 所示。

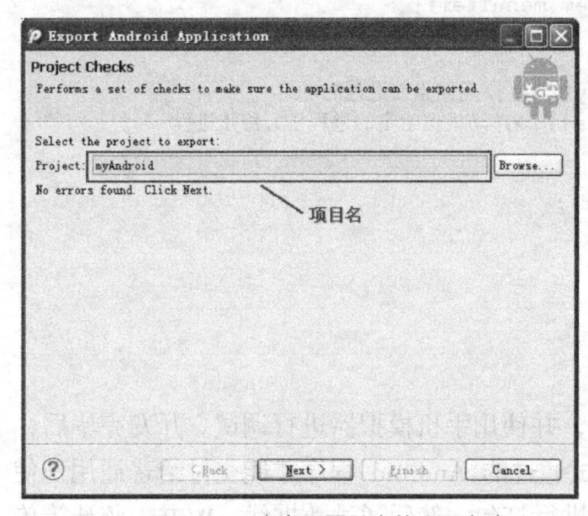

图 2-30　确定需要导出的项目名称

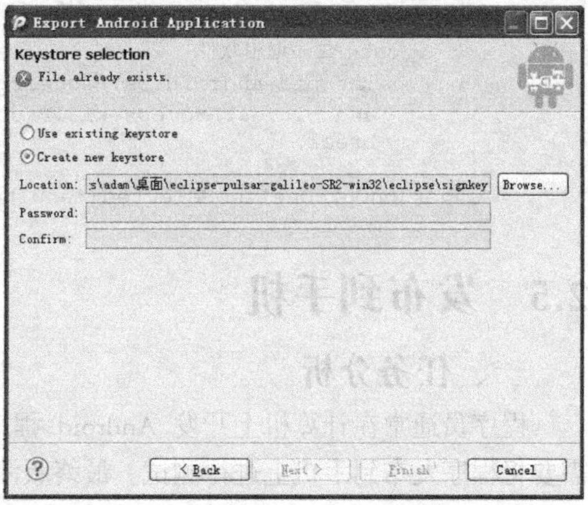

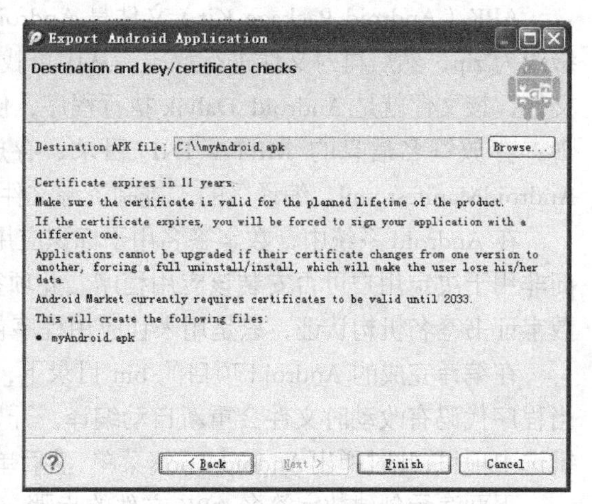

图 2-31　选择密钥库

④ 填写密钥的相关信息，分别是：别名、密码、确认密码、有效期（年）、姓名、组织单元、组织、城市、省、国家代码，如图 2-32 所示。

⑤ 确认需要签名的 APK 文件，以及签名后所存放的目录，并单击【Finish】按钮完成操作，如图 2-33 所示。

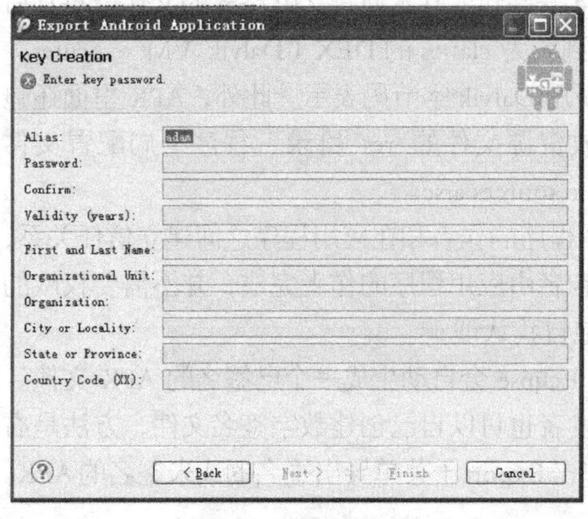

图 2-32 填写密钥信息

图 2-33　目标 APK

生成 APK 文件后，可以通过数据线将其复制到手机上，然后在手机上单击该 APK 文件，即可以完成软件在手机上的安装。也可以借助于豌豆夹、360 手机助手、腾讯手机管家等软件安装和卸载手机应用程序。

2.6　完整项目实施

标准个人身高计算器由一个 main.xml 文件和 HeightCalculatorActivity 类实现,其完整代码如下。

（1）标准个人身高程序界面布局 XML：【main.xml】。

```
<?xml version="1.0" encoding="utf-8"?>
<LinearLayout
xmlns:android="http://schemas.android.com/apk/res/android"
    android:orientation="vertical"
    android:layout_width="fill_parent"
    android:layout_height="fill_parent" >
    <!-- 标题 -->
    <LinearLayout
    android:layout_width="fill_parent"
    android:layout_height="wrap_content"
    android:orientation="horizontal"
    android:gravity="center_horizontal">
    <TextView android:layout_width="wrap_content"
    android:layout_height="wrap_content"
    android:text="个人标准身高计算器"
     android:textSize="22.0dip"
    android:textStyle="bold"
    android:layout_marginTop="20.0dip"/>
</LinearLayout>

<!--体重 输入框 -->
<LinearLayout
    android:layout_width="wrap_content"
    android:layout_height="wrap_content"
    android:orientation="horizontal"
    android:layout_marginTop="10.0dip"
    >
    <TextView android:layout_width="120.0dip"
    android:layout_height="wrap_content"
    android:text="请输入你的体重:"
    android:layout_marginLeft="5.0dip"
    />
      <EditText android:layout_width="150.0dip"
        android:id="@+id/weight"
        android:layout_height="wrap_content"
        android:inputType="number" />
    <TextView android:layout_width="wrap_content"
            android:layout_height="wrap_content"
             android:text="kg"
    />
   </LinearLayout>
   <!--性别选择控件-->
     <LinearLayout
    android:layout_width="wrap_content"
    android:layout_height="wrap_content"
    android:orientation="horizontal"
    >
    <TextView android:layout_width="120.0dip"
    android:layout_height="wrap_content"
    android:text="请选择你的性别:"
```

```
android:layout_marginLeft="5.0dip"
/>
 <CheckBox
        android:id="@+id/man"
        android:layout_width="wrap_content"
        android:layout_height="wrap_content"
         android:orientation="vertical">
 <RadioButton android:id="@+id/man"
      android:layout_width="wrap_content"
      android:layout_height="wrap_content"
      android:checked="true"
      android:text="男"></RadioButton>
 <RadioButton android:id="@+id/woman"
      android:layout_width="wrap_content"
      android:layout_height="wrap_content"
      android:text="女"></RadioButton>
</RadioGroup>
  </LinearLayout>

  <!-- 计算事件按钮 -->
  <LinearLayout
   android:layout_width="fill_parent"
   android:layout_height="wrap_content"
   android:orientation="horizontal"
   android:gravity="center_horizontal"
   >
   <Button
   android:layout_marginTop="20.0dip"
   android:layout_width="200.0dip"
   android:layout_height="wrap_content"
   android:id="@+id/calculator"
   android:text=" 运          算 " />
  </LinearLayout>

   <!-- 结果显示 -->
   <LinearLayout
   android:layout_width="fill_parent"
   android:layout_height="wrap_content"
   android:orientation="horizontal"
   android:gravity="center_horizontal"
   >
  <TextView  android:layout_width="wrap_content"
          android:layout_height="wrap_content"
          android:id="@+id/result"
          android:layout_marginTop="10.0dip"/>
  </LinearLayout>
</LinearLayout>
```

（2）标准个人身高程序实现类:【HeightCalculatorActivity.java】。

```
package com.demo.pr2;
import android.app.Activity;
import android.app.AlertDialog;
import android.content.DialogInterface;
import android.os.Bundle;
import android.view.Menu;
import android.view.MenuItem;
import android.view.View;
import android.view.View.OnClickListener;
import android.widget.Button;
import android.widget.CheckBox;
import android.widget.EditText;
import android.widget.TextView;
/********************(C)COPYRIGHT 2012********************
 * 项目名称：个人标准身高计算器。
```

```
 * 项目目标：
 *(1)开发输入界面。
 *(2)进行事件处理。
 *(3)处理计算结果。
 *(4)发布到手机。
*******************************************************************/
public class HeightCalculatorActivity extends Activity {
    /** Called when the activity is first created. */
    //计算按钮
    private  Button  calculatorButton;
    //体重输入框
    private  EditText weightEditText;
    //男性选择框
    private  CheckBox manCheckBox;
    //女性选择框
    private  CheckBox womanCheckBox;
    //显示结果
    private  TextView resultTextView;

    @Override
    public void onCreate(Bundle savedInstanceState){
        super.onCreate(savedInstanceState);
        //设置页面布局
        setContentView(R.layout.main);
        //从 main.xml 页面布局中获得对应的 UI 控件
        calculatorButton=(Button)findViewById(R.id.calculator);
        weightEditText=(EditText)findViewById(R.id.weight);
        manCheckBox=(CheckBox)findViewById(R.id.man);
        womanCheckBox=(CheckBox)findViewById(R.id.woman);
        resultTextView=(TextView)findViewById(R.id.result);
    }
@Override
protected void onStart() {
     super.onStart();
     //注册事件
     registerEvent();
}
/**
* 注册事件
*/
private  void  registerEvent()
{
     //注册按钮事件
     calculatorButton.setOnClickListener(new OnClickListener() {
         @Override
         public void onClick(View v){
         //判断是否已填写体重数据
         if(!weightEditText.getText().toString().trim().equals(""))
         {
                  //判断是否已选择性别
              if(manCheckBox.isChecked()||womanCheckBox.isChecked())
                  {
                        Double weight=Double.parseDouble(weightEditText
                        .getText().toString());
                          StringBuffer  sb=new StringBuffer();
                          sb.append("------评估结果----- \n");
                          if(manCheckBox.isChecked())
                          {
                               sb.append("男性标准身高：");
                               //执行运算
                               double result=evaluateHeight(weight,"男");
                               sb.append((int)result+"(厘米)");
                          }
                          if(womanCheckBox.isChecked())
                          {
```

```
                                sb.append("女性标准身高：");
                                //执行运算
                                double result=evaluateHeight(weight,"女");
                                sb.append((int)result+"(厘米)");
                            }
                            //输出页面显示结果
                            resultTextView.setText(sb.toString());
                    }else
                    {
                        showMessage("请选择性别！");
                    }
            }else
            {
                showMessage("请输入体重！");
            }
          }
    });
 }
 /**
 *计算处理执行代码事件
 */
 private double  evaluateHeight(double weight,String sex)
 {
  double height;
  if(sex=="男"){
       height=170-(62-weight)/0.6;
  }else{
      height =158-(52-weight)/0.5;
      }
   return height;
 }
 /**
 * 消息提示
 * @param message
 */
 private  void  showMessage(String message)
 {
     //提示框
     AlertDialog  alert=new AlertDialog.Builder(this).create();
     alert.setTitle("系统信息");
     alert.setMessage(message);
     alert.setButton("确定",new DialogInterface.OnClickListener() {
      public void onClick(DialogInterface dialog,int whichButton){
              //可在此方法内编写按下确定按钮的处理代码，但在本项目中不需要编写处理代码
          }
     });
     alert.show();// 显示窗口
 }
 **
 * 创建菜单
 */
 public boolean onCreateOptionsMenu(Menu menu){
     menu.add(Menu.NONE,1,Menu.NONE,"退出");
     return super.onCreateOptionsMenu(menu);
 }
/**
 * 菜单事件
 */
 public boolean onOptionsItemSelected(MenuItem item){
  // TODO Auto-generated method stub
  switch(item.getItemId()){
     case 1://退出
        finish();
        break;
     }
```

```
        return super.onOptionsItemSelected(item);
    }
}
```

2.7 实训项目

一、用户登录界面

1．实训目的与要求

掌握用户界面中文本框和按钮事件的使用。

2．实训内容

（1）编写登录界面，提供账号和密码两个文本输入框。

（2）在界面上提供【退出】和【提交】按钮：

按下【退出】按钮，整个应用程序【退出】；

按下【提交】按钮，如果账号输入为test，密码为123，则在界面上显示“登录成功”，否则显示“登录失败”。

3．思考

① 如何将输入框在手机屏幕上居中显示？

② 如果比较两个字符串是否相等？

二、调查问卷程序

1．实训目的与要求

学会利用XML布局技术开发手机应用程序，掌握在Android中使用菜单功能。

2．实训内容

（1）编写调查问卷界面，信息包括：姓名、出身日期（使用日期类）、性别、院系（使用下拉列表框）、电话号码、对饭堂是否满意、建议等。

（2）在界面上提供【退出】和【提交】按钮。

按下【退出】按钮，则退出整个应用程序。

按下【提交】按钮，有两种情况：如果用户没有写全录入信息，则打开一个信息提示框，提示没有录入的信息；如果用户所有信息录入完整，则将界面中用户录入的信息用一个TextView类显示出来，表示该用户填写的调查问卷结果。

3．思考

（1）如何通过XML文件来定义菜单？

（2）如何利用其他布局方式（例如，相对布局）来实现本任务？

PART3 项目三 开发手机通讯录

本项目工作情景的目标是让学生掌握 Android 的数据管理技术。主要的工作任务划分为①添加联系人记录、②修改联系人记录、③查找号码记录、④查看联系人记录、⑤删除号码记录、⑥对外共享数据、⑦设计主界面。本项目主要涉及的关键技术包括：SQLite 数据库的使用、Activity 类的切换、ListView 和 Adapter 结合显示数据、ContentProvider 共享数据等。

3.1 Android 的数据存储技术

无论使用何种平台或开发环境，也不管开发哪种类型的应用程序，数据处理都是核心。一个对数据存储有良好支持的开发平台将会对应用程序的开发发挥良好的促进作用。

应用程序的数据存储方式主要分为 3 类：文件存储、数据库存储和网络存储。其中文件存储可以自行定义数据格式，使用较为灵活；数据库存储在管理大量数据时较为方便，性能较高，不仅能够对数据进行查询、删除、增加、修改操作，还可以进行加密、加锁、跨应用和跨平台等操作；网络存储则用于实时数据的处理，如在运输、科研、勘探、航空、移动办公等场景下，实时将采集到的数据通过网络传输到数据处理中心。

对 Android 平台而言，其存储方式同样是包括文件存储、数据库存储和网络存储 3 种。但从开发者具体使用的角度来划分，有如下的 5 种。

① 使用 SharedPreferences 存储数据：通过 XML 文件将一些简单的配置信息存储到设备中。只能在同一个包内使用，不能在不同的包之间使用。

② 文件存储数据：在 Android 中读取/写入文件，与 Java 中实现 I/O 的程序完全一样，提供了 openFileInput()和 openFileOutput()方法来写入和读取设备上的文件。

③ SQLite 数据库存储数据：SQLite 是 Android 自带的一个标准数据库，支持 SQL 语句，是一个轻量级的嵌入式数据库（将在本项目中做重点介绍）。

④ 使用 ContentProvider 存储数据：主要用于在应用程序之间进行数据交换，从而能够让其他应用读取或者保存某个 ContentProvider 的各种数据类型。

⑤ Internet 网络存储数据：通过网络提供的存储空间来上传（存储）和下载（获取）在网络空间中的数据。

一、使用 SharedPreferences 存储数据

SharedPreferences 方式是 Android 读取外部数据最简单的方法，常用于将用户个性化设置的字体、颜色、位置等参数信息保存在文件中，适用于存储数据量较少的场合。这种方式用来存储关键字（Key）和值（Value）成对映射的数据结构（key，value），其中 value 值只能是 int、long、boolean、string 和 float 五种基本数据类型。事实上该方式相当于一个 HashMap，不同之处在于 HashMap 的 value 值可以是任何对象，而 SharedPreferences 中的值只能存储基本数据类型。使用 SharedPreferences 存储数据主要应用 SharedPreferences 和 SharedPreferences.Editor 接口，下面分别进行介绍。

SharedPreferences 接口的主要方法如下。

① boolean contains（String key）：检查在数据文件中是否已经包含有关键字 key 的数据。

② SharedPreferences.Editor edit()：获取 Editor 对象。Editor 对象的作用是存储 key-value 键值对数据，可对数据进行修改，并确保参数值在提交存储后保持状态一致。

③ boolean getBoolean（String key，boolean defValue）：从参数数据中检索出关键字 key 所对应的 boolean 数值类型。参数 key 为需要检索的关键字；参数 defValue 为在数据文件中查找不到相应的关键字时返回的默认数值。

④ float getFloat(String key，float defValue)：从数据文件中检索出关键字 key 所对应的 float 数值类型。

⑤ int getInt（String key，int defValue）：从数据文件中检索出关键字 key 所对应的 int 类型数值。

⑥ long getLong(String key，long defValue)：从数据文件中检索出关键字 key 所对应的 long 类型数值。

⑦ String getString（String key，String defValue）：从数据文件中检索出关键字 key 所对应的 String 类型的数值。

SharedPreferences.Editor 接口的主要方法如下。

① void apply()：将 Editor 中对参数进行修改的数据向 SharedPreferences 对象提交。

② SharedPreferences.Editor clear()：删除 SharedPreferences 对象中的所有数据。

③ boolean commit()：将 Editor 中对参数进行修改的数据向 SharedPreferences 对象提交。

④ SharedPreferences.Editor putBoolean（String key，boolean value）：对关键字 key 设置 boolean 类型的数值 value。

⑤ SharedPreferences.Editor putFloat（String key，float value）：对关键字 key 设置 float 类型的数值 value。

⑥ SharedPreferences.Editor putInt（String key，int value）：对关键字 key 设置 int 类型的数值 value。

⑦ SharedPreferences.Editor putLong（String key，long value）：对关键字 key 设置 long 类型的数值 value。

⑧ SharedPreferences.Editor putString（String key，String value）：对关键字 key 设置 String 类型的数值 value。

⑨ SharedPreferences.Editor remove（String key）：从参数中删除关键字为 key 的数值。

下面举例说明实现通过 SharedPreferences 对登录用户的用户名与密码进行读取和写入的两个方法。可以将这两个方法放入两个菜单选项中进行调用测试，也可以放在项目二实训项

目一的用户登陆界面中进行测试：用户首次登陆时，保存其密码和账号；在下一次登陆时，则检查输入的账号、密码是否与之前保存的账号、密码相匹配。

```
void ReadSharedPreferences(){
    String strName,strPassword;
    SharedPreferences  user=getSharedPreferences("user",0);
    strName=user.getString("NAME","");
    strPassword=user.getString("PASSWORD","");
}

void WriteSharedPreferences(String strName,String strPassword){
    SharedPreferences  user=getSharedPreferences("user",0);
    Editor editor=user.edit();
    editor.putString("NAME",strName);
    editor.putString("PASSWORD",strPassword);
    editor.commit();
}
```

代码分析如下。

代码中的 getSharedPreferences（String name，int mode）方法是 Activity 类的方法，用于获取 SharedPreferences 对象。其中，参数 name 表示保存参数信息的文件名，如果该名字的文件不存在，则在执行提交操作后自动新建；参数 mode 表示操作模式，体现了对文件操作的权限，其值在 Context 类中定义。

- MODE_PRIVATE 或者 0：是默认操作，表示文件为私有数据，只能被调用的应用程序访问。在该模式下，写入的内容会覆盖原文件的内容。
- MODE_APPEND 或者 32768：为追加模式，会检查文件是否存在，存在就向文件中追加内容，否则就创建新文件。
- MODE_WORLD_READABLE 或者 1：表示创建的文件可以被其他应用读取。
- MODE_WORLD_WRITEABLE 或者 2：表示创建的文件可以被其他应用写入。

如果希望创建的文件既可以被其他应用读取又可以被其他应用写入，则：

```
int mode= Context.MODE_WORLD_READABLE+Context.MODE_WORLD_WRITEABLE;
SharedPreferences sharedPreferences=context.getSharedPreferences("color",mode);
```

如果要读取配置文件信息，只需要直接使用 SharedPreferences 对象的 getXXX()方法即可。而如果要写入配置信息，则必须先调用 SharedPreferences 对象的 edit()方法，使其处于可编辑状态，然后再调用 putXXX()方法写入配置信息，最后调用 commit()方法提交更改后的配置文件。实际上，SharedPreferences 采用 XML 格式将数据存储到设备中，可通过 DDMS 中的 File Explorer 在“/data/data/<package name>/shares_prefs”下进行查看。以上述数据存储结果为例，可以看到一个名为 user.xml 的文件，打开后有如下数据：

```
<?xml version="1.0" encoding="UTF-8"?>
<map>
  <string name="NAME">test </string>
  <string name="PASSWORD">123 </string>
</map>
```

二、文件存储数据

Android 文件读取或者写入介质分两类：一类是 ROM（只读内存）文件存储，另一类是 SD 卡文件存储。两者之间的差别在于，SD 卡存储的内容可以共享给任何应用程序，而 ROM 存储数据则由应用程序本身独有，其他应用程序无法访问。

1．ROM 存储数据

使用 Activity 类的 openFileInput()和 openFileOutput()方法来操作设备上的文件，创建的文件默认存放在“/data/data/<package name>/files”目录下，如在包名为【com.company.business】

的程序中创建一个【date.txt】文件，存放路径将是【/data/data/com.company.business/files/date.txt】。在默认状态下，文件不能在不同的程序之间共享，这两个方法只支持读取该应用目录下的文件，若读取非自身目录下的文件将会抛出 FileNotFoundException 异常。

FileInputStream openFileInput(String name): 读取名字为 name 的文件，返回 FileInputStream 流。因为文件存放在默认的路径下，因此不需要在参数 name 中指定文件的所在路径。

FileOutputStream openFileOutput(String name, int mode): 向名字为 name 的文件中写入数据，返回 FileOutputStream 流。如果文件不存在，Android 将会自动创建。和 getSharedPreferences 方法的 mode 参数类似，这里的参数 mode 取值为 0 或者 MODE_PRIVATE 时，表示所创建的文件为私有数据，只能被调用的程序存取；取值为 MODE_APPEND 时，表示将数据追加到已有的文件中；取值为 MODE_WORLD_READABLE 或 MODE_WORLD_WRITEABLE 时，可以控制其他的应用程序是否有读取或者写入文件的权限。

关键的实现代码如下，建议将这两个方法放入两个菜单选项中进行调用，以查看程序效果。

（1）读取文件代码。

```
public void readFile(String sFileNmae) {
     try {
         FileInputStream fis = openFileInput(sFileNmae);
         // 将字节流转换成字符流
         InputStreamReader inReader = new InputStreamReader(fis);
         // 转换成带缓存的 bufferedReader
         BufferedReader bufferedReader = new BufferedReader(inReader);
         String s;
         while ((s = bufferedReader.readLine()) != null) {
             Toast.makeText(this, s, Toast.LENGTH_LONG).show();
         }
         fis.close(); // 关闭输入流
     } catch (Exception ex) {
         ex.printStackTrace();
     }
}
```

（2）写入文件代码。

```
public void writeFile(String sFileName) {
       String s = "Hello";
       s = s + "\n" + "Nice to meet you!";
       try {
          // 定义一个文件字节输出流，名字为 sFileName
          FileOutputStream fos = openFileOutput(sFileName, 0);
          // 将文件字节输出流转换成文件字符输出流
          OutputStreamWriter outWriter = new OutputStreamWriter(fos);
          // 再将文件字符输出流转成缓存字符输出流
          BufferedWriter bufferedWriter = new BufferedWriter(outWriter);
          // 使用 write 方法将信息写入文件
          bufferedWriter.write(s);
          bufferedWriter.flush();
          fos.close();
       } catch (Exception ex) {
          ex.printStackTrace();
       }
}
```

2．SD 卡存储数据

SD 卡存储数据的操作实际就是 J2SE 的 I/O 操作，数据可以在任何应用程序间共享。在 SD 卡中若找不到对应的文件将会抛出 FileNotFoundException 异常，创建的文件存放在用户指定的 SD 卡目录下。

进行 SDK 操作，需要在配置文件 AndroidManifest.xml 中添加如下权限声明：

```
<uses-permission android:name="android.permission.WRITE_EXTERNAL_STORAGE"/>
<uses-permission android:name="android.permission.MOUNT_UNMOUNT_FILESYSTEMS"/>
```

获取 SD 卡的根目录可以通过

android.os.Environment.getExternalStorageDirectory().getPath()方法来获取。下面列出关键的实现代码。

（1）读取文件代码。

```
public void readSD(String sName){
    String sPath=android.os.Environment.getExternalStorageDirectory().getPath();
    String  PATH=sPath+"/"+sName;
    Toast.makeText(this, PATH, Toast.LENGTH_SHORT).show();
            //建立一个文件对象
            File   file=new File(PATH);
            //接下来的操作类似 java 的 I/O 文件操作
            int length =(int)file.length();
            byte[] b=new byte[length]; //使用字节数组存储读取出来的数据
            try {
                FileInputStream fis=new FileInputStream(file);
                fis.read(b,0,length);
                 String data="";
                for (byte element: b )
                {
                 data+=element;
                }
                Toast.makeText(this, data, Toast.LENGTH_SHORT).show();
                fis.close();
            } catch(FileNotFoundException e){
                e.printStackTrace();
            } catch(IOException e){
              e.printStackTrace();
            }
    }
```

（2）将数据写入文件代码，参数 sName 表示文件名，data 表示写入文件的数据。

```
public void writeSD(String sName,byte data[]){
String  PATH=android.os.Environment.getExternalStorageDirectory().getPath()+"/"+ sName;
        Toast.makeText(this, PATH, Toast.LENGTH_SHORT).show();
                //建立一个文件对象
                File   file=new File(PATH);
                //接下来的操作类似 Java 的 I/O 文件操作
                try{
                    FileOutputStream fos=new FileOutputStream(file);
                    fos.write(data);
                    fos.flush();
                    fos.close();
                }catch(Exception ex){
                    Toast.makeText(this, "出现异常了", Toast.LENGTH_SHORT).show();
                    ex.printStackTrace();
                }

    }
```

三、SQLite 数据库存储数据

计算机应用程序常用的数据库有 SQL Server、My SQL、Oracle、DB2 等。SQLite 则是一款著名的应用于嵌入式设备的开源轻量级数据库，占用的系统资源非常少，在嵌入式设备中，只需要 350 KB 的内存，这也是 Android 等手持设备系统采用 SQLite 数据库的重要原因之一。目前 SQLite 的最新版本为 3.7.13，可以在官方网站 http://www.sqlite.org 获得相应源代码和文档。SQLite 数据库有如下优点。

① 独立性：SQLite 数据库的核心引擎不需要依赖第三方软件。

② 隔离性：SQLite 数据库中的所有信息（如表、视图、触发器等）都包含在一个文件中，

便于管理和维护。

③ 跨平台：SQLite 目前支持大部分操作系统，例如，Unix（Linux、Mac OS-X、Android、iOS）、Windows（Win32、WinCE、WinRT）等平台。

④ 多语言接口：SQLite 数据库支持多语言编程接口，例如，C、PHP、Perl、Java、C#、Python 等。

⑤ 安全性：SQLite 数据库支持 ACID 事务，通过数据库级的独占性和共享锁来实现独立事务处理。这就意味着多个进程可以在同一时间从同一数据库读取数据，但只能有一个可以写入数据。

⑥ 轻量级：SQLite 大致由 3 万行 C 代码完成，只需要一个较小的动态库，就可以享有其全部功能。

⑦ 零配置：无需安装和管理配置。

Android 集成了 SQLite 数据库，所以每个 Android 应用程序都可以使用 SQLite 数据库。对于熟悉 SQL 的开发人员来说，在 Android 开发中使用 SQLite 较为简单。不同的是 Android 没有 JDBC 链接数据库的概念，因为这样会耗费很多系统资源，为此 Android 系统提供了一些链接数据库的 API，在开发 SQLite 应用程序时，程序员应先学会使用这些 API。Android 系统下每个程序的数据存放在“/data/data/（package name）/”目录下，数据库则存放在“.../dababases/”目录下，即“/data/data/（package name）/databases/”下，当然，也可以指定新的存储目录。

SQLite 数据库的一个比较重要的概念是数据库版本。在 SQLite 的开发中常需要指定数据库的版本号。这是因为在移动开发中，随着应用程序的升级，手机用户中可能有多个旧版本并存，这些旧版本都可能需要升级到最新版本，但要执行的升级操作可能不一样，尤其在保留用户旧数据的情况下。

SQLite 数据库的使用方法将在本项目后面的任务中进行详细的介绍。

四、使用 ContentProvider 对外共享数据

一个应用程序可以创建自己的数据，这些数据对该应用程序来说是私有的，外界无法看到，因此也无从得知数据是通过数据库、文件还是网络存储的。对此，可以为外界提供一套与程序中的数据打交道的标准及统一的接口，以便进行添加、删除、查询、修改等操作。

在本章的第 3.7 节中会详细介绍如何使用 ContentProvider 实现对外共享数据。

五、Internet 网络存储数据

网络存储相对于以上四种存储方式显得不够经济，这是因为移动终端的网络稳定性和产生的流量费用对用户来说是一种负担。对于重要的实时数据，或需要发送给远端服务器处理的数据，可以使用网络进行实时发送，例如，移动办公、实时天气预报、即时通信软件 QQ 等应用程序产生的数据。

Android 网络存储使用标准的 Web 协议，编写方法与 Java 标准版实现网络应用程序的方法类似。先建立一个 Web 服务器端，负责接收 Andorid 提交的数据，然后返回 XML 数据或 JSON 格式的数据。Android 的客户端则负责向服务器发送相应的请求并接收相应的数据，过程代码示例如下。

```
String path="http://www.myweb.com/images/icon.png";//网址
URL url=new URL(path);
HttpURLConnection conn =(HttpURLConnection)url.openConnection();
```

```
conn.setConnectTimeout(5 * 1000);          //设置超时
conn.setRequestMethod("GET");
InputStream  in=conn.getInputStream();  //获取数据流
ByteArrayOutputStream bytestream=new ByteArrayOutputStream();
int ch;
while((ch=in.read())!= -1){
  bytestream.write(ch);
}
byte data[]=bytestream.toByteArray(); //得到的服务器端数据
```

3.2 添加联系人记录

一、任务分析

本任务需要实现的效果如图 3-1 所示。

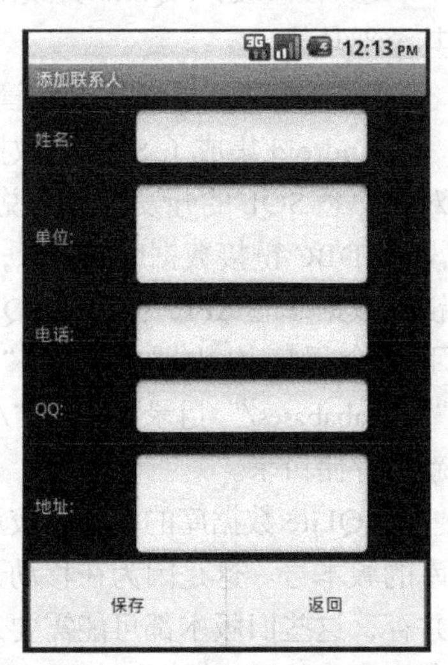

图 3-1 添加联系人记录界面

手机通讯录用于记录联系人的姓名、单位、电话号码、QQ、地址等信息，是手机常用且必备的功能。添加联系人记录的模块主要是为用户提供录入联系人基本信息的界面，并将用户录入的数据保存起来。要完成本次任务，需要思考如下问题。

（1）在手机上如何存储应用程序的数据？

（2）如何把界面上的数据保存到手机上？

二、相关知识

1. SQLiteDatabase 类

SQLiteDatabase 类是一个最终类，用于管理 SQLite 数据库，类中有创建、删除、执行 SQL 命令以及执行其他常见的数据库管理任务的命令。SQLite 除了提供 execSQL() 和 rawQuery()等直接对 SQL 语句进行解析的方法外，还针对 INSERT、UPDATE、DELETE 和 SELECT 等操作专门定义了相关的方法，其主要方法介绍如下。

（1）void beginTransaction()：以独占模式开始事务。执行该方法后，如果代码在后面不执行 setTransactionSuccessful()方法，对数据库的操作将会回退。

（2）int delete（String table，String whereClause，String[] whereArgs）：删除数据库中的某些行。其中参数 table 表示需要删除记录的表；参数 whereClause 表示删除数据的条件，如果取值为 null 则表示删除所有的记录；参数 whereArgs 表示 where 语句中表达式的“?”占位参数列表，参数只能为 String 类型。

（3）boolean deleteDatabase（File file）：删除数据库。参数 file 表示数据库文件的路径。

（4）void endTransaction()：结束事务。

（5）void execSQL（String sql）：执行一条不是 Select 或者其他有数据返回的 SQL 语句，使用参数 sql 表示要执行的 SQL 语句。运行该方法不会返回任何数值，建议熟悉 SQL 的开发者使用该方法。

（6）long getMaximumSize()：返回数据库可以扩展到的最大空间。

（7）String getPath()：返回数据库文件所在的路径。

（8）int getVersion()：返回数据库的版本。

（9）long insert（String table，String nullColumnHack，ContentValues values）：向数据库中

插入一行数据。其中，参数 table 表示需要插入记录的表名。参数 nullColumnHack 为可选参数，如果取值为 null，则 values 取值不能为空，这是因为 SQL 不允许在没有给列命名的情况下，插入一个完全空的行；如果该参数取值不是 null，那么其所提供的列名表示如果 value 值为空，则向该列插入一个空值。参数 values 表示要插入行数据的 ContentValues 对象，即列名和列值的映射。

（10）boolean isOpen()：判断数据库当前是否已经被打开。

（11）boolean isReadOnly()：判断数据库当前是否以只读的方式被打开。

（12）SQLiteDatabase openDatabase（String path，SQLiteDatabase.CursorFactory factory，int flags，DatabaseErrorHandler errorHandler）：根据指定的方式打开数据库。其中，参数 path 表示打开或者创建的数据库文件；参数 factory 表示一个可选的工厂类，用于创建一个查询时需要使用的游标对象，默认取值为 null；参数 flags 表示数据库访问的方式，取值有 OPEN_READWRITE、OPEN_READONLY、CREATE_IF_NECESSARY 或者 NO_LOCALIZED_COLLATORS；参数 errorHandler 用于当 SQLite 报告数据库发生毁坏时进行错误处理。

（13）Cursor query（String table，String[] columns，String selection，String[] selectionArgs，String groupBy，String having，String orderBy，String limit）：查询给定的表，并把查询结果以 Cursor 游标对象的形式返回。

① 参数 table 表示查询的表名。

② 参数 columns 表示返回的列。使用字符串数组表示，若取值为 null 则返回所有的列。

③ 参数 selection 指定需要返回哪些行的 where 条件语句（此处不需要包括 SQL 的关键字 where），若取值为 null 则表示返回所有的行。

④ 参数 selectionArgs 表示 where 语句中表达式的？占位参数列表，参数只能为 String 类型。

⑤ 参数 groupBy 表示对结果集进行分组的 group by 语句（此处不需要包括 SQL 的 group by 关键字），若取值为 null 将不对结果集进行分组。

⑥ 参数 having 表示对分组结果集设置条件的 having 语句（此处不需要包括 SQL 的 having 关键字）。必须是配合 groupBy 参数使用，若取值为 null 将不对分组结果集设置条件。

⑦ 参数 orderBy 表示对结果集进行排序的 order by 语句（此处不需要包括 SQL 的 order by 关键字）。若取值为 null 将对结果集使用默认的排序，通常情况下是不排序的。

⑧ 参数 limit 限制查询返回的行数，若取值为 null 将不限制返回的行数。

（14）Cursor rawQuery（String sql，String[] selectionArgs）：执行一条 SQL 查询语句，并把查询结果以 Cursor 对象的形式返回。其中，参数 sql 表示需要执行的 SQL 语句字符串；参数 selectionArgs 表示 SQL 语句中表达式的“?”占位参数列表，参数只能为 String 类型。

（15）void setTransactionSuccessful()：将当前的事务标记为成功。

（16）void setVersion（int version）：设置数据库的版本号。参数 version 表示新的数据库版本号。

（17）int update（String table，ContentValues values，String whereClause，String[] whereArgs）：更新数据库中的行数据。其中参数 table 表示需要插入记录的表名；参数 values 表示要更新行数据的 ContentValues 对象，即列名和列值的映射；参数 whereClause 指定符合更新条件行的 SQL Where 语句，若取值为 null 则更新所有的行。

2. SQLiteOpenHelper 类

Android 提供 SQLiteOpenHelper 类来辅助创建一个 SQLite 数据库，通过继承 SQLiteOpenHelper 类可以轻松地创建数据库，并支持数据库的版本更新管理。这两项功能是 SQLiteDatabase 类所欠缺的，因此在一般情况下，将 SQLiteOpenHelper 和 SQLiteDatabase 两个类结合起来使用。

SQLiteOpenHelper 是一个抽象类，根据开发应用程序的需要，封装了创建和更新数据库使用的逻辑，是实现 SQLite 数据库的一个关键类。编程时，应定义一个 SQLiteOpenHelper 的子类，至少应实现 onCreate（SQLiteDatabase db）和 onUpgrade（SQLiteDatabase db，int oldVersion，int newVersion）两个方法，下面对该类进行详细的介绍。

SQLiteOpenHelper 类有两种构造方法，格式分别如下。

① SQLiteOpenHelper(Context context，String name，SQLiteDatabase.CursorFactory factory，int version)。其中，参数 context 表示上下文；参数 name 表示数据库文件的名字，对于在内存中的数据库取值为 null；参数 factory 表示可选的工厂类，用于创建一个查询时用到的游标对象，默认取值为 null；参数 version 表示数据库的版本，取值从 1 开始。

② SQLiteOpenHelper(Context context，String name，SQLiteDatabase.CursorFactory factory，int version，DatabaseErrorHandler errorHandler)。前四个参数的含义和第一个构造函数一样，第五个参数 errorHandler 用于当 SQLite 报告一个数据库发生毁坏错误时进行错误处理。

SQLiteOpenHelper 类的常用方法介绍如下。

① void close()：关闭任何打开的数据库对象。

② String getDatabaseName()：返回已经被打开的 SQLite 数据库的名字。

③ SQLiteDatabase getReadableDatabase()：创建或者打开一个数据库，返回一个 SQLite 数据库对象。当数据库的磁盘空间已满时，将以只读的方式访问数据库。

④ SQLiteDatabase getWritableDatabase()：创建或者打开一个数据库，返回一个 SQLite 数据库对象。当数据库的磁盘空间已满时，执行该方法会报错。

⑤ void onCreate（SQLiteDatabase db）：当数据库首次被创建时调用该方法，用于创建表和初始化表中的数值。如果数据库之前已经被创建，则该方法将不会被执行。参数 db 表示 SQLite 数据库对象，在方法中根据需要对该对象填充表和初始化数据。

⑥ void onDowngrade（SQLiteDatabase db，int oldVersion，int newVersion）：当数据库需要降级时调用该方法。其中，参数 db 表示 SQLite 数据库对象；参数 oldVersion 表示旧的数据库版本号；参数 newVersion 表示新的数据库版本号。

⑦ void onOpen（SQLiteDatabase db）：数据库被打开时该方法被自动调用。例如，执行 getReadableDatabase()或 getWritableDatabase()时，若数据库已存在，则调用 onOpen 方法，否则调用 onCreate 方法。

⑧ onUpgrade（SQLiteDatabase db，int oldVersion，int newVersion）：当数据库需要升级时调用该方法。其中，参数 db 表示 SQLite 数据库对象；参数 oldVersion 表示旧的数据库版本号；参数 newVersion 表示新的数据库版本号。

更新数据库的版本是通过重新构造 SQLiteOpenHelper 对象来实现的。假设数据库原来版本是 2，下面为示例代码：

```
SQLiteOpenHelper databaseHelper=new SQLiteOpenHelper(Context context,String string,
null,1);
SQLiteDatabase db=databaseHelper.getReadableDatabase();
```

由于 SQLiteOpenHelper 构造函数的最后一个参数是 1，跟原来的版本不同，而且是把版本降级了，所以系统就会调用 onDowngrade(SQLiteDatabase db, int oldVersion, int newVersion)方法来处理降级事件。

若 SQLiteOpenHelper 构造函数的最后一个参数是 3，那么就比原来的版本要高，所以就会调用 onUpgrade（SQLiteDatabase db，int oldVersion，int newVersion）方法来处理升级。

下面的示例代码展示了如何继承 SQLiteOpenHelper 类来创建一个名为 My_DB.db 的数据库。

```
public class MyDataBase extends SQLiteOpenHelper{
   private static String DB_NAME="My_DB.db";  //数据库名称
   private static int DB_VERSION=1;  //版本号
   private  SQLiteDatabase db; //数据库操作对象
   public  MyDataBase(Context context)
   {
      super(context,DB_NAME,null,DB_VERSION);
      //以写方式打开数据库,该方法是获得 SQLite 数据库的关键。
      //后面对数据库的操作就可以直接使用 SQLiteDatabase 类中的方法
      db=getWritableDatabase();
   }
   //创建数据库后，该方法被调用
   @Override
   public void onCreate(SQLiteDatabase db){
   }
   //每次成功打开数据库后首先被执行
   @Override
   public void onOpen(SQLiteDatabase db){
      super.onOpen(db);
   }
   //升级数据库版本，该方法被调用
   @Override
   public void onUpgrade(SQLiteDatabase db,int oldVersion,
                int newVersion){
      // TODO 更改数据库版本的操作
   }
   public void createTable(String sql){
        db.execSQL(sql);// 创建表
     }
   public void insertData(String sql){
    //execSQL 只能用于没有返回结果的 SQL 语句，例如:insert,delete,create 这样的 SQL 语句
    db.execSQL(sql);// 向表插入数据
   }
}
```

在 Activity 类中用 MyDataBase 类来生成一个对象，即可创建一个数据库：

MyDataBase database= new MyDataBase (this);

可以在“/data/data/<pachage name>/database/”目录下找到数据库文件 My_DB.db，接着可以利用 SQL 语句创建一个表。

```
sql = "CREATE TABLE IF NOT EXISTS person (_id int primary key,name VARCHAR," +
                                          "age int)";
   database.createTable(sql);
```

在表创建好之后，可以向表中插入数据：

```
sql="insert into person values(1,'小方',18)";
   database.insertData(sql);
   sql="insert into person values(2,'小东',20)";
   database.insertData(sql);
```

3. ContentValues 类

ContentValues 类和 Hashtable 类较为相似，它用于存储一组键值对，可以被 ContentResolver 类处理，但是它存储的键值对中的键是一个 String 类型，而值都是基本类型。ContentValues

类作为一个重要的参数在 SQLiteDatabase 中的 insert、update 等方法中使用。下面对该类进行详细的介绍。

ContentValues 类有 3 种构造方法，格式分别如下。

① ContentValues()：使用默认的初始大小创建一个空集。

② ContentValues（int size）：使用指定的初始大小 size 值创建一个空集。

③ ContentValues（ContentValues from）：复制给定的集合 from，用于创建一组集合数值。

ContentValues 类的常用方法介绍如下。

① void clear()：清空集合中的所有数值。

② boolean containsKey（String key）：如果集合中包含了指定的关键字 key，则返回 true，否则返回 false。

③ Object get（String key）：返回关键字 key 对应的数值，返回数值类型为 Object，通常还需要进行强制类型转换。

④ void put（String key，Integer value）：将一个整形数值加入到集合中，其中参数 key 表示集合中的关键字；参数 value 表示要添加的数据。ContentValues 类还有很多 put 方法，主要的区别是第二个参数为其他数据类型，例如，put（String key，Byte value）、put（String key，Float value）、put（String key，Short value）、put（String key，byte[] value）、put（String key，String value）、put（String key，Double value）、put（String key，Long value）、put（String key，Boolean value）。

⑤ void remove（String key）：将某个关键字 key 的数值从集合中删除。

⑥ int size ()：返回集合中数值的个数。

创建 ContentValues 的一般步骤是首先定义一个 ContentValues 对象，然后使用 put 方法将值放入集合，下面为示例代码。

```
ContentValues values= new ContentValues();
values.put("name","小明");
values.put("age",17);
```

4．Toast 类

Toast 类用于向用户显示帮助或者提示信息。与 Dialog 不同的是，Toast 类显示的信息在短暂地显示一段时间后会自动消失。信息可以是简单的文本，也可以是复杂的图片及其他内容（显示一个 View）。

使用 Toast 构造函数创建 Toast 对象较为麻烦，最便捷的生成 Toast 对象的方法是直接调用其静态方法。

① makeText（Context context，int resId，int duration）：创建一个标准的 Toast 对象，只包含一个来自资源文件的文字视图。其中，参数 context 表示上下文；参数 resId 表示字符串资源文件的 ID；参数 duration 表示显示消息的时间长度，取值为 LENGTH_SHORT 或者 LENGTH_LONG，分别表示短时间或者长时间显示视图或文本提示。

② makeText（Context context，CharSequence text，int duration）：创建一个标准的 Toast 对象，只包含一个文字视图。参数 text 表示消息的文字内容，可用字符串表示。

调用上述两个方法之一创建 Toast 对象之后，再调用该对象的 show()方法，就可以打开 Toast 的信息提示。

当需要调整 Toast 在屏幕上的显示位置时，可以使用 setGravity（int gravity，int xOffset，int yOffset）方法。其中参数 gravity 表示显示信息的起点位置，常用的取值有 Gravity. CENTER_

HORIZONTAL、Gravity.CENTER_VERTICAL、Gravity.CENTER、Gravity.TOP、Gravity.LEFT、Gravity.RIGHT、Gravity. BOTTOM；参数 xOffset 表示水平位移，大于 0 则向右移，小于 0 则向左移；参数 yOffset 表示垂直位移，大于 0 则向下移，小于 0 则向上移。如 Toast 对象.setGravity（Gravity.CENTER_VERTICAL，0，0）表示把 Toast 对象定位在屏幕的正中间。

当需要设置 Toast 在屏幕上显示的持续时间时，可以使用 setDuration（int duartion）方法。其中，参数 duartion 表示时间，单位是毫秒（ms），LONG_DELAY 为 3500 ms，SHORT_DELAY 为 2000 ms。需要注意的是，参数 duartion 若取其他数值，并不会使显示时间更长或者更短。

当需要修改 Toast 显示的文本内容时，可以使用 setText（int resId）或者 setText（CharSequence s）方法。

当需要设置 Toast 在屏幕中的显示视图时，可以使用 setView（View view）方法。其中，参数 view 表示视图。该方法一般用于在 Toast 中显示自定义的视图。

图 3-2 Toast 程序主界面

下面的示例用于说明创建 Toast 的三种方法，程序的主界面效果如图 3-2 所示。

（1）定义项目的布局 XML 文件，命名为【toast_demo.xml】，在界面上以垂直方式显示 3 个按钮，代码如下。

```
<LinearLayout xmlns:android="http://schemas.android.com/apk/res/android"
    xmlns:tools="http://schemas.android.com/tools"
    android:id="@+id/LinearLayout1"
    android:layout_width="fill_parent"
    android:layout_height="fill_parent"
    android:orientation="vertical" >
    <Button
        android:id="@+id/button1"
        android:layout_width="fill_parent"
        android:layout_height="wrap_content"
        android:text="Toast 直接输出文字" />
    <Button
        android:id="@+id/button2"
        android:layout_width="fill_parent"
        android:layout_height="wrap_content"
        android:text="Toast 输出 View 内容" />
    <Button
        android:id="@+id/button3"
        android:layout_width="fill_parent"
        android:layout_height="wrap_content"
        android:text="Toast 输出自定义 View" />
</LinearLayout>
```

（2）定义一个 Toast 调用的信息显示布局文件【toastinfo.xml】，在文件中定义一个 Image View 控件和一个 TextView 控件。

```
<?xml version="1.0" encoding="utf-8"?>
<LinearLayout xmlns:android="http://schemas.android.com/apk/res/android"
    android:layout_width="match_parent"
    android:layout_height="match_parent"
    android:orientation="vertical" >
    <ImageView
        android:id="@+id/imageView1"
        android:layout_width="wrap_content"
        android:layout_height="wrap_content"
```

```
      android:src="@drawable/icon" />
   <TextView
      android:id="@+id/textView1"
      android:layout_width="wrap_content"
      android:layout_height="wrap_content"
      android:text="我是Android小图标" />
</LinearLayout>
```

（3）编写 Android 程序。

```
public class ToastDemo extends Activity {

   @Override
   public void onCreate(Bundle savedInstanceState){
      super.onCreate(savedInstanceState);
      setContentView(R.layout.toast_demo);
      Button button1=(Button)this.findViewById(R.id.button1);
      Button button2=(Button)this.findViewById(R.id.button2);
      Button button3=(Button)this.findViewById(R.id.button3);
      button1.setOnClickListener(showToast1);
      button2.setOnClickListener(showToast2);
      button3.setOnClickListener(showToast3);
   }
   OnClickListener showToast1=new OnClickListener(){
      public void onClick(View v){
         Toast.makeText(ToastDemo.this,"直接输出信息",Toast.LENGTH_SHORT).show();
      }
   };
   OnClickListener showToast2=new OnClickListener(){
      public void onClick(View v){
         //获取LayoutInflater对象
         LayoutInflater
      li=(LayoutInflater)getSystemService(Context.LAYOUT_INFLATER_SERVICE);
    //获取一个View对象
        View view=li.inflate(R.layout.toastinfo,null);
         Toast toast=new Toast(ToastDemo.this);
         toast.setView(view);
         toast.show();
      }
   };
   OnClickListener showToast3=new OnClickListener(){
      public void onClick(View v){
        //1.创建Toast
        Toast toast=Toast.makeText(ToastDemo.this,"图文显示",Toast.LENGTH_LONG);
        //2.创建Layout,并设置为水平布局
         LinearLayout mLayout=new LinearLayout(ToastDemo.this);
       mLayout.setOrientation(LinearLayout.VERTICAL);
        ImageView mImage=new ImageView(ToastDemo.this); //创建显示图像的
        ImageView mImage.setImageResource(R.drawable.icon);
        View toastView=toast.getView(); //获取显示文字的Toast View
        mLayout.addView(mImage); //将图片添加到Layout
        mLayout.addView(toastView);
        //3.设置Toast显示的View,参数为上面生成的Layout
         toast.setView(mLayout);
        toast.show();
    }
   };
}
```

代码补充说明：上述代码使用到了 LayoutInflater 类。在实际开发中 LayoutInflater 类较为重要，作用类似于 findViewById()。不同点是 LayoutInflater 用来解析 res/layout/下的 XML 布局文件，并进行实例化。对于一个没有被载入或者想要动态载入的界面，都可以使用 LayoutInflater.inflate()来加载；而 findViewById()则是用于查找 XML 布局文件下的具体 widget 控件（如 Button、TextView 等）。对于一个已经载入的界面，可在 Activity 程序中直接调用

findViewById()方法获得其中的界面元素。

三、任务实施

1．创建联系人录入界面

（1）创建项目：创建一个 Android Project，建项目名称为【MyContacts】，包名为【com.demo.pr3】，选择 Android SDK 2.2 版本，并创建一个 Activity，将其命名为【MyContactsActivity】。

（2）设计用户输入界面，提供姓名、单位、电话、QQ 和地址 5 项信息供用户录入，每项录入信息使用一个 LinearLayout 嵌套一个 TextView 和 EditText 子元素来表示，然后再将这些 LinearLayout 作为上一级 LinearLayout 的子元素。以下是布局页面的 XML 代码。

```
<?xml version="1.0" encoding="utf-8"?>
<LinearLayout
    xmlns:android="http://schemas.android.com/apk/res/android"
    android:orientation="vertical"
    android:layout_width="fill_parent"
    android:layout_height="fill_parent"
    >
   <!--姓名-->
   <LinearLayout
    android:layout_width="wrap_content"
    android:layout_height="wrap_content"
    android:orientation="horizontal"
    android:layout_marginTop="10.0dip"
    >
        <TextView android:layout_width="80.0dip"
        android:layout_height="wrap_content"
        android:text="姓名:"
        android:layout_marginLeft="5.0dip"
        />
        <EditText android:layout_width="190.0dip"
        android:id="@+id/name"
        android:layout_height="wrap_content" />
   </LinearLayout>
   <!--单位-->
   <LinearLayout
    android:layout_width="wrap_content"
    android:layout_height="wrap_content"
    android:orientation="horizontal"
    android:layout_marginTop="10.0dip"
   >
        <TextView android:layout_width="80.0dip"
        android:layout_height="wrap_content"
        android:text="单位:"
        android:layout_marginLeft="5.0dip"
        />
        <EditText android:layout_width="220.0dip"
         android:layout_height="wrap_content"
         android:id="@+id/danwei"
         android:minLines="2"
     />
   </LinearLayout>
   <!--手机号码-->
   <LinearLayout
    android:layout_width="wrap_content"
    android:layout_height="wrap_content"
    android:orientation="horizontal"
    android:layout_marginTop="10.0dip"
   >
        <TextView android:layout_width="80.0dip"
```

```
        android:layout_height="wrap_content"
        android:text="电话:"
        android:layout_marginLeft="5.0dip"
        />
        <EditText android:layout_width="190.0dip"
         android:id="@+id/mobile"
        android:layout_height="wrap_content"
        android:inputType="number" />
   </LinearLayout>
    <!--qq-->
   <LinearLayout
    android:layout_width="wrap_content"
    android:layout_height="wrap_content"
    android:orientation="horizontal"
    android:layout_marginTop="10.0dip"
   >
       <TextView android:layout_width="80.0dip"
        android:layout_height="wrap_content"
        android:text="QQ:"
        android:layout_marginLeft="5.0dip"
        />
       <EditText android:layout_width="190.0dip"
             android:layout_height="wrap_content"
                 android:id="@+id/qq"
     />
   </LinearLayout>
   <!--地址-->
   <LinearLayout
    android:layout_width="wrap_content"
    android:layout_height="wrap_content"
    android:orientation="horizontal"
    android:layout_marginTop="10.0dip"
   >
        <TextView android:layout_width="80.0dip"
        android:layout_height="wrap_content"
        android:text="地址:"
        android:layout_marginLeft="5.0dip"
        />
        <EditText android:layout_width="220.0dip"
        android:id="@+id/address"
        android:layout_height="wrap_content"
        android:minLines="2"
      />
   </LinearLayout>
</LinearLayout>
```

2．创建数据库的操作类

（1）创建一个 SQLiteOpenHelper 的子类 MyDB，用于对数据库进行管理，主要包括创建表，并对表数据进行增加、删除、修改和查找，以及打开和关闭数据库等基础功能操作。

首先定义该类的主要数据成员：

```
private static String DB_NAME="My_DB.db";  //数据库名称
private static int DB_VERSION=2;  //版本号
private  SQLiteDatabase db; //数据库操作对象
```

下面介绍该类主要方法的实现。

① 构造函数。

```
public  MyDB(Context context)
{
   //在子类的构造函数实现过程中,一般首先调用父类的构造函数
   super(context,DB_NAME,null,DB_VERSION);
```

```
    db=getWritableDatabase();
}
```

② 打开 SQLite 数据库连接。

```
public SQLiteDatabase openConnection ()
{
    db=getWritableDatabase();
    return  db;
}
```

③ 关闭 SQLite 数据库连接操作。

```
public void closeConnection ()
{
    db.close();
}
```

④ 创建表。

```
public  boolean creatTable(String  createTableSql)
{
      try{
           openConnection();
           db.execSQL(createTableSql);
         }catch(Exception ex){
           ex.printStackTrace();
           return false;
       }finally
       {
           closeConnection();
       }
       return true;
}
```

⑤ 添加数据：…部分省略的代码与 creatTable 方法一样。

```
public  boolean  save(String tableName,
ContentValues values)
 {      ...
       db.insert(tableName,null,values);
       ...
 }
```

⑥ 修改数据：…部分省略的代码与 creatTable 方法一样。

```
public  boolean  update(String table,
ContentValues values,String whereClause,String []whereArgs)
{      ...
       db.update(table,values,whereClause,whereArgs);
       ...
 }
```

⑦ 删除数据：…部分省略的代码与 creatTable 方法一样。

```
public  boolean  delete(String table,
String deleteSql,String obj[])
{  ...
   db.delete(table,deleteSql,obj);
...
}
```

⑧ 查询操作：除了 return 操作之外，…部分省略的代码与 creatTable 方法一样。

```
public  Cursor  find(String findSql,String obj[])
{   ...
    Cursor cursor=db.rawQuery(findSql,obj);
    return  cursor;
    ...
}
```

⑨ 判断表是否存在：对表进行查询，如果出错表示数据库中不存在所要查询的表。

```
public boolean isTableExits(String tablename)
{
   try{
      openConnection();
```

```
    String str="select count(*)xcount  from  "
          +tablename;
    db.rawQuery(str,null).close();
  }catch(Exception ex)
  {
        return false;
  }
  return true;
}
```

（2）创建一个表示联系人信息的类【User.java】。

该辅助类的作用主要是采用面向对象的方法管理联系人的数据，以便于对联系人表中的字段统一进行设置和获取操作。

```
public  class User {
  ...
  private  String name; //用户名
  private  String moblie;//手机号码
  private  String danwei;//单位
  private  String qq;  //QQ
  private  String address;//地址
  private  int id_DB=-1;//表主键 ID
  ...
}
```

（3）创建 ContactsTable.java 操作类。

ContactsTable 类用于封装联系人表中数据的操作，例如将界面上的数据保存到数据库中，主要实现是通过调用 MyDB 类的方法。

① 定义类的成员变量。

```
private   final static String TABLENAME="contactsTable";//表名
private   MyDB  db;//定义 SQLiteOpenHelper 数据库管理对象
```

② 在构造方法中创建表。

```
public    ContactsTable(Context context)
{
  db=new  MyDB(context);
  if(!db.isTableExits(TABLENAME))
  {   //定义 SQL 语句
     String  createTableSql="CREATE TABLE IF NOT EXISTS " +
     TABLENAME+"(id_DB  integer   " +
     "primary key  AUTOINCREMENT," +
     User.NAME    +"  VARCHAR," +
     User.MOBLIE   +"  VARCHAR,"+
     User.QQ   +"  VARCHAR,"+
     User.DANWEI   +"  VARCHAR,"+
     User.ADDRESS+ " VARCHAR)";
     //创建表
     db.creatTable(createTableSql);
  }
}
```

③ 创建添加数据到数据库的方法，将联系人数据放入到一个 ContentValues 对象中。

```
public  boolean  addData(User user)
{
  ContentValues values=new ContentValues();
  values.put(User.NAME,user.getName());
  values.put(User.MOBLIE,user.getMoblie());
  values.put(User.DANWEI,user.getDanwei());
  values.put(User.QQ,user.getQq());
  values.put(User.ADDRESS,user.getAddress());
  returndb.save(TABLENAME,values);
}
```

3．将界面上数据保持到数据库

创建一个显示添加数据界面的 Activity 类 AddContactsActivity。首先根据界面要显示的控件，相应地定义该类的主要数据成员。

```
private  EditText nameEditText;     //姓名输入框
private  EditText mobileEditText;   //手机输入框
private  EditText qqEditText;       //qq输入框
private  EditText danweiEditText;   //单位输入框
private  EditText addressEditText;  //地址输入框
```

下面介绍该类主要方法的实现。

（1）从布局文件中获得界面上的控件对象。

```
nameEditText=(EditText)findViewById(R.id.name);
 mobileEditText=(EditText)findViewById(R.id.mobile);
 danweiEditText=(EditText)findViewById(R.id.danwei);
 qqEditText=(EditText)findViewById(R.id.qq);
 addressEditText=(EditText)findViewById(R.id.address);
```

（2）设置界面上返回和保存的菜单。

```
public boolean onCreateOptionsMenu(Menu menu){
      menu.add(Menu.NONE,1,Menu.NONE,"保存");
      menu.add(Menu.NONE,2,Menu.NONE,"返回");
      return super.onCreateOptionsMenu(menu);
}
```

（3）实现保存和返回的处理操作，将输入界面中的信息先保存在User对象中，然后再通过前面定义的ContactsTable类addData方法将数据保存到SQLite数据库的表中。

```
public boolean onOptionsItemSelected(MenuItem item){
    switch(item.getItemId()){
      case 1://保存
      if(!nameEditText.getText().toString().equals(""))
      {
              User user=new User();
              user.setName(nameEditText.getText().toString());
              user.setMoblie(mobileEditText.getText().toString());
              user.setDanwei(danweiEditText.getText().toString());
              user.setQq(qqEditText.getText().toString());
              user.setAddress(addressEditText.getText().toString());
              ContactsTable ct=
                  new ContactsTable(AddContactsActivity.this);
          if(ct.addData(user))
          {
           Toast.makeText(AddContactsActivity.this,"添加成功!
               ",Toast.LENGTH_SHORT).show();
          }else
          {
           Toast.makeText(AddContactsActivity.this,"添加失败!
               ",Toast.LENGTH_SHORT).show();
          }
       }else
           {
           Toast.makeText(AddContactsActivity.this,"请先输入数据!
                ",Toast.LENGTH_SHORT).show();
           }
           break;
     case 2://返回
        finish();
        break;
      default:
        break;
    }
   return super.onOptionsItemSelected(item);
}
```

3.3 修改联系人记录

一、任务分析

本任务需要实现的效果如图 3-3 所示。

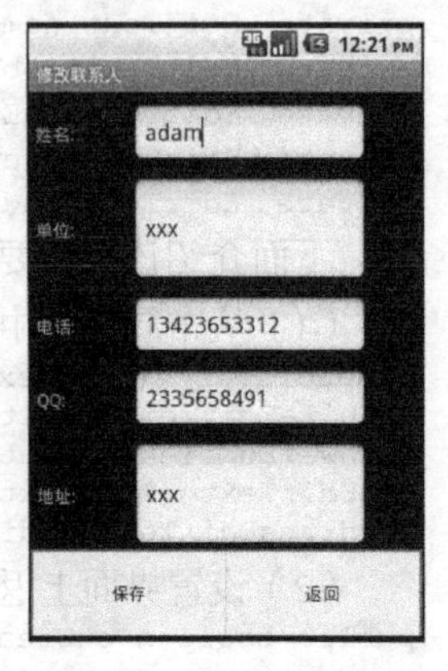

图 3-3 修改联系人记录

对联系人信息进行修改，首先需要将联系人的信息从 SQLite 数据库读出并显示在界面上。当用户修改数据之后，再将更新后的记录写回到存储器当中。要完成本次任务，需要思考以下三个问题。

（1）在 Android 中如何切换两个 Activity 程序？

（2）如何把界面上修改后的记录更新到存储器中？

（3）如何知道用户单击修改的是哪一条记录？

二、相关知识

1. Cursor 类

Cursor 是一个游标类，可以在一组数据库查询结果的集合上前后移动，进而检索每一行的数据。下面对该类进行详细的介绍。

（1）void close()：关闭游标，释放所有的资源，并使游标变为无效。

（2）int getColumnCount()：返回列的个数。

（3）int getColumnIndex（String columnName）：根据字段的列名返回其索引值如果不存在则返回-1。其中参数 columnName 表示字段的列名。

（4）String getColumnName（int columnIndex）：根据字段的索引值，返回字段的列名。

（5）String[] getColumnNames()：返回一组表示所有字段列名的字符串数组。

（6）int getCount()：返回游标中行的个数。

（7）double getDouble（int columnIndex）：将请求的列数值以 double 类型返回，参数 columnIndex 表示列的索引值。

（8）float getFloat（int columnIndex）：将请求的列数值以 float 类型返回，参数 columnIndex 表示列的索引值。

（9）int getInt（int columnIndex）：将请求的列数值以 int 类型返回，参数 columnIndex 表示列的索引值。

（10）long getLong(int columnIndex)：将请求的列数值以 Long 类型返回，参数 columnIndex 表示列的索引值。

（11）int getPosition()：返回游标的当前位置。

（12）short getShort（int columnIndex）：将请求的列数值以 short 类型返回，参数 columnIndex 表示列的索引值。

（13）String getString（int columnIndex）：将请求的列数值以 String 类型返回，参数 columnIndex 表示列的索引值。

（14）boolean isAfterLast()：判断游标是否指向最后一行之后的位置。

（15）boolean isBeforeFirst()：判断游标是否指向第一行前面的位置。

（16）boolean isClosed()：判断游标是否被关闭。

（17）boolean isFirst()：判断游标是否指向第一行。

（18）boolean isLast()：判断游标是否指向最后一行。

（19）boolean isNull（int columnIndex）：如果指定的列数值是 null，则返回 true。参数 columnIndex 表示列的索引值。

（20）boolean move（int offset）：将游标从当前位置向前或者向后移动到一个新的位置。参数 offset 表示一个相对的移动距离。

（21）boolean moveToFirst()：将游标移到第一行。

（22）boolean moveToLast()：将游标移到最后一行。

（23）boolean moveToNext ()：将游标移到下一行，如果游标当前已经在最后一行，执行该方法将会返回 false。

（24）boolean moveToPosition（int position）：将游标从当前位置向前或者向后移动到一个新的位置。参数 position 表示一个绝对的目的位置，取值范围为大于等于-1，小于等于集合中记录的个数。

下面举例说明如何使用 Cursor 类来对数据库的查询结果数据进行遍历，其中 db 是一个 SQLiteDatabase 类对象，person 是需要查询的个人信息表。

```
String sName;
Int iAge;
Cursor cursor=db. rawQuery("select * from student",null);
while(cursor.moveToNext()){
   //获取游标当前行的 name 列数值
   sName=cursor.getString(cursor.getColumnIndex("name"));
   //获取游标当前行的 age 列数值
   iAge=cursor.getInt(cursor.getColumnIndex("age"));
   …
}
```

2．Bundle 类

Bundle 类可以用于不同 Activity 之间的数据传递，将数据打包到 Intent 对象中，辅助 Intent 对象携带数据。Bundle 类与 Java 的集合类 Map 类似，用于存放 key-value 键值对形式的值，提供了各种常用类型的 put×××()/get×××() 方法，例如，putString()/getString()、putInt()/getInt()。其中，put××× ()用于向 Bundle 对象放入某种类型的数据，get××× ()方法用于从 Bundle 对象获取某种类型的数据，操作与 Intent 对象的 putExtra()方法相对应。Bundle 内部实际使用 HashMap<String，Object>类型的变量来存放 put××× ()方法放入的值。

3．Activity 的切换

一个 Android 应用程序可以有多个 Activity 类，因此常常需要处理各 Activity 类之间的切换。例如用户登录 Activity 在验证账号和密码成功之后，需要打开应用程序的主界面 Activity。

Intent 的作用就是在程序运行时连接两个不同的组件，为它们之间传递数值，可以理解为意图。Android 的 Activity、Service 和 BroadcastReceiver 三种基本组件，都是由 Intent 的运行绑定机制激活的，这三种组件在传递 Intent 的实现上各不相同。Intent 在运行时将各个组件相互结合在一起。可以把 Intent 看作是请求从其他组件执行行动的使者，不管该组件是否属于该应用程序。

Intent 调用目标组件有两种方式：一种是显式 Intent，即明确指出目标组件；另一种是隐式 Intent，即只是指出调用组件的特征，由系统决定调用哪个组件。以 Activity 为例，在 AndroidManifest.xml 文件中，使用<intent-filter>元素的子元素< action android:name>定义待选的 Activity 会被哪些动作调用。

对于 Activity 和 Service，Intent 定义要执行的动作，如：查看或发送某种类型的数据，并可以指定需要操作的数据的URI。例如，一个Intent可以为一个Activity传送要求，要求其显示图像或打开网页。在某些情况下，可以启动一个Activity去接收请求、进行处理并将结果返回Intent中。例如，发出一个Intent让用户可以选择某个人的通讯录，并让这个Intent返回一个指向所选择通讯录的URI。

对于BroadcastReceiver，Intent只简单定义一个广播公告，如广播显示设备电池电量低，只需包括一个已知的显示“电池电量低”的动作字符串。

对于其他类型的组件，ContentProvider并不是被Intent激活的，而是由ContentResolver激活的。ContentResolver和ContentProvider一起处理共享数据的访问事务。

下面是激活每种类型组件的不同方法。

① 开启一个Activity:通过传递一个Intent对象到startActivity() 或者startActivityForResult()方法中（注：当希望Activity返回结果时调用后一个方法。

② 开启一个Service:通过传递一个Intent对象到startService ()方法或者bindService()方法。

③ 开启一个Broadcast:通过传递一个Intent对象到sendBroadcast()、sendOrderedBroadcast()或者sendStickyBroadcast()方法。

④ 在ContentResolver上调用query()方法，对ContentProvider执行查询操作。

本节主要介绍Intent在Activity中的使用方法。Intent是从一个Activity到达另一个Activity的引路者，包含当前Activity、目标Activity、分类、Activity之间切换所需的动作、传送的数据、标志位和附加的信息等。在使用Intent时，可根据需要来填写这些数据。

Intent类有六种构造方法，格式分别如下，其中构造方法①～④是隐式的，构造方法⑤、⑥种是显式的。

① Intent()：创建一个空的Intent。

② Intent（Intent o）：复制一个已有的Intent对象。参数o表示已有的Intent对象。

③ Intent（String action）：根据给定的动作创建一个Intent对象。其中，参数action表示Intent触发动作的名字。Android系统提供了很多标准的Activity动作和Broadcast动作。例如，android.intent.action.MAIN表示主程序入口，android.intent.action.CALL_BUTTON表示打开系统应用中的拨号界面，ACTION_AIRPLANE_MODE_CHANGED表示用户切换飞行模式的Broadcast动作。详细的标准Activity动作和Broadcast动作定义可以查看官方文档（http://developer.android.com/reference/android/content/Intent.html）。

④ Intent（String action，Uri uri）：根据给定的动作和数据网址来创建一个Intent对象。其中参数action表示Intent触发动作的名字；uri表示动作处理的数据所在的位置。Uri类的使用在下一节进行介绍。

⑤ Intent（Context packageContext，Class<?> cls）：为特定的组件建立一个Intent对象。其中，参数packageContext表示实现当前类的应用上下文；参数cls表示用于Intent的组件类。Contex表示一个开发上下文的接口，代表某种开发环境。有些组件或者控件在应用时需要知道它们所在的环境或上下文信息。

⑥ Intent（String action，Uri uri，Context packageContext，Class<?> cls）：根据特定的动作和数据，为特定的组件建立一个Intent对象。

Intent类的常用方法介绍如下。

（1）Intent提供了各种putExtra()方法，作用是向Intent对象添加不同类型的数据。putExtra()方法的参数数量有两种形式：一种是putExtra（String name，datatype value），它带有两个参数，其中参数 name 表示数据的名字，value 表示数据值，datatype 表示某种数据类型或者数组，如 putExtra（String name，int value）、putExtra（String name，int[] value）；另一种是只带有一个参数的，如 putExtras（Bundle extras）和 putExtras（Intent src）。

（2）Bundle getExtras()：获得 Intent 携带的一组扩充数据，即其之前在其他组件通过 putExtra()方法添加的数据。

（3）Uri getData()：返回 Intent 正在操作数据的 Uri。

（4）Intent setClass（Context packageContext，Class<?> cls）：通过指定 Class 对象的方式设置需要跳转的目标组件。其中，参数 packageContext 表示实现当前类的应用上下文，一般的取值形式是：当前 Activity 的类名.this；参数 cls 表示要跳转到的目的 Activity，一般的取值形式是：Activity 类名.class。

在一个应用程序内往往会通过 Intent 对象来指定需要跳转的目标 Activity，并调用 startActivity（Intent intent）或者 startActivityForResult（Intent intent，int requestCode）方法来启动该 Activity。这两个方法的区别是：前者调用一个新的 Activity，不需要接收其返回的结果；而后者除了调用新的 Activity 之外，还将其处理结果返回到前一个执行的 Activity 的 onActivityResult（int requestCode，int resultCode，Intent data）方法中。参数 requestCode 表示请求码，用于识别调用的 Activity。

设置完 Intent 对象后，若通过 startActivity 或者 startActivityForResult 方法启动新的 Activity 时，设置的值也会一同传递到新启动的 Activity 中。在新的 Activity 中，可以通过在 oncreate 方法内调用 getIntent.getExtra()方法获得一个 Bunble 对象，并调用相应的 Get 方法获取之前 Intent 对象利用 putExtra 方法写入的值。

如果通过 putExtra()方法来传递一个复杂的数据，那么该数据必须是可序列化的。例如，要传递一个 Person 对象的实例，则该类必须实现 Serializable 接口。

```
class Person implements Serializable {
      String name="小明" ;
      int  age= 17;
  }
Person person=new Person();
putExtra(person);
```

说明

使用 putExtra()方法传递一个实现了 Serializable 接口的类的实例对象时，这个类中所有的成员也必须是可序列化的，否则会抛出异常。

一个简单 Intent 实现如下。

```
Intent intent=new Intent();                    //创建一个 Intent 对象
intent.setClass(activity1.this,activity2.class); //描述起点 activity1 和目标 activity2
startActivity(intent);                         //从 activity1 跳转到 activity2
```

下面举例说明在 Intent 中附加数据的两种方法。

（1）先将数据写入到 Bundle 对象中，然后再将 Bundle 对象传入 Intent 对象。

```
Intent intent=new Intent();
Bundle bundle=new Bundle();//该类用作携带数据
bundle.putString("name","小明");
bundle.putInt("age",17);
intent.putExtras(bundle);//为 intent 追加额外的数据,intent 原来已经具有的数据不会丢失,但与 key 同名的数据会被替换
```

（2）直接把数据逐个添加到 Intent 对象中。

```
Intent intent=new Intent();
```

```
intent.putExtra("name","小明");
intent.putExtra("age",17);
```

启动当前 Activity 的 Intent 对象是通过 getIntent()方法，而 Bundle 对象则可以通过 Activity 类的 getIntent().getExtras()方法返回。

下面举例说明 Activity 之间如何切换并相互传递数据，效果如图 3-4 所示。在示例中有两个 Activity 程序，分别是 IntentDemo 和 IntentDemo2，其中 IntentDemo 中有 1 个 EditText 用于接收用户输入的 1 个字符串，要求把该字符串转换成大写状态，但把转换的工作交给 IntentDemo2 来做，并把转换成大写的字符串再返回给 IntentDemo，由其负责显示。

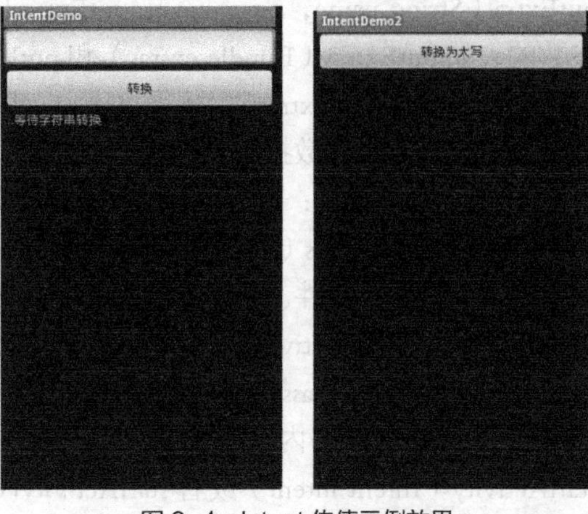

图 3-4　Intent 传值示例效果

（1）构建 IntentDemo 所需的界面布局，文件名为【intent_demo.xml】。

```
<LinearLayout xmlns:android="http://schemas.android.com/apk/res/android"
    xmlns:tools="http://schemas.android.com/tools"
    android:id="@+id/LinearLayout1"
    android:layout_width="match_parent"
    android:layout_height="match_parent"
    android:orientation="vertical" >
    <EditText
        android:id="@+id/editText1"
        android:layout_width="match_parent"
        android:layout_height="wrap_content"
        android:ems="10" >
        <requestFocus />
    </EditText>
    <Button
        android:id="@+id/button1"
        android:layout_width="match_parent"
        android:layout_height="wrap_content"
        android:text="转换" />
    <TextView
        android:id="@+id/textView1"
        android:layout_width="wrap_content"
        android:layout_height="wrap_content"
        android:text="...等待字符串转换" />
</LinearLayout>
```

（2）编写 IntentDemo 程序代码。

```
package com.example.project3;

public class IntentDemo extends Activity {
    private EditText eText;
    private Button button;
    private TextView tText;

    @Override
    public void onCreate(Bundle savedInstanceState){
        super.onCreate(savedInstanceState);
        setContentView(R.layout.intent_demo);
        eText =(EditText)this.findViewById(R.id.editText1);
        button =(Button)this.findViewById(R.id.button1);
        tText =(TextView)this.findViewById(R.id.textView1);
        button.setOnClickListener(new OnClickListener() {
            public void onClick(View v){
```

```
                    Bundle bundle=new Bundle();
                    bundle.putString("value",eText.getText().toString());
                    Intent intent=new Intent(IntentDemo.this,IntentDemo2.class);
                    intent.putExtras(bundle);
                    startActivityForResult(intent,10);
                }
        });
    }

    @Override
    protected void onActivityResult(int requestCode,int resultCode,Intent data){
        switch(requestCode){
        case 10:
            Bundle bundle=data.getExtras();
            tText.setText(bundle.getString("result"));
        }

    }
}
```

（3）构建 IntentDemo2 所需界面的布局，文件名为【intent_demo2.xml】。

```
<LinearLayout xmlns:android="http://schemas.android.com/apk/res/android"
   xmlns:tools="http://schemas.android.com/tools"
   android:id="@+id/LinearLayout1"
   android:layout_width="match_parent"
   android:layout_height="match_parent"
   android:orientation="vertical" >
   <Button
      android:id="@+id/button1"
      android:layout_width="match_parent"
      android:layout_height="wrap_content"
      android:text="转换为大写" />
</LinearLayout>
```

（4）编写 IntentDemo2 程序代码。

```
package com.example.project3;
public class IntentDemo2 extends Activity {
   private Button button;
   String value;
   @Override
   public void onCreate(Bundle savedInstanceState){
      super.onCreate(savedInstanceState);
      setContentView(R.layout.intent_demo2);
      button=(Button)this.findViewById(R.id.button1);
      Intent intent=this.getIntent();
      Bundle bundle=intent.getExtras();
      value=bundle.getString("value");
      value=value.toUpperCase();
      button.setOnClickListener(new OnClickListener(){
      public void onClick(View v){
           Intent intent=new Intent();
           intent.putExtra("result",value);
           IntentDemo2.this.setResult(RESULT_OK,intent);
           IntentDemo2.this.finish();
      }

      }
      );
   }
}
```

3．URI 类

URI（Universal Resource Identifier）称为通用资源标识符，表示需要操作的数据，特别是包含了数据所在的位置。Android 的每一种资源都可以使用 URI 表示。URI 包含主机名、片段标识符、相对 URI 等部分。一般情况下，开发者并不需要自己定义 URI，只需要调用即可。

URI 一般具有如下形式：

```
<scheme name> : scheme-specific-part
```

scheme name：命名 URI 名称空间的标识符，例如，http、file、tel、smsto、mailto、content 等。

scheme-specific-part：冒号把 scheme name 与 scheme-specific-part 分开，scheme-specific-part 的语法和语义（意思）由 URI 的名称空间决定。

以 ContentProvidr 为例，URI 主要包含了两部分信息：需要操作的 ContentProvider 和要进行操作的 ContentProvider 中的数据，如图 3-5 所示。

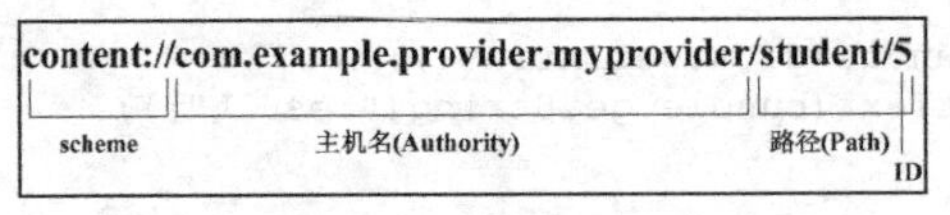

图 3-5　URI 结构图

（1）scheme name：ContentProvider（内容提供者）的 scheme name 已经由 Android 规定，命名为 content://。

（2）主机名（Authority）：用于唯一标识 ContentProvider，外部调用者可以根据此标识来找到它。如果把 ContentProvider 看作是一个网站，那么主机名就是其域名。

（3）路径（Path）：用来表示要操作的数据，路径的构建应根据业务而定。

- 要操作 student 表中 ID 为 5 的记录，构建的路径为/student/5。
- 要操作 student 表中 ID 为 5 的记录的 name 字段，构建的路径为 student/5/name。
- 要操作 student 表中的所有记录，构建的路径为/student。
- 要操作×××表中的记录，可以构建路径为/×××。当然要操作的数据不一定来自数据库，也可以是文件、XML 或网络等其他存储方式。如要操作 XML 文件中 student 节点下的 name 节点，可以构建路径：/student/name。

URI 类的主要作用是解析标识符。如果要把一个字符串转换成 URI，可以使用 URI 类中的 parse()方法：

```
Uri uri=Uri.parse("content://com.ljq.provider.personprovider/person")
```

4．修改表记录的 SQL 语法

继续使用 3.2 节中 MyDataBase 类定义的对象 database

```
sql="update person set name='小花' where name='小东'";
database.db.execSQL(sql);
```

三、任务实施

本任务将要创建一个新的 Activity 类 UpdateContactsActivity，它负责更新通讯录，并对 3.2 节中的类进行修改，具体步骤如下。

（1）设计联系人信息修改界面。该界面与添加联系人记录的界面类似，不同之处在于前者的文本输入框显示了从数据库中读取的联系人信息，后者的所有文本框都为空。

（2）在 ContactsTable.java 类中实现根据联系人 ID 来获取联系人信息的方法。

```
public User getUserByID(int id)
{
     Cursor cursor=null;
     try {
       cursor=db.find("select * from "+TABLENAME +"  where  "
            +"id_DB=?",new String[]{id+""});
       User temp=new User();
    //游标一开始指向-1 位置,moveToNext 方法将游标移动到了下一行,即第一行
```

```
        cursor.moveToNext();
        temp.setId_DB(cursor.getInt(cursor.getColumnIndex("id_DB")));
        temp.setName(cursor.getString(cursor.getColumnIndex(User.NAME)));
temp.setMoblie(cursor.getString(cursor.getColumnIndex(User.MOBLIE)));
        temp.setDanwei(cursor.getString(cursor.getColumnIndex(User.DANWEI)));
        temp.setQq(cursor.getString(cursor.getColumnIndex(User.QQ)));
        temp.setAddress(cursor.getString(cursor.getColumnIndex(User.ADDRESS)));
        return temp;
     } catch(Exception e){
        e.printStackTrace();
     } finally {
        if(cursor != null){
            cursor.close();
        }
        db.closeConnection();
     }
     return null;
}
```

（3）将联系人数据赋值到用户界面。

```
//将要修改的联系人数据赋值到用户界面进行显示
public void onCreate(Bundle savedInstanceState){
   super.onCreate(savedInstanceState);
   ... //获得界面上的控件类，此处代码省略
  //获取 Activity 传过来的数据
  Bundle localBundle=getIntent().getExtras();
  //参数 id 表示用户单击需要修改的记录值的主键
  int id=localBundle.getInt("user_ID");
  ContactsTable ct=new ContactsTable(this);
  //getUserByID()以主键 id 作为查询条件，调用 SQL 语句查询出符合条件的联系人数据
  user =ct.getUserByID(id);
  //将要修改的联系人数据赋值到用户界面进行显示
  nameEditText.setText(user.getName());
  mobileEditText.setText(user.getMoblie());
  qqEditText.setText(user.getQq());
  danweiEditText.setText(user.getDanwei());
  addressEditText.setText(user.getAddress());
  ...
}
```

（4）在 ContactsTable.java 类中添加更新 updateUser()方法。

```
public  boolean updateUser(User user)
{
    ContentValues values=new ContentValues();
    values.put(User.NAME,user.getName());
    values.put(User.MOBLIE,user.getMoblie());
    values.put(User.DANWEI,user.getDanwei());
    values.put(User.QQ,user.getQq());
    values.put(User.ADDRESS,user.getAddress());
    //将界面上修改的数据重新写入到 SQLite 数据库中
    return db.update(TABLENAME,
         values," id_DB=? ",new String[]{user.getId_DB()+""});
}
```

（5）在 UpdateContactsActivity 上设置【返回】和【保存】按钮。

```
public boolean onCreateOptionsMenu(Menu menu){
     menu.add(Menu.NONE,1,Menu.NONE,"保存");
     menu.add(Menu.NONE,2,Menu.NONE,"返回");
     return super.onCreateOptionsMenu(menu);
}
```

（6）实现【保存】按钮函数，调用（4）中的 updateUser 方法将输入界面中的信息更新到记录存储当中。

```
public boolean onOptionsItemSelected(MenuItem item){
     switch(item.getItemId()){
     case 1://保存
     if(!nameEditText.getText().toString().equals(""))
```

```
    {
    user.setName(nameEditText.getText().toString());
    user.setMoblie(mobileEditText.getText().toString());
    user.setDanwei(danweiEditText.getText().toString());
    user.setQq(qqEditText.getText().toString());
    user.setAddress(addressEditText.getText().toString());
    ContactsTable ct=
                new ContactsTable(UpdateContactsActivity.this);
      //修改数据库联系人信息
      if(ct.updateUser(user))
      {
         Toast.makeText(UpdateContactsActivity.this,"修改成功!
              ",Toast.LENGTH_SHORT).show();
      }else
      {
         Toast.makeText(UpdateContactsActivity.this,"修改失败! ",
              Toast.LENGTH_SHORT). show();
       }
      }else
      {
        Toast.makeText(UpdateContactsActivity.this,"数据不能为空!
            ",Toast.LENGTH_SHORT).show();
      }
      break;
      case 2://返回
         finish();
         break;
      default:
       break;
  }
  return super.onOptionsItemSelected(item);
}
```

3.4　查找号码记录

一、任务分析

本任务需要实现的效果如图 3-6 所示。

当有很多条联系人数据存储在数据库中时，为了方便用户快速找到所需信息，需要提供一项查找功能。查询联系人界面应提供一个文本框，由用户输入联系人的部分信息后，单击【查询】按钮，界面显示出符合条件的联系人。要完成本次任务，需要思考如下两个问题。

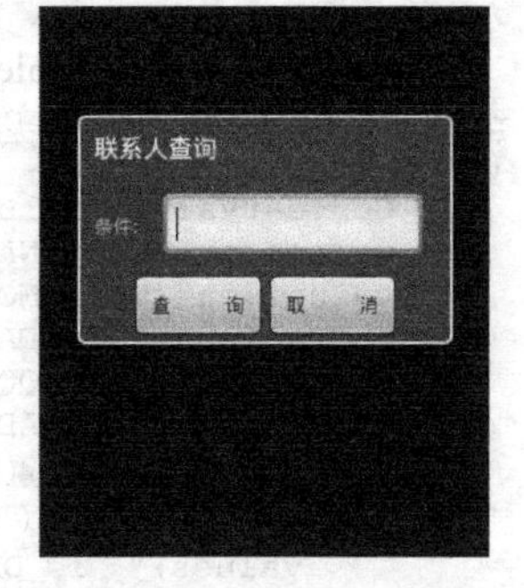

图 3-6　查询联系人记录

（1）如何把查询条件传递给 SQLite 数据库并进行模糊查询?

（2）如何把查询结果的每条记录显示在界面上?

二、相关知识

1. Dialog 类

对话框一般是出现在当前 Activity 之上的小窗口。处于下层的 Activity 失去焦点，对话框与用户进行交互，一般用于提示信息。对话框在程序中不是必备的，但是合适的对话框能够增加应用的友好性。常见的对话框情景有：用户登录、网络正在下载、下载成功或失败的提示、电池电量耗尽等。

Dialog 类是其他类型对话框的父类，可以继承 Dialog 自定义对话框，在任务实施部分将会演示如何定义一个查找对话框。

Dialog 类有两种构造方法，格式分别如下。

① Dialog（Context context）：使用默认的框架样式创建一个对话框，参数 context 表示上下文。

② Dialog（Context context，int theme）：使用定制的样式创建一个对话框，参数 theme 表示描述窗口主题的样式资源。可以在项目的“res/values”目录下定义 XML 样式文件。

Dialog 作为一个可以显示的对话框，很多实现方法和 Activity 相同，在此不再一一赘述，下面主要对其常用的方法进行介绍。

① void cancel()：退出对话框。

② void dismiss ()：退出对话框，功能和 cancel()相同。

③ void hide()：隐藏但不退出对话框。

④ void onCreate（Bundle savedInstanceState）：在对话框创建时被系统调用，利用该方法初始化对话框的设置，包括调用 setContentView（View）。

⑤ void setContentView（View view）：设置对话框的视图，参数 view 表示需要显示的内容。

⑥ void setContentView（int layoutResID）：设置对话框的视图，参数 layoutResID 表示资源布局文件的 ID。

⑦ void setTitle（int titleId）：设置对话框的标题，参数 titleId 表示标题文本的资源标识符。

⑧ void setTitle（CharSequence title）：设置对话框的标题，参数 title 表示标题文本的内容。

⑨ void show ()：启动对话框并显示在屏幕上。

Dialog 有 5 种不同类型的对话框的子类，其中警告对话框 AlertDialog 已经在项目二中进行了介绍，这里介绍另外 4 种类型对话框。

① 字符选择对话框（Character PickerDialog）：让用户选择设定的字符。例如图 3-7 所示的数字 0～9 和字母 A～F 就是通过程序定义并用对话框显示出来的。

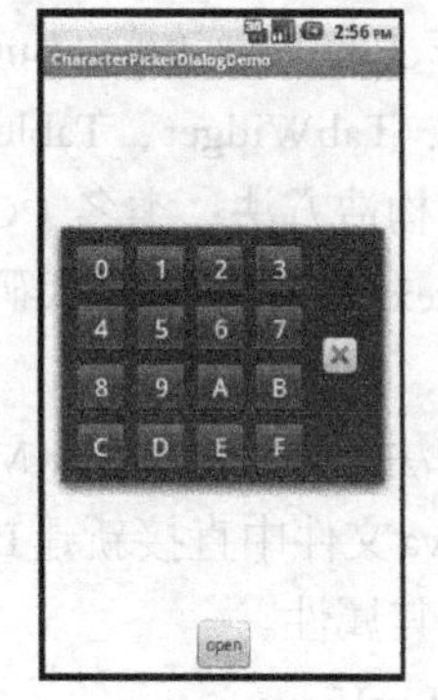

图 3-7 字符选择对话框示例 图 3-8 日期选择对话框示例

② 进度对话框（Progress Dialog）：显示一个圆形进度条或者一个长条形进度条。由于 ProgressDialog 是 AlertDialog 的扩展，所以 Progress Dialog 也支持按钮事件。ProgressDialog 类提供静态方法 ProgressDialog show（Context context，CharSequence title，Char Sequence message，boolean indeterminate）用于显示进度对话框。其中，参数 context 表示上下文；参数 title 表示进度对话框的标题；参数 message 表示对话框的内容；参数 indeterminate 表示进度是否不确定，如果取值为 true，表示不确定。

图 3-9 时间选择对话框示例 图 3-10 圆形进度条对话框示例

③ 日期选择对话框（Date Picker Dialog）：让用户选择一个日期，界面示例如图 3-8 所示。

④ 时间选择对话框（Time Picker Dialog）：让用户选择一个时间，界面示例如图 3-9 所示。

圆形进度条显示对话框的示例如下，要求实现图 3-10 所示的效果。

下面为进度对话框显示的示例：

```
public class ProgressDialogDemo extends Activity {
ProgressDialog dialog;
   @Override
   public void onCreate(Bundle savedInstanceState){
       super.onCreate(savedInstanceState);
       dialog=ProgressDialog.show(ProgressDialogDemo.this,"加载","加载中,请稍候..",true);
       //使用一个线程,在 4 秒后移除进度对话框
       new Thread(){
          public void run(){
           try{
            sleep(4000);

           }catch(Exception e){
            e.printStackTrace();
           }finally{
              dialog.dismiss();
           }
          }
       }.start();
   }
}
```

2. ListView 类

ListView 是比较常用的控件，以垂直列表的形式展示具体内容。内容可以是文字和图片，并且能够根据数据的数量自适应显示。本任务在显示联系人名单时就用到了 ListView。

除此之外，常用的 View 控件有：TextView、ImageView、ProgressBar、AutoComplete TextView、Button、CheckBox、EditText、ImageButton、ImageSwitcher、RadioButton、RadioGroup、SeekBar、Spinner、TabHost、TabWidget、TableLayout、WebView 等。

这些控件有一个类似的构造方法：类名（Context context）。例如 ListView 可以使用构造方法 ListView（Context context），若在 Java 代码中生成一个 View 实例，就可以用这个万金油式的构造方法。

创建 ListView 有 3 种方法：第 1 种是在 XML 文件中使用 ListView 控件；第 2 种是直接继承 ListActivity；第 3 种是在 Java 文件中直接新建 ListView 对象。建议在 XML 文件中使用 ListView 控件，这样可以较好地控制其属性。

在 Activity 中构造一个 ListView：

```
ListView listView=new ListView(this);
```

下面介绍 ListView 两个比较重要的方法。

（1）void setAdapter（ListAdapter adapter）方法：向 ListView 对象添加适配器。参数 adapter 表示适配器。这样当数据发生变化之后，ListView 会自动更新显示。

（2）void setOnItemClickListener（AdapterView.OnItemClickListener listener））：对适配器中的数据项添加监听器，当数据项被单击时，将触发相应的操作。参数 listener 表示监听器。下面是一般的编写方法，其中 listView 为 ListView 对象，onItemClick 方法的 position 参数表示用户在列表中单击数据的位置，数值从 0 开始。

```
listView.setOnItemClickListener(new OnItemClickListener() {
    public void onItemClick(AdapterView<?> parent, View view, int position, long id){
            // 在函数中编写数据项被单击后的处理代码
            …
    }
});
```

3．Adapter 类

Adapter 在 Android 应用程序中起着非常重要的作用，提供对数据的访问权限，也负责为每一项数据产生一个对应的 View。Adapter 的应用非常广泛，可看作是数据源和 ListView、Spinner、Gallery、GridView 等 UI 显示控件之间的桥梁。当数据源发生变化时，Adapter 可以使 UI 控件的数据自动更新显示。图 3-11 展示了 Adapter 在数据和 UI 控件之间的作用。比较常用的基础数据适配器有 BaseAdapter、ArrayAdapter、SimpleAdapter、SimpleCursorAdapter、CursorAdapter 等。其中 ArrayAdapter、CursorAdapter、SimpleAdapter 等还直接提供了 getFiter()方法对数据进行过滤的功能，例如用户在文本框中输入部分字符，控件可以只显示匹配的内容。

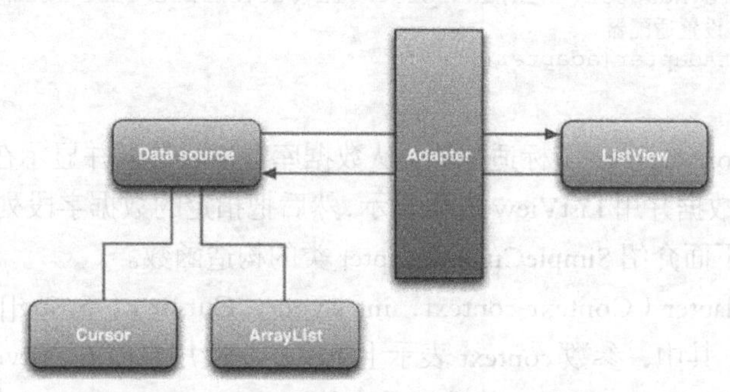

图 3-11　Adpater 在数据和 UI 控件之间的作用

（1）ArrayAdapter：数组适配器，使用起来较为简单，但只能用于显示文字。ArrayAdapter 常用的构造函数是 ArrayAdapter（Context context，int textViewResourceId，T[] objects），可创建一个用于装配数据的数组装配器。其中，参数 context 表示上下文，一般用当前的 Actvity 表示；参数 textViewResourceId 表示布局文件，该布局文件的作用是描述数组中每一条数据的显示布局(如定义一个 TextView），既可以使用 Android 内部提供的布局文件（如 android.R.layout.simple_list_item_1，android.R.layout.simple_expandable_ list_item_1，详细定义可以在 SDK 安装目录下查看），也可以使用开发者自己定义的布局文件；objects 表示要显示的数据数组。listView 会根据这 3 个参数，遍历数组中的每一条数据，每读出一条，则将其显示到第 2 个参数 textViewResourceId 对应的布局中，这样就形成了我们看到的 ListView 的显示效果。

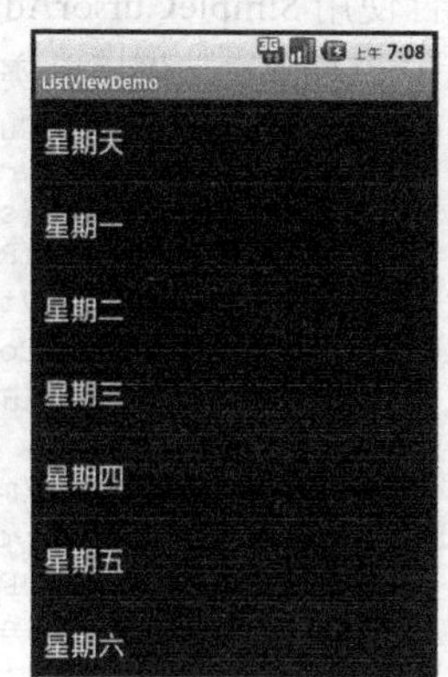

图 3-12　程序效果

下面举例进行说明，实现如图 3-12 所示的效果。

① 定义布局文件【listviewdemo.xml】，添加一个 ListView，代码如下。

```
<RelativeLayout xmlns:android="http://schemas.android.com/apk/res/android"
   xmlns:tools="http://schemas.android.com/tools"
   android:layout_width="match_parent"
   android:layout_height="match_parent" >
   <ListView
      android:id="@+id/list_view"
      android:layout_width="match_parent"
      android:layout_height="wrap_content"
      android:layout_alignParentLeft="true"
      android:layout_alignParentTop="true" >
   </ListView>
</RelativeLayout>
```

② ListViewDemo 程序代码如下。

```
public class ArrayAdapterDemo extends Activity {
   @Override
   public void onCreate(Bundle savedInstanceState){
      super.onCreate(savedInstanceState);
      //显示名为【listviewdemo】的布局文件
      setContentView(R.layout.listviewdemo);
      ListView listView =(ListView)findViewById(R.id.list_view);
      //用数组来表示列表项需要显示的数据
      String[] weeks={"星期天","星期一","星期二","星期三","星期四","星期五","星期六"};
      //建立数组适配器，android.R.layout.simple_list_item_1 为系统自带的布局
      ArrayAdapter adapter=
           new ArrayAdapter(this,android.R.layout.simple_list_item_1,weeks);
      //为 ListView 设置适配器
      listView.setAdapter(adapter);
   }
}
```

（2）SimpleCursorAdapter：游标适配器，从数据库中读取数据并显示在列表上。其原理是：从 Cursor 游标取出数据并用 ListView 进行显示，然后把指定的数据字段列映射到 TextView 或者 ImageView 中。下面介绍 SimpleCursorAdapter 类的构造函数。

SimpleCursorAdapter（Context context，int layout，Cursor c，String[] from，int[] to）：创建一个游标适配器。其中，参数 context 表示上下文，一般用当前的 Actvity 表示；参数 layout 表示布局文件，作用与 ArrayAdapter 构造函数中的 textViewResourceId 一样；参数 c 表示数据库游标，如果游标不可用，则取值为 null；参数 from 表示绑定到显示在 UI 上的数据字段列表；参数 to 表示对应显示参数 from 列的控件。从构造函数来看，以数据库作为数据源时才适合使用 SimpleCursorAdapter。

下面的示例用以实现将系统通讯录的联系人信息显示在列表中。

```
public class SimpleCursorAdapterDemo extends Activity {
  public void onCreate(Bundle savedInstanceState){
     super.onCreate(savedInstanceState);
     setContentView(R.layout.listviewdemo);
     ListView mListView =(ListView)findViewById(R.id.list_view);
Cursor mCursor= getContentResolver().query(People.CONTENT_URI,null,null,null,null);
     //将 Cursor 对象交由 Activity 管理,这样 Cursor 的生命周期和 Activity 便能够自动同步,省去手动管理
Cursor 的烦恼
startManagingCursor(mCursor);
     //在 android.R.layout.simple_expandable_list_item_1 文件中定义了 android.R.id.text1
     SimpleCursorAdapter mAdapter=
         new   SimpleCursorAdapter(this,android.R.layout.simple_expandable_list_item_1,
mCursor,new String[]{People.NAME},new int[]{android.R.id.text1});
      mListView.setAdapter(mAdapter);
   }
}
```

代码说明如下。

① getContentResolver()返回一个 ContentResolver 对象,用于访问某个 ContentProvider。

② 用 query(Uri uri,String[] projection,String selection,String[] selectionArgs,String sortOrder)方法获得一个指定 URI 的 Cursor 对象。其中参数 projection 表示返回的列，参数 selection 表示 SQL 的 where 查询条件，参数 selectionArgs 表示 where 语句中表达式的？占位参数列表，参数 sortOrder 表示要进行升序或降序的字段。

注意还需要在【AndroidMainfest.xml】文件中把读取通讯录的权限打开。

```
<uses-permission android:name="android.permission.READ_CONTACTS" />
```

（3）SimpleAdapter：具有较强的扩展性，能自定义出各种效果。不仅可以添加 ImageView（图片），还可以添加 Button（按钮）、CheckBox（复选框）等。SimpleAdapter 类的构造函数是：SimpleAdapter（Context context，List<? extends Map<String，?>> data，int resource，String[] from，int[] to）。其中，参数 context 表示上下文，一般用当前的 Actvity 表示；参数 data 表示由 Map 构成的 List，List 集合中的每个数据由一个 Map 集合组成，而每个 Map 集合的数据对应显示在 ListView 的每一行。要求 Map 集合的第一个关键字必须是 String 类型；参数 resource 表示布局文件，作用与 ArrayAdapter 构造函数中的 textViewResourceId 一样；参数 from 表示绑定到 UI 上的数据字段列表；参数 to 表示对应显示参数 from 列的控件。

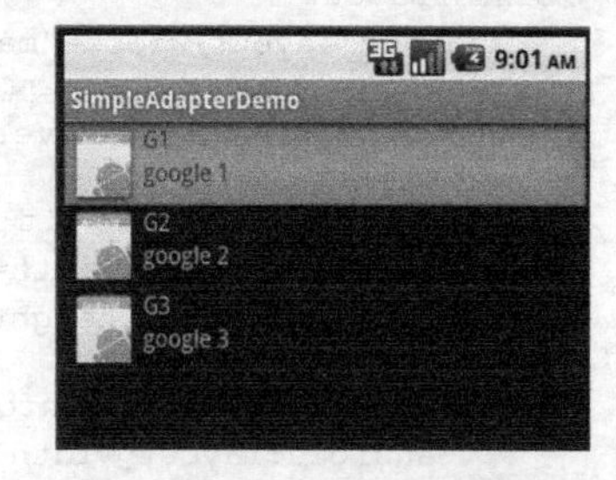

图 3-13　SimpleAdapter 示例实现界面

下面的示例通过直接继承 ListActivity 类，利用 SimpleAdapter 创建一个能够显示图片和文字的程序界面，如图 3-13 所示。ListActivity 类其实就是含有一个 ListView 控件的 Activity 类。如前面两个例子，若直接在普通的 Activity 中加入一个 ListView 也完全可以取代 ListActivity。

```
public class SimpleAdapterDemo extends ListActivity {
    public void onCreate(Bundle savedInstanceState){
      super.onCreate(savedInstanceState);
     SimpleAdapter adapter=new SimpleAdapter(this,getData(),R.layout.simpleadapter_demo,
           new String[]{"imge","title","info"},
           new int[]{R.id.imageView,R.id.titleView,R.id.infoView});
      setListAdapter(adapter);
    }

   private List<Map<String,Object>> getData() {
      List<Map<String,Object>> list=new ArrayList<Map<String,Object>>();
      Map<String,Object> map=new HashMap<String,Object>();
      map.put("title","G1");
      map.put("info","google 1");
      map.put("imge",R.drawable.icon);
      list.add(map);

      map=new HashMap<String,Object>();
      map.put("title","G2");
      map.put("info","google 2");
      map.put("imge",R.drawable.icon);
      list.add(map);

      map=new HashMap<String,Object>();
      map.put("title","G3");
      map.put("info","google 3");
      map.put("imge",R.drawable.icon);
      list.add(map);

      return list;
    }
}
```

代码中的<...>部分是用于限制集合的数据类型。出于简化代码的目的，也可以将 getData() 方法中的<...>代码全部删除掉，即：<Map<String,Object>>，<String,Object>。从上面的代码可以看出，要达到显示效果需要创建一个名为【simpleadapter_demo.xml】的布局文件，其包含一个用于显示图片的 ImageView 控件和两个显示文本的 TextView 控件。

```
<LinearLayout xmlns:android="http://schemas.android.com/apk/res/android"
   xmlns:tools="http://schemas.android.com/tools"
   android:id="@+id/LinearLayout1"
   android:layout_width="match_parent"
```

```
    android:layout_height="match_parent"
    android:orientation="horizontal" >
    <ImageView
        android:id="@+id/imageView"
        android:layout_width="wrap_content"
        android:layout_height="wrap_content" />
    <LinearLayout
        android:layout_width="match_parent"
        android:layout_height="wrap_content"
        android:orientation="vertical" >
        <TextView
            android:id="@+id/titleView"
            android:layout_width="wrap_content"
            android:layout_height="wrap_content" />
        <TextView
            android:id="@+id/infoView"
            android:layout_width="wrap_content"
            android:layout_height="wrap_content" />
    </LinearLayout>
</LinearLayout>
```

（4）BaseAdapter：通过继承 BaseAdapter 子类，可以在列表上添加处理事件，例如按钮单击处理。自定义 BaseAdapter 的子类能够灵活实现各种适配器数据显示效果，因此也能够实现前面 ArrayAdapter、SimpleCursorAdapter、SimpleAdapter 的效果。

定义一个 BaseAdapter 的子类，需要实现如下方法。

① int getCount()：返回适配器所表示的数据项数量。

② Object getItem（int position）：返回数据集中指定位置的数据项。其中，参数 position 表示位置。

③ long getItemId（int position）：返回列表中指定位置的行 ID。其中，参数 position 表示位置。

④ View getView（int position，View convertView，ViewGroup parent）：返回一个显示数据集指定位置数据的视图。参数 position 表示某个位置的数据；参数 convertView 用于显示每一数据项的视图；参数 parent 表示父视图，例如 Spinner、ListView、GridView 等，即显示最终会被附加到的父级视图。需要注意的是：当调用这个方法时，参数 convertView 是循环再用的。因此在编写该方法的实现代码时需要判断 convertView 对象是否为空，若为空，则需要为它实例化一个视图对象并配置相关的显示属性。getView 方法非常重要，用于定义数据如何显示在 ListView 中。

ListView 在开始绘制的时候，系统会首先调用 BaseAdapter 适配器的 getCount()方法，根据其返回值得到 ListView 的长度，然后根据这个长度，调用 getView()逐一绘制每一行。若 getCount()返回值为 0 时，列表将不显示数值。

当视图关联的数据集有变动时（在本项目中，将表示用户信息的数组 users 作为显示的数据源，见 3.9 项目的完整代码），可调用 BaseAdapter 类的 notifyDataSetChanged()方法通知 ListView 通过视图刷新数据，这样就避免了人工重新检索和更新数据显示的繁琐。

下面举例说明如何定义 BaseAdapter 适配器：

```
private class MyAdapter extends BaseAdapter{
    private String data[];//data 为一组需要显示的数据，用数组表示
    private Context context;
    public MyBaseAdapter(Context context,String data[]){
            this.data=data;
            this.context=context;
}
```

```
@Override
     public int getCount() {
        return data.length;
     }
     @Override
     public Object getItem(int position){
           //返回 position 位置处的数据
           return data[position];
     }
     @Override
     public long getItemId(int position){
           return position;
     }
     //下面代码处理数组显示，效果等同于 ArrayAdapter
     @Override
     public View getView(int position,View convertView,ViewGroup parent){
           //getApplicationContext()用于获取应用程序上下文，也可以使用 context
           TextView mTextView=new TextView(getApplicationContext());
           mTextView.setText("A BaseAdapter Demo");
           mTextView.setTextColor(Color. YELLOW);
           return mTextView;
     }
}
```

在 Actvity 程序中定义 MyAdapter 对象，通过 ListView 对象 setAdapter 方法去调用：

```
MyAdapter mybaseadapter=new MyAdapter (this,{ "周末","周一","周二","周三","周四","周五"});
listView.setAdapter(mybaseadapter);
```

相对于 ArrayAdapter、SimpleAdapter 等其他适配器，BaseAdapter 能够更灵活地去配置数据在 ListView 中的显示。尤其是在 ListView 中每一行的数据显示风格可能存在差异的情况，更加需要用 BaseAdapter 来进行判断，并设置每一行的显示风格。下面对它的使用步骤进行总结，以便于开发者更好地理解它的使用方法。

（1）定义一个 BaseAdapter 的子类，然后通过 Android 自带的纠错功能“Add unimplemented methods”来自动生成所需要实现的方法。系统自动会生成四个方法“getCount()”、“getItem()”、“getItemId()”、“getView()”。此外，开发者自己要另外创建一个构造方法，将需要显示的数据（其实就是数据源）以集合的方式传递进来，因此对应在子类中定义一个集合对象，用于接收构造方法传递过来的数据源。例如，

```
private class MyAdapter extends BaseAdapter{
     private String data[];//data 为一组需要显示的数据,用数组表示
     private Context context;
     public MyBaseAdapter(Context context,String data[]){
            this.data=data;// 利用构造方法将数据从外部传递到内部变量
            this.context=context;
        }
 }
```

（2）分别实现系统自动生成的四个方法，其中核心需要实现的方法是 getCount()和 getView()。getCount()方法是返回集合中数据的个数。getItem()返回对应位置（其实就是某一行）的数据、getItemId()返回对应数据的 ID、getView()返回对应数据的显示风格。

（3）方法 getView(int position, View convertView, ViewGroup parent)决定了 ListView 中每行数据的显示，因此需要在该方法编写具体的数据显示方式，主要有下面四个步骤。

① 首先判断参数 convertView 是否为空，若为空，则生成一个布局 LayoutInflater 对象，加载列表项布局。

```
if(convertView==null)
    {  //context 是上下文，可以通过构造方法传递进来。file_item 为列表项布局文件。
       convertView =context.getLayoutInflater().inflate(R.layout.file_item, null);
    }
```

② 通过 convertView.findViewById()方法获得列表项布局文件的各种控件。

③ 根据项目本身数据显示的需要，对第②步得到的控件进行赋值或者属性设置。其中可能用到判断语句，即根据不同的数据取值，显示不同的风格控件。

④ 在方法的最后返回 convertView 对象：return convertView;。

（4）上面的步骤完成了一个 BaseAdapter 子类的定义，在使用中，只需使用该子类定义一个对象，然后将这个对象作为参数赋值到 ListView.setAdapter()中即可。

4．查找记录的 SQL 语法

向 3.2 节的 MyDataBase 类中添加查询表数据的方法 selectData：

```
public Vector selectData(String sql){
        //Vector 表示一个集合，可以理解为动态数组，放在集合中的数据类型既可以普通类型，也可以是对象
        String s="";
        Vector result=new Vector();
        //查询 Sql 语句，将结果返回到一个游标对象
        Cursor cursor=db.rawQuery(sql, null);
   //通过循环，将游标中的数据取出
        while(cursor.moveToNext()!=false){
             s="姓名："+cursor.getString(cursor.getColumnIndex("name"));
             s=s+",年龄："+cursor.getInt(cursor.getColumnIndex("age"));
             //add 方法向集合添加对象
             result.add(s);
        }
        return result;
  }
```

继续使用 3.2 节中 MyDataBase 类定义的对象 database，显示查询结果：

```
sql=" select * from person";
Vector vector=database.selectData(sql);
for(int i=0;i<vector.size();i++){
    //get 方法就是根据索引从集合中取出某个对象（一般要做强制类型转换）
    Toast.makeText(this, (String)vector.get(i), Toast.LENGTH_SHORT).show();
}
```

三、任务实施

（1）定义一个 XML 布局页面【find.xml】作为查找界面，为用户提供输入搜索信息的文本框，并设置【查询】和【取消】按钮。

```
<?xml version="1.0" encoding="utf-8"?>
<LinearLayout
   xmlns:android="http://schemas.android.com/apk/res/android"
   android:orientation="vertical"
   android:layout_width="fill_parent"
   android:layout_height="fill_parent"
   >
  <LinearLayout
   android:layout_width="wrap_content"
   android:layout_height="wrap_content"
   android:orientation="horizontal"
  >
      <TextView android:layout_width="50.0dip"
      android:layout_height="wrap_content"
      android:text="条件:"
      android:layout_marginLeft="10.0dip"
      />
      <EditText android:layout_width="190.0dip"
       android:id="@+id/value"
       android:layout_height="wrap_content"
       android:layout_marginRight="20.0dip" />
  </LinearLayout>
  <LinearLayout
   android:layout_width="wrap_content"
```

```
  android:layout_height="wrap_content"
  android:orientation="horizontal"
  android:layout_marginTop="10.0dip"
  android:layout_gravity="center_horizontal"
 >
  <Button
     android:layout_width="wrap_content"
     android:layout_height="wrap_content"
     android:id="@+id/find"
     android:text=" 查          询 " />
     <Button
     android:layout_width="wrap_content"
     android:layout_height="wrap_content"
     android:id="@+id/cancel"
     android:text=" 取          消 " />
  </LinearLayout>
</LinearLayout>
```

（2）在 ContactsTable.java 类中添加查询 findUserByKey（String key）方法。该方法使用符号“%”来模糊匹配联系人的用户名、手机号、QQ 号等信息，使用一个游标 Cursor 对象表示查找结果，然后通过循环调用取出游标中的数值。

```
public  User[] findUserByKey(String key)
{
      //定义一个专门存放 User 类型数据的 Vector 向量类对象
           Vector<User> v=new Vector<User>();
      Cursor cursor=null;
      try {
           cursor=db.find("select * from "+TABLENAME +"   where  "
             +User.NAME+" like '%"+key+"%' " +
             " or "+User.MOBLIE+" like '%"+key+"%' " +
             " or "+User.QQ+" like  '%"+key+"%' "
             ,null);
        while(cursor.moveToNext()){
          User temp=new User();
          temp.setId_DB(cursor.getInt(
             cursor.getColumnIndex("id_DB")));
          temp.setName(cursor.getString(
                 cursor.getColumnIndex(User.NAME)));
          temp.setMoblie(cursor.getString(
             cursor.getColumnIndex(User.MOBLIE)));
          temp.setDanwei(cursor.getString(
               cursor.getColumnIndex(User.DANWEI)));
          temp.setQq(cursor.getString(
               cursor.getColumnIndex(User.QQ)));
          temp.setAddress(cursor.getString(
               cursor.getColumnIndex(User.ADDRESS)));
          //将符合查询条件的 User 对象添加到向量集合中
          v.add(temp);
        }
      } catch(Exception e){
          e.printStackTrace();
      } finally {
          if(cursor != null){
              cursor.close();
           }
           db.closeConnection();
      }
      if(v.size() > 0){
          //将 vector 中的数据，强制以 User 数组形式返回
          return v.toArray(new User[] {});
      }else
      {
        User[]  users=new User[1];
        User  user=new User();
        user.setName("无结果");
```

```
        users[0]=user;
        return users;
    }
}
```

（3）定义用户查找对话框类，继承于 Dialog 类，显示用户布局页面【find.xml】，并处理【查找】和【退出】按钮的相关事件。

```
public  class FindDialog extends Dialog{
    public FindDialog(Context context){
        super(context);

    }
    protected void onCreate(Bundle savedInstanceState){
        super.onCreate(savedInstanceState);
        setContentView(R.layout.find);
        setTitle("联系人查询");
        Button find=(Button)findViewById(R.id.find);
        Button cancel=(Button)findViewById(R.id.cancel);
        find.setOnClickListener(new View.OnClickListener() {
          @Override
              public void onClick(View v){
              EditText value=(EditText)findViewById(R.id.value);
        ContactsTable ct=newContactsTable(MyContactsActivity.this);
      //查询结果，这里出于示例目的，将 users 设置为局部变量，在完整项目中它应是全局变量
          User users=ct.findUserByKey(value.getText().toString());
              for(int i=0;i<users.length;i++)
              {
                System.out.println("姓名是"+users[i].getName()+
                             ",电话是" +users[i].getMoblie());
                  }
              dismiss();
            }
        });
        cancel.setOnClickListener(new View.OnClickListener() {

            @Override
            public void onClick(View v){
                // TODO Auto-generated method stub
                dismiss();
                }
          });
    }
}
```

3.5 查看联系人记录

一、任务分析

本任务需要实现的效果如图 3-14 所示。

本次任务的实现方法和 3.3 节具有一定的相似性，区别在于 3.3 节任务的作用是可以修改联系人记录，而本任务只是将联系人的信息显示在界面上，不允许用户修改联系人的信息。要完成本次任务，需要思考如何在界面上只显示信息？

图 3-14 查看记录

二、任务实施

（1）设计用户显示界面，用于显示联系人的元素都使用 TextView 类，这样用户便无法对数据进行修改。

```
<?xml version="1.0" encoding="utf-8"?>
<LinearLayout
   xmlns:android="http://schemas.android.com/apk/res/android"
```

```
  android:orientation="vertical"
  android:layout_width="fill_parent"
  android:layout_height="fill_parent"
  >
  <!--姓名-->
  <TextView
      android:id="@+id/name"
      android:layout_width="fill_parent"
      android:layout_height="wrap_content"
      android:text="姓名:"
      android:textSize="22.0dip"
      android:layout_marginLeft="5.0dip"
  />
  <!--单位-->
  <TextView
      android:id="@+id/danwei"
      android:layout_width="fill_parent"
      android:layout_height="wrap_content"
      android:text="单位:"
      android:textSize="22.0dip"
      android:layout_marginLeft="5.0dip"
   />

  <!--手机号码-->
  <TextView
      android:id="@+id/mobile"
      android:layout_width="fill_parent"
      android:layout_height="wrap_content"
      android:text="电话:"
      android:textSize="22.0dip"
      android:layout_marginLeft="5.0dip"
  />
  <!--qq-->
   <TextView
      android:id="@+id/qq"
      android:layout_width="fill_parent"
      android:layout_height="wrap_content"
      android:text="QQ:"
      android:textSize="22.0dip"
      android:layout_marginLeft="5.0dip"
      />
   <!--地址-->
   <TextView
      android:id="@+id/address"
      android:layout_width="fill_parent"
      android:layout_height="wrap_content"
      android:text="地址:"
      android:textSize="22.0dip"
      android:layout_marginLeft="5.0dip"
  />
</LinearLayout>
```

（2）将联系人数据赋值到用户界面。

```
public void onCreate(Bundle savedInstanceState){
      super.onCreate(savedInstanceState);
      setContentView(R.layout.message);
      setTitle("联系人信息");

      //从布局文件中获得对应的控件
      nameTextView=(TextView)findViewById(R.id.name);
      mobileTextView=(TextView)findViewById(R.id.mobile);
      danweiTextView=(TextView)findViewById(R.id.danwei);
      qqTextView=(TextView)findViewById(R.id.qq);
      addressTextView=(TextView)findViewById(R.id.address);
```

```
    //通过 Intent 将联系人数据赋值到用户界面进行显示，数据来自于其他的 Activity
    Bundle localBundle=getIntent().getExtras();
    int id=localBundle.getInt("user_ID");
    ContactsTable ct=new ContactsTable(this);
    user =ct.getUserByID(id);
    nameTextView.setText("姓名:"+user.getName());
    mobileTextView.setText("电话:"+user.getMoblie());
    qqTextView.setText("QQ:"+user.getQq());
    danweiTextView.setText("单位:"+user.getDanwei());
    addressTextView.setText("地址:"+user.getAddress());
}
```

3.6　删除号码记录

一、任务分析

为用户单击菜单上的“删除联系人”操作时，应提示用户“是否确定删除选定的记录”，如果用户选择【确定】，则将该记录删除；如果用户选择【取消】，则不执行删除操作。要完成本次任务，需要思考如下两个问题。

（1）如果在界面上为用户提供删除操作的信息提示？

（2）如何将一条记录从 SQLite 中删除？

二、相关知识

删除一条记录使用 SQLiteDatabase 类的 delete（String table，String whereClause，String[] whereArgs）方法。从参数可以看出，删除是以记录的 where 条件参数作为依据的。示例如下。

```
//删除 student 表中 name 值为“小明”的记录
String sName="小明";
db.delete("student","name=?",new String[]{sName };
或者：db.delete("student","name="+ sName,null);
//删除 student 表中 name 值不是“小明”且年龄大于 18 的记录
String sName="小明";
int iAge=18;
db.delete("student","name<>? and age>?",new String[]{sName,iAge };
或：db.delete("student","name<>"+ sName+"and age>"+iAge,null);
```

三、任务实施

（1）在 ContactsTable.java 类中添加 deleteByUser（User user）方法，该方法根据联系人表中的主键 ID 来删除联系人。

```
public  boolean  deleteByUser(User user)
{
    return db.delete(TABLENAME,"id_DB=?",new String[]{user.getId_DB()+""});
}
```

（2）利用 AlertDialog 定义一个显示窗口，让用户选择是否确定删除某个联系人记录。如果用户选择【确定】，则删除该联系人；否则，返回原有联系人显示列表。

```
  Builder alert=new AlertDialog.Builder(this);
   alert.setTitle("系统信息");
   alert.setMessage("是否删除联系人？");
   alert.setPositiveButton("是", new DialogInterface.OnClickListener() {
       public void onClick(DialogInterface dialog,int whichButton){
           ContactsTable ct=new ContactsTable(MyContactsActivity.this);
           //删除联系人信息
           if(ct.deleteByUser(users[selecteItem]))
           {
              Toast.makeText(MyContactsActivity.this,"删除成功！",
                                  Toast.LENGTH_SHORT).show();
```

```
            //打开联系人记录显示列表界面,显示更新后的联系人记录,代码省略
            …
        }else
        {
            Toast.makeText(MyContactsActivity.this,"删除失败!",
                                Toast.LENGTH_SHORT).show();
        }
    }
});
alert.setNegativeButton("否",new DialogInterface.OnClickListener() {
    public void onClick(DialogInterface dialog,int whichButton){

    }
});
alert.show();
```

3.7 对外共享数据

一、任务分析

Android 系统本身已经自带了一个通讯录，本次任务是将本项目创建的通讯录数据和 Android 系统的通讯录共享，实现将本项目创建的通讯录数据导入到系统通讯录中。要完成本次任务，需要思考如下两个问题。

（1）如何理解 Android 的系统通讯录？

（2）如何访问系统通讯录，并对其进行操作？

二、相关知识

1. ContentProvider 类

通过 SQLiteDatabase 创建的数据库只能被其创建者使用，其他的应用不能访问，所以 Android 提供了 ContentProvider 来实现对外数据的共享。一个程序如果允许其程序操作自己的数据，就可以定义自己的 ContentProvider，然后在【AndroidManifest.xml】中注册，其他 Application 可以通过 URI 有选择性地操作程序中的数据。若使用其他方法也可以对外共享数据，但数据访问方式会因数据存储方式而不同，如文件存储、SharedPreferences 数据存储、SQLite 数据存储等访问数据的方法均不相同。

当外部应用需要对 ContentProvider 中的数据进行添加、删除、修改和查询操作时，可以使用 ContentResolver 类来完成。ContentResolver 类的作用是能够访问 ContentProvider，提供了 insert、delete、query、update 等主要接口。要获取 ContentResolver 对象，可以使用 Activity 提供的 getContentResolver()方法，即：

```
ContentResolver contentresolver=getContentResolver();
```

下面对 ContentResolver 的主要方法进行介绍。

（1）Uri insert（Uri url，ContentValues values）：向给定 URI 的表中插入一行数据。

（2）Cursor query（Uri uri，String[] projection，String selection，String[] selectionArgs，String sortOrder）：根据给定的 URI 数据，进行数据查询，查询结果集用游标对象表示。其中，参数 uri 表示数据的 URI；参数 projection 表示返回的列；参数 selection 表示指定需要返回哪些行的 where 语句（不包括 SQL 的关键字 where），若取值为 null 则表示返回所有的行；参数 selectionArgs 表示 where 语句中表达式的“?”占位参数列表，参数只能为 String 类型；参数 sortOrder 表示对结果集进行排序的 order by 语句（不包括 SQL 的 order by 关键字），若取值为 null 将对结果集使用默认的排序，通常不进行排序。

（3）update（Uri uri，ContentValues values，String where，String[] selectionArgs）：根据给定的 URI 更新行数据。其中，参数 uri 表示数据的 URI；参数 values 表示要插入行数据的 ContentValues 对象，即列名和列值的映射；参数 where 表示指定需要更新哪些行的 where 语句（不包括 SQL 的关键字 where），若取值为 null 则表示更新所有的行；参数 selectionArgs 表示 where 语句中表达式的"?"占位参数列表，参数只能为 String 类型。

（4）delete（Uri url，String where，String[] selectionArgs）：删除给定 URI 的记录。其中，参数 url 表示数据的 URI；参数 where 表示删除数据的条件 where 语句（不包括 SQL 的关键字 where），如果取值为 null 则表示删除所有的记录；参数 whereArgs 表示 where 语句中表达式的"?"占位参数列表，参数只能为 String 类型。

Android 中的通讯录数据即通过 ContentProvider 实现数据共享。实际上，在 Android 系统中已经存在很多共享的 URI。可以使用 ContentResolver 来操作不同表之间的数据，如 MediaStore.Images.Media.INTERNAL_ CONTENT_URI（或 content://media/internal/images）表示访问设备上存储的所有图片。

当应用程序需要通过 ContentProvider 对外共享数据时，第一步需要继承 ContentProvider 并重写下面的方法。

```
public class MyProvider extends ContentProvider{
  public boolean onCreate()
  //数据插入
  public Uri insert(Uri uri,ContentValues values)
  //数据删除
  public int delete(Uri uri,String selection,String[] selectionArgs)
  //数据更新
  public int update(Uri uri,ContentValues values,String selection,String[] selectionArgs)
  //数据查询
  public Cursor query(Uri uri,String[] projection,String selection,String[] selectionArgs,
String sortOrder)
  //返回数据的 MIME 类型
  public String getType(Uri uri)
}
```

为了能让其他应用程序找到该 ContentProvider，第二步需要在【AndroidManifest.xml】中使用<provider>对该 ContentProvider 进行配置。采用 authorities（主机名/域名）对 ContentProvider 进行唯一标识。

```
<manifest.... >
  <application android:icon="@drawable/icon" android:label="@string/app_name">
    <provider android:name=".MyProvider"
       android:authorities="com.example.providers.myprovider"/>
    </application>
</manifest>
```

其中，android:name 需要设置为 ContentProvider 的子类名；android:authorities 则指定了在 content://样式的 URI 中标识 ContentProvider 的字符串。另外，可以设置 android:readPermission 和 android:writePermission 两个属性，来分别指定对 ContentProvider 中数据进行读、写操作的权限。也可以在 ContentProvider 类的 onCreate()方法中调用 setReadPermission()和 setWritePermission()方法来动态指定权限。

2．Android 系统通讯录

Android 系统的通讯录已经通过 ContentProvider 封装。我们可以通过 SDK 提供的 URI 和字段来对其进行增加、删除、修改、查询等操作。对系统通讯录进行操作，需要在【AndroidManifest.xml】文件中添加如下的读写权限。

```
<uses-permission android:name="android.permission.WRITE_CONTACTS"></uses-permission>
<uses-permission android:name="android.permission.READ_CONTACTS"></uses-permission>
```

ContactsContract 类主要用于表示通讯录提供者和应用程序之间的约定，包括支持的 URI 和列的定义。其中 ContactsContract.Data.CONTENT_URI（或 content://com.android.contacts/contacts）表示系统联系人的 URI。

ContentProvider 自己管理一个 SQLite 数据库文件，名为【contacts2.db】。该文件的路径为 “/data/data/com.android.providers.contacts/databases/contacts2.db”。从手机模拟器 DDMS 的操作界面图中可以看出，除了 Android 的通讯录信息放在 “/databases” 目录之外，Android 系统的其他 ContentProvider 也放在相应的目录下，如图 3-15 所示。

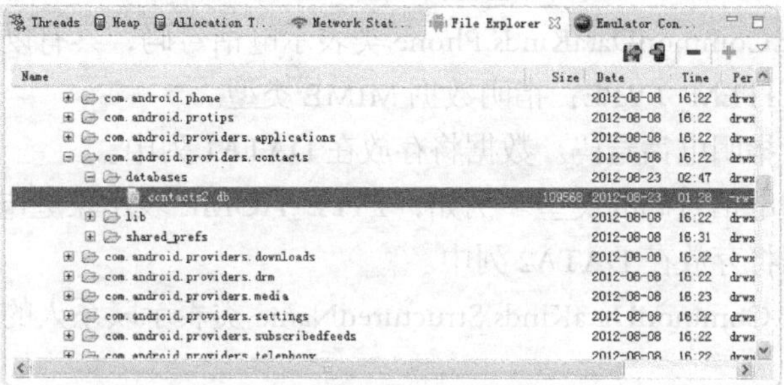

图 3-15 数据库文件浏览

利用【File Explorer】选项卡中右上角的【Pull a file from the device】按钮，可以将 contacts 2.db 数据库文件从手机模拟器中导出到计算机上。

SQLite 的官方网站 http://www.sqlite.org/cvstrac/wiki?p=ManagementTools 列出了很多查看 SQLite 数据库的软件。这里以 SQLite Administrator(http://sqliteadmin.orbmu2k.de/)为例进行介绍。需要注意的是，SQLite Administrator 不能够读取存放在中文目录下的 SQLite 数据库，因此需要将 contacts2.db 放在没有中文的目录下。

在手机模拟器的系统通讯录程序上创建两个联系人信息，名字分别为 adam 和 peter，电话号码分别为 386588888 和 356288888，图 3-16 所示为打开的 contacts2.db 的 data 表信息。

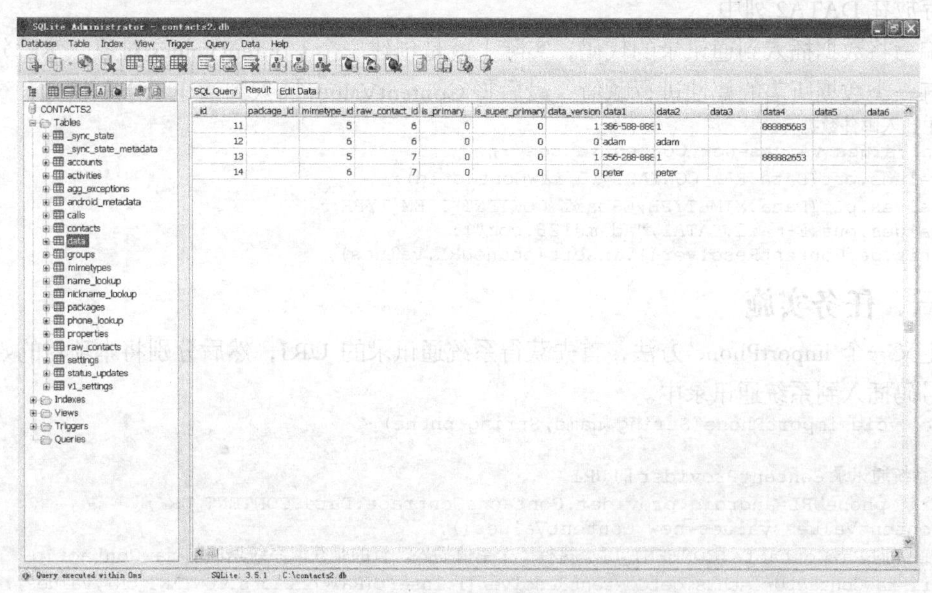

图 3-16 contacts2.db 的 data 表信息

针对 data 表，Android 提供了 ContactsContract.Data 类进行管理。该类表示的是一个通用表，可以保存任何类型的联系人数据，如电话号码和 E-mail 地址。每类数据存储在一个给定的行中，并指定行的 MIMETYPE 值。MIMETYPE 的取值确定了表中 DATA1～DATA15 列的含义。DATA1 列是索引列，通常用于存放在查询条件中最常用的数据。例如某一行表示邮件地址，那么 DATA1 应该存放邮件地址，而 DATA2 等其他列则用来存放邮件地址类型的辅助信息。DATA15 一般用于存放二进制数据。

ContactsContract.CommonDataKinds 类中定义了一些数据类型的种类，例如，ContactsContract.CommonDataKinds.Phone、ContactsContract.CommonDataKinds.StructuredName、ContactsContract.Common DataKinds.E-mail，下面分别进行介绍。

ContactsContract.CommonDataKinds.Phone 类表示电话号码，具有以下重要属性。

① CONTENT_ITEM_TYPE：指明数据 MIME 类型。

② NUMBER：指明电话号码。数据将存放在 DATA1 列中。

③ TYPE：指明电话号码的类型。例如，TYPE_HOME 表示家庭电话；TYPE_MOBILE 表示移动电话。数据将存放在 DATA2 列中。

ContactsContract.CommonDataKinds.StructuredName 类表示联系人的名字，具有以下的重要属性。

① CONTENT_ITEM_TYPE：数据 MIME 类型。

② GIVEN_NAME：指明联系人名字。数据放在 DATA2 列中。

③ FAMILY_NAME：指明联系人姓氏。数据放在 DATA3 列中。

④ PREFIX：尊称前缀，例如，Mr、Ms、Dr。数据放在 DATA4 列中。

ContactsContract.CommonDataKinds.E-mail 表示 E-mail 信息，具有以下重要属性。

① CONTENT_ITEM_TYPE：指明数据 MIME 类型。

② ADDRESS：指明 E-mail 地址，存放在 DATA1 列中。注意 ADDRESS 名在 Android 3.0 版本后才有定义，在此版本前的版本中可以直接使用 DATA1。

③ TYPE：指明邮件类型。例如，TYPE_HOME 表示家庭邮件、TYPE_WORK 工作邮件。存放在 DATA2 列中。

编写将数据插入到通讯录的代码，实际上就是创建一个 ContentValues 对象，然后分别为上述每一类数据所需的属性进行赋值，最后将 ContentValues 插入到通讯录中。例如，

```
//向表中插入邮件数据
ContentValues values=new ContentValues();
    values.put(Data.RAW_CONTACT_ID,rawContactId);
    values.put(Data.MIMETYPE,E-mail.CONTENT_ITEM_TYPE);
    values.put(E-mail.DATA1,"adam@126.com");
    this.getContentResolver().insert(phoneURL,values);
```

三、任务实施

定义一个 importPhone 方法，首先获得系统通讯录的 URI，然后分别将本项目的姓名和电话号码插入到系统通讯录中。

```
public  void importPhone(String name,String phone)
{
    //系统通讯录 ContentProvider 的 URI
    Uri  phoneURL=android.provider.ContactsContract.Data.CONTENT_URI;
    ContentValues values=new ContentValues();
    //首先向 RawContacts.CONTENT_URI 执行一个空值插入，目的是获取系统返回的 rawContactId
    Uri rawContactUri=this.getContentResolver().insert(RawContacts.CONTENT_URI,values);
    long rawContactId=ContentUris.parseId(rawContactUri);
```

```
    //向 data 表插入姓名
    values.clear();
    values.put(Data.RAW_CONTACT_ID,rawContactId);
    values.put(Data.MIMETYPE,StructuredName.CONTENT_ITEM_TYPE);
    values.put(StructuredName.GIVEN_NAME,name);
    this.getContentResolver().insert(phoneURL,values);
    //向 data 表插入电话号码
    values.clear();
    values.put(Data.RAW_CONTACT_ID,rawContactId);
    values.put(Data.MIMETYPE,Phone.CONTENT_ITEM_TYPE);
    values.put(Phone.NUMBER,phone);
    values.put(Phone.TYPE,Phone.TYPE_MOBILE);
    this.getContentResolver().insert(phoneURL,values);
}
```

3.8 设计主界面

一、任务分析

本任务需要实现的效果如图 3-17 所示。

用户在打开手机联系人程序时，主界面将以列表形式显示出当前存储器中的联系人信息。手机通讯录程序有添加、编辑、删除、查看信息、查询和导入到手机电话簿、退出 7 项功能。将这些操作集成在主界面上，可以方便用户随时进行各种操作。要完成本次任务，需要思考如下两个问题。

（1）如何将存储器中的主要信息显示在主界面上？

（2）如何把添加、编辑、删除、查看信息、查询和导入到手机电话簿、退出 7 项功能较好地集成在主界面上？

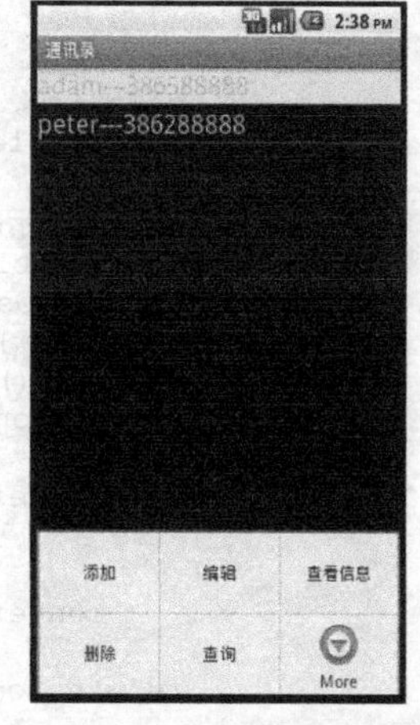

图 3-17 示例实现主界面

二、任务实施

（1）设计用户显示界面，主界面由 ListView 和菜单组成。

```
<?xml version="1.0" encoding="utf-8"?>
<LinearLayout
    xmlns:android="http://schemas.android.com/apk/res/android"
    android:orientation="vertical"
    android:layout_width="fill_parent"
    android:layout_height="fill_parent"
    >
    <ListView
        android:id="@+id/listView"
        android:layout_weight="1.0"
        android:layout_width="fill_parent"
        android:layout_height="fill_parent"
    />
</LinearLayout>
```

（2）将存储器中的联系人信息显示在列表（ListView）上，将【添加】、【编辑】、【删除】、【查看信息】、【查询】、【导入到手机电话簿】和【退出】等功能作为主界面的菜单项。

```
public class MyContactsActivity extends Activity {
    private  ListView listView; //显示结果列表
    private BaseAdapter  listViewAdapter; //ListView 列表适配器
    private  User users[]; //通讯录用户
    private  int selecteItem=0; //当前选择的数据项
    @Override
    public void onCreate(Bundle savedInstanceState){
        super.onCreate(savedInstanceState);
        setContentView(R.layout.main);
```

```
        setTitle("通讯录");
        listView =(ListView)findViewById(R.id.listView);
        loadContacts();
    }
    /**
     * 加载联系人列表
     */
    private void  loadContacts()
    {
      //获取所有通讯录联系人
      ContactsTable ct=new ContactsTable(this);
      users=ct.getAllUser();
      //采用匿名内部类方式定义BaseAdapter适配器对象
     listViewAdapter=new BaseAdapter() {
            @Override
            public View getView(int position,
                    View convertView,ViewGroup parent){
             if(convertView==null)
             {
               TextView textView=new TextView(MyContactsActivity.this);
              textView.setTextSize(22);
              convertView=textView;
             }
             //适配器显示的数据源是users数组
             String moblie=users[position].getMoblie()==null?""
              :users[position].getMoblie();
             ((TextView)convertView).setText(users[position]
              .getName()+"---"+moblie);
             if(position==selecteItem)
             {
                 convertView.setBackgroundColor(Color.YELLOW);
             }else
             {
                 convertView.setBackgroundColor(0);
             }
              return convertView;
            }
            @Override
            public long getItemId(int position){
               return position;
            }
            @Override
            public Object getItem(int position){
               return users[position];
            }
            @Override
            public int getCount() {
               return users.length;
            }
        };
        //设置listView控件的适配器
        listView.setAdapter(listViewAdapter);
        listView.setOnItemClickListener(new OnItemClickListener() {
         @Override
         public   void   onItemClick(AdapterView<?>   arg0,View   arg1,int   arg2,long
arg3){                          selecteItem=arg2; //记录单击列的位置
             listViewAdapter.notifyDataSetChanged();//刷新列表
         }

     });
    }

    /**
  * 创建菜单
  */
public boolean onCreateOptionsMenu(Menu menu){
    menu.add(Menu.NONE,1,Menu.NONE,"添加");
```

```
    menu.add(Menu.NONE,2,Menu.NONE,"编辑");
    menu.add(Menu.NONE,3,Menu.NONE,"查看信息");
    menu.add(Menu.NONE,4,Menu.NONE,"删除");
    menu.add(Menu.NONE,5,Menu.NONE,"查询");
    menu.add(Menu.NONE,6,Menu.NONE,"导入到手机电话簿");
    menu.add(Menu.NONE,7,Menu.NONE,"退出");
    return super.onCreateOptionsMenu(menu);
   }
  /**
   * 菜单事件
   */
public boolean onOptionsItemSelected(MenuItem item){
        // TODO Auto-generated method stub
        switch(item.getItemId()){
      case 1://添加
        //添加新的一条记录,在 3.2 节已经实现
         break;
      case 2://编辑
        //编辑选中的记录,在 3.3 节已经实现
         break;
      case 3://查看信息
         //查看选中的记录,在 3.5 节已经实现
         break;
      case 4://删除
          //删除选中的记录,在 3.6 节已经实现
          break;
      case 5://查询
          //打开查询界面,在 3.4 节已经实现
          break;
      case 6:// 导入到手机电话簿
          //导入到手机电话簿功能,在 3.7 节已经实现
          break;
      case 7://退出
          System.exit(0)
          break;
      default:
           break;
      }
      return super.onOptionsItemSelected(item);
    }
    @Override
    protected void onResume() {
        // TODO Auto-generated method stub
        super.onResume();
    }
}
```

3.9　完整项目实施

手机通讯录程序由多个类和 XML 布局文件组成，下面分别对每个类的实现进行详细介绍。

（1）Android 项目配置文件：【AndroidManifest.xml】。

```
<?xml version="1.0" encoding="utf-8"?>
<manifest xmlns:android="http://schemas.android.com/apk/res/android"
     package="com.demo.pr3"
     android:versionCode="1"
     android:versionName="1.0">
   <uses-sdk android:minSdkVersion="8" />
   <application android:icon="@drawable/icon"
              android:label="@string/app_name">
     <activity android:name=".MyContactsActivity"
             android:label="@string/app_name">
        <intent-filter>
           <action android:name="android.intent.action.MAIN" />
```

```
        <category android:name="android.intent.category.LAUNCHER"/>
      </intent-filter>
    </activity>
    <!-- 增加号码录入 Activity -->
    <activity android:name=".AddContactsActivity"></activity>
    <!-- 修改号码录入 Activity -->
    <activity android:name=".UpdateContactsActivity"></activity>
    <!-- 查询号码录入 Activity -->
    <activity android:name=".ContactsMessageActivity"></activity>
  </application>
  <!-- 添加导出手机电话簿所需权限 -->
  <uses-permission android:name="android.permission.READ_CONTACTS" />
  <uses-permission android:name="android.permission.WRITE_CONTACTS"
  />
</manifest>
```

（2）手机通讯录程序主界面类：【MyContactsActivity.java】。

```
package com.demo.pr3;
import android.app.Activity;
import android.app.AlertDialog;
import android.app.Dialog;
import android.app.AlertDialog.Builder;
import android.content.ContentUris;
import android.content.ContentValues;
import android.content.Context;
import android.content.DialogInterface;
import android.content.Intent;
import android.graphics.Color;
import android.net.Uri;
import android.os.Bundle;
import android.provider.ContactsContract.RawContacts;
import android.provider.ContactsContract.CommonDataKinds.Phone;
import android.provider.ContactsContract.CommonDataKinds.*;
import android.provider.ContactsContract.Contacts.Data;
import android.view.Menu;
import android.view.MenuItem;
import android.view.View;
import android.view.ViewGroup;
import android.widget.AdapterView;
import android.widget.BaseAdapter;
import android.widget.Button;
import android.widget.EditText;
import android.widget.ListView;
import android.widget.TextView;
import android.widget.Toast;
import android.widget.AdapterView.OnItemClickListener;

import com.demo.pr3.datax.ContactsTable;
import com.demo.pr3.datax.User;
/********************(C)COPYRIGHT 2012********************
*主界面
**********************************************************/
public class MyContactsActivity extends Activity {
     private  ListView listView; //显示结果列表
     private BaseAdapter  listViewAdapter; //ListView 列表适配器
     private  User users[];//通讯录用户
     private  int selecteItem=0; //当前选择的数据项

    @Override
    public void onCreate(Bundle savedInstanceState){
       super.onCreate(savedInstanceState);
       setContentView(R.layout.main);
       setTitle("通讯录");
       listView =(ListView)findViewById(R.id.listView);
       loadContacts();
    }
```

```
/**
 * 加载联系人列表
 */
private void  loadContacts()
{
  //获取所有通讯录联系人
 ContactsTable ct=new ContactsTable(this);
 users=ct.getAllUser();
 //为 listView 列表创建适配器
 listViewAdapter=new BaseAdapter() {
        @Override
        public View getView(int position,View convertView,ViewGroup parent){
         if(convertView==null)
         {
             TextView textView=new TextView(MyContactsActivity.this);
             textView.setTextSize(22);
             convertView=textView;
         }
         String moblie=users[position].getMoblie()==null?""
                   :users[position].getMoblie();
         ((TextView)convertView).setText(users[position]
                 .getName()+"---"+moblie);
         if(position==selecteItem)
         {
              convertView.setBackgroundColor(Color.YELLOW);
         }else
         {
              convertView.setBackgroundColor(0);
         }
          return convertView;
        }
        @Override
        public long getItemId(int position){
           return position;
        }
        @Override
        public Object getItem(int position){
           return users[position];
        }
        @Override
        public int getCount() {
           return users.length;
        }
    };
    //设置 listView 控件的适配器
    listView.setAdapter(listViewAdapter);
    listView.setOnItemClickListener(new OnItemClickListener() {
    @Override
    public void onItemClick(AdapterView<?> arg0,View arg1,int arg2,long arg3){
                 selecteItem=arg2; //记录单击列的位置
         listViewAdapter.notifyDataSetChanged();//刷新列表
      }
    });
}
/**
  * 创建菜单
  */
 public boolean onCreateOptionsMenu(Menu menu){
     menu.add(Menu.NONE,1,Menu.NONE,"添加");
     menu.add(Menu.NONE,2,Menu.NONE,"编辑");
     menu.add(Menu.NONE,3,Menu.NONE,"查看信息");
     menu.add(Menu.NONE,4,Menu.NONE,"删除");
     menu.add(Menu.NONE,5,Menu.NONE,"查询");
     menu.add(Menu.NONE,6,Menu.NONE,"导入到手机电话簿");
     menu.add(Menu.NONE,7,Menu.NONE,"退出");
     return super.onCreateOptionsMenu(menu);
```

```
   }
   /**
    * 菜单事件
   */
public boolean onOptionsItemSelected(MenuItem item){
  String sHint;
  switch(item.getItemId()){
   case 1://添加
    Intent intent=new Intent(MyContactsActivity.this, AddContactsActivity.class);
    startActivity(intent);
    break;
   case 2://编辑
        //根据数据库ID判断当前记录是否可以操作
    if(users[selecteItem].getId_DB()>0){
        intent=new Intent(MyContactsActivity.this, UpdateContactsActivity.class);
        intent.putExtra("user_ID",users[selecteItem].getId_DB());
        startActivity(intent);
     }else{

        sHint="无结果记录,无法操作!";
             Toast.makeText(this,hint , Toast.LENGTH_SHORT).show();

     }
       break;
     case 3://查看信息
       if(users[selecteItem].getId_DB()>0)
       {
       intent=new Intent(MyContactsActivity.this, ContactsMessageActivity.class);
              intent.putExtra("user_ID",users[selecteItem].getId_DB());
              startActivity(intent);
        }else
        {
          sHint="无结果记录,无法操作!";
              Toast.makeText(this, sHint , Toast.LENGTH_SHORT).show();
        }
         break;
        case 4://删除
          if(users[selecteItem].getId_DB()>0)
          {
            delete();
          }else
          {
             sHint="无结果记录,无法操作!";
                  Toast.makeText(this, sHint,Toast.LENGTH_SHORT).show();
          }
          break;
        case 5://查询
           new FindDialog(this).show();
           break;
        case 6://导入到手机电话簿
          if(users[selecteItem].getId_DB()>0)
          {
            importPhone(users[selecteItem].getName(),
                          users[selecteItem].getMoblie());
  sHint="已经成功导入'"+users[selecteItem].getName()+"'导入到手机电话簿!";
Toast.makeText(this, sHint , Toast.LENGTH_SHORT).show();
          }else
          {
               sHint="无结果记录,无法操作!";
Toast.makeText(this, sHint, Toast.LENGTH_SHORT).show();
          }
          break;
        case 7://退出
            finish();
            break;
```

```
        default:
            break;
        }
            return super.onOptionsItemSelected(item);
}
@Override
protected void onResume() {
        super.onResume();
        //重新加载数据
        ContactsTable ct=new ContactsTable(this);
        users=ct.getAllUser();
        //刷新数据显示列表
        listViewAdapter.notifyDataSetChanged();
}
/**
* 查询
*/
public  class FindDialog extends Dialog{
        public FindDialog(Context context){
            super(context);
        }
        protected void onCreate(Bundle savedInstanceState){
            super.onCreate(savedInstanceState);
            setContentView(R.layout.find);
            setTitle("联系人查询");
            Button find=(Button)findViewById(R.id.find);
            Button cancel=(Button)findViewById(R.id.cancel);
            find.setOnClickListener(new View.OnClickListener() {
                @Override
                public void onClick(View v){
                    EditText value=(EditText)findViewById(R.id.value);
                    ContactsTable ct=new
                        ContactsTable(MyContactsActivity.this);
                    users=ct.findUserByKey(value.getText().toString());
                    for(int i=0;i<users.length;i++)
                    {
                     System.out.println("姓名是"+users[i].getName()+
                                ",电话是" +users[i].getMoblie());
                    }
                    listViewAdapter.notifyDataSetChanged();
                  selecteItem=0;
                    dismiss();
                }
            });
            cancel.setOnClickListener(new View.OnClickListener() {
                @Override
                public void onClick(View v){
                    dismiss();
                }
             });
        }
    }
    /**
     * 删除联系人
     */
    public  void delete()
    {
        Builder alert=new AlertDialog.Builder(this);
        alert.setTitle("系统信息");
        alert.setMessage("是否删除联系人？");
        alert.setPositiveButton("是",
        new DialogInterface.OnClickListener() {
          public void onClick(DialogInterface dialog,int whichButton){
            ContactsTable ct=new ContactsTable(MyContactsActivity.this);
            //删除联系人信息
            if(ct.deleteByUser(users[selecteItem]))
            {
```

```
                //重新获取数据
                users=ct.getAllUser();
                //刷新列表
                listViewAdapter.notifyDataSetChanged();
                selecteItem=0;
            Toast.makeText(MyContactsActivity.this,"删除成功！",Toast.LENGTH_SHORT).
show();
            }else
            {
            Toast.makeText(MyContactsActivity.this,"删除失败！",Toast.LENGTH_SHORT).
show();
            }
        }
    });
    alert.setNegativeButton("否",
    new DialogInterface.OnClickListener() {
        public void onClick(DialogInterface dialog,int whichButton)
    {

        }
    });
    alert.show();
}
///导入到手机电话簿
public  void importPhone(String name,String phone)
{
    //系统通讯录ContentProvider的URI
    Uri  phoneURL=android.provider.ContactsContract.Data.CONTENT_URI;
    ContentValues values=new ContentValues();
    //首先向RawContacts.CONTENT_URI执行一个空值插入，目的是获取系统返回的rawContactId
 Uri rawContactUri=this.getContentResolver().
 insert(RawContacts.CONTENT_URI,values);
 long rawContactId=ContentUris.parseId(rawContactUri);
 //向data表插入姓名
 values.clear();
 values.put(Data.RAW_CONTACT_ID,rawContactId);
 values.put(Data.MIMETYPE,StructuredName.CONTENT_ITEM_TYPE);
 values.put(StructuredName.GIVEN_NAME,name);
 this.getContentResolver().insert(phoneURL,values);
 //向data表插入电话号码
 values.clear();
 values.put(Data.RAW_CONTACT_ID,rawContactId);
 values.put(Data.MIMETYPE,Phone.CONTENT_ITEM_TYPE);
 values.put(Phone.NUMBER,phone);
 values.put(Phone.TYPE,Phone.TYPE_MOBILE);
 this.getContentResolver().insert(phoneURL,values);
 }
}
```

（3）手机通讯录程序数据库作类：【MyDB.java】。

```
package com.demo.pr3.datax;
import android.content.ContentValues;
import android.content.Context;
import android.database.Cursor;
import android.database.sqlite.SQLiteDatabase;
import android.database.sqlite.SQLiteOpenHelper;

/**
 *SQLite数据库操作管理
 */
public class MyDB extends SQLiteOpenHelper{
    private static String DB_NAME="My_DB.db";  //数据库名称
    private static int DB_VERSION=2;  //版本号
        private  SQLiteDatabase db; //数据库操作对象
        public  MyDB(Context context)
  {
```

```
        super(context,DB_NAME,null,DB_VERSION);
        db=getWritableDatabase();
    }
    @Override
    public void onCreate(SQLiteDatabase db){
        // TODO 创建数据库后，对数据库的操作
    }

    @Override
    public void onOpen(SQLiteDatabase db){
        // TODO 每次成功打开数据库后首先被执行
        super.onOpen(db);
    }

    @Override
    public void onUpgrade(SQLiteDatabase db,
            int oldVersion,int newVersion){
        // TODO 更改数据库版本的操作
    }
    /***
     * 执行 SQLite 数据库链接
     */
     public SQLiteDatabase openConnection (){
        if(!db.isOpen())
        {
            //以读写方式获取 SQLiteDatabase
            db=getWritableDatabase();
         }
         return db;
     }
     /***
     * 关闭 SQLite 数据库链接操作
     */
     public void closeConnection (){
         try{
            if(db!=null&&db.isOpen())
               db.close();
         }catch(Exception e)
         {
            e.printStackTrace();
         }
     }
     /**
     * 创建表
     */
     public  boolean creatTable(String  createTableSql)
     {
        try{
           openConnection();
           db.execSQL(createTableSql);
        }catch(Exception ex)
        {
           ex.printStackTrace();
           return false;
        }finally
        {
            closeConnection();
        }
        return true;
     }
     /**
     * 添加操作
     * @param tableName   表名
     * @param values      集合对象表示要插入表中的记录
     */
     public  boolean  save(String tableName,ContentValues values)
     {
```

```
    try{
        openConnection();
        db.insert(tableName,null,values);

     }catch(Exception ex)
     {
         ex.printStackTrace();
         return  false;
     }finally
     {
          closeConnection();
     }
     return true;
}
/**
 * 更新操作
 */
public  boolean  update(String table,
        ContentValues values,String whereClause,String []whereArgs)
{
    try{
       openConnection();
       db.update(table,values,whereClause,whereArgs);
    }catch(Exception ex)
    {
       ex.printStackTrace();
       return  false;
    }finally
    {
        closeConnection();
    }
   return true;
}
/**
 * 删除
 * @param   deleteSql   对应删除的条件
 * @param   obj[]       对应删除条件的值
 */
public  boolean  delete(String table,String deleteSql,String obj[])
{
     try{
         openConnection();
         db.delete(table,deleteSql,obj);

      }catch(Exception ex)
      {
          ex.printStackTrace();
          return  false;
      }finally
      {
          closeConnection();
      }
      return true;
}
/**
* 查询操作
* @param findSql  对应查询字段，如:
*    select * from person limit wher name=?
* @param obj      对应查询字段？占位参数的取值，如:
*    new String[]
*    String.valueOf(sName)}
*/
public  Cursor  find(String findSql,String obj[])
{

    try{
        openConnection();
```

```
            Cursor cursor=db.rawQuery(findSql,obj);
            return  cursor;

        }catch(Exception ex)
        {
            ex.printStackTrace();
             return null;
        }
    }
    /**
    * 判断表是否存在
    */
    public boolean isTableExits(String tablename){
        try{
            openConnection();
            String str="select count(*)xcount  from  " +tablename;
            db.rawQuery(str,null).close();
        }catch(Exception ex)
        {
            return false;
        }finally
        {
            closeConnection();
        }
        return true;
    }
}
```

（4）手机通讯录程序添加、编辑，查看信息操作类：【ContactsTable.java】。

```
package com.demo.pr3.datax;

import java.util.Vector;
import android.content.ContentValues;
import android.content.Context;
import android.database.Cursor;
/**
 * 联系人表操作
 */
public class ContactsTable {
private   final static String TABLENAME="contactsTable";//表名
private   MyDB   db;//数据库管理对象
public    ContactsTable(Context context)
{
    db=new  MyDB(context);
    if(!db.isTableExits(TABLENAME))
    {
        String   createTableSql="CREATE TABLE IF NOT EXISTS " +
        TABLENAME+"(id_DB  integer   " +
        "primary key  AUTOINCREMENT," +
        User.NAME    +"  VARCHAR," +
        User.MOBLIE   +"  VARCHAR,"+
        User.QQ   +"  VARCHAR,"+
        User.DANWEI   +"  VARCHAR,"+
        User.ADDRESS+ " VARCHAR)";
        //创建表
         db.creatTable(createTableSql);
    }
}
/**
 * 添加数据到联系人表
*/
public  boolean  addData(User user)
{

     ContentValues values=new ContentValues();
     values.put(User.NAME,user.getName());
     values.put(User.MOBLIE,user.getMoblie());
     values.put(User.DANWEI,user.getDanwei());
```

```
    values.put(User.QQ,user.getQq());
    values.put(User.ADDRESS,user.getAddress());
    return   db.save(TABLENAME,values);
}

/**
* 获取联系人表数据
*/
public  User[] getAllUser()
{
    Vector<User> v=new Vector<User>();
        Cursor cursor=null;
        try {
          cursor=db.find("select * from "+TABLENAME,null);
          while(cursor.moveToNext()){
           User temp=new User();
            temp.setId_DB(cursor.getInt(cursor.getColumnIndex("id_DB")));
            temp.setName(cursor.getString(cursor.getColumnIndex(User.NAME)));
            temp.setMoblie(cursor.getString(cursor.getColumnIndex(User.MOBLIE)));
            temp.setDanwei(cursor.getString(cursor.getColumnIndex(User.DANWEI)));
            temp.setQq(cursor.getString(cursor.getColumnIndex(User.QQ)));
           temp.setAddress(cursor.getString(cursor.getColumnIndex(User.ADDRESS)));
            v.add(temp);
          }
        } catch(Exception e){
            e.printStackTrace();
        } finally {
            if(cursor != null){
                cursor.close();
            }
            db.closeConnection();
        }
        if(v.size() > 0){
            return v.toArray(new User[] {});
        }else
        {
          User[]  users=new User[1];
          User  user=new User();
          user.setName("无结果");
          users[0]=user;
          return users;
        }
}
/**
 * 根据数据库改变主键 ID 来获取联系人
*/
public User getUserByID(int id)
{
    Cursor cursor=null;
    try {
        cursor=db.find("select * from "+TABLENAME +"   where  "
             +"id_DB=?",new String[]{id+""});
        User temp=new User();
        cursor.moveToNext();
        temp.setId_DB(cursor.getInt(cursor.getColumnIndex("id_DB")));
        temp.setName(cursor.getString(cursor.getColumnIndex(User.NAME)));
        temp.setMoblie(cursor.getString(cursor.getColumnIndex(User.MOBLIE)));
        temp.setDanwei(cursor.getString(cursor.getColumnIndex(User.DANWEI)));
        temp.setQq(cursor.getString(cursor.getColumnIndex(User.QQ)));
        temp.setAddress(cursor.getString(cursor.getColumnIndex(User.ADDRESS)));
        return temp;
    } catch(Exception e){
        e.printStackTrace();
    } finally {
        if(cursor != null){
            cursor.close();
        }
        db.closeConnection();
```

```
        }
        return null;
}
/**
* @return
*/
public  User[] findUserByKey(String key)
{
    Vector<User> v=new Vector<User>();
    Cursor cursor=null;
    try {
       cursor=db.find("select * from "+TABLENAME +"  where "
               +User.NAME+" like '%"+key+"%' " +
               " or "+User.MOBLIE+" like '%"+key+"%' " +
               " or "+User.QQ+" like  '%"+key+"%' "
               ,null);
       while(cursor.moveToNext()){
          User temp=new User();
           temp.setId_DB(cursor.getInt(cursor.getColumnIndex("id_DB")));
           temp.setName(cursor.getString(cursor.getColumnIndex(User.NAME)));
           temp.setMoblie(cursor.getString(cursor.getColumnIndex(User.MOBLIE)));
           temp.setDanwei(cursor.getString(cursor.getColumnIndex(User.DANWEI)));
           temp.setQq(cursor.getString(cursor.getColumnIndex(User.QQ)));
           temp.setAddress(cursor.getString(cursor.getColumnIndex(User.ADDRESS)));
           v.add(temp);
          }
    } catch(Exception e){
          e.printStackTrace();
    } finally {
          if(cursor != null){
              cursor.close();
          }
          db.closeConnection();
    }
    if(v.size() > 0){
       return v.toArray(new User[] {});
    }else
    {
        User[]  users=new User[1];
        User  user=new User();
        user.setName("无结果");
        users[0]=user;
        return users;
    }
}
/**
 * 修改联系人信息
 */
public  boolean updateUser(User user)
{
  ContentValues values=new ContentValues();
  values.put(User.NAME,user.getName());
  values.put(User.MOBLIE,user.getMoblie());
  values.put(User.DANWEI,user.getDanwei());
  values.put(User.QQ,user.getQq());
  values.put(User.ADDRESS,user.getAddress());
  return db.update(TABLENAME, values," id_DB=? ",new String[]{user.getId_DB()+""});
}
/**
 * 删除联系人
*/
public  boolean  deleteByUser(User user)
{
     return db.delete(TABLENAME, "  id_DB=?",new String[]{user.getId_DB()+""});
  }
}
```

（5）手机通讯录程序辅助类：【User.java】。

```
package com.demo.pr3.datax;
/**
 * 联系人信息类
 */
public   class User {

      public  final static String  NAME="name";
      public  final static String  MOBLIE="mobile";
      public  final static String  DANWEI="danwei";
      public  final static String  QQ="qq";
      public  final static String  ADDRESS="address";
      private  String name; //用户名
      private  String moblie;//手机号码
      private  String danwei;//单位
      private  String qq;  //QQ
      private  String address;//地址
      private  int id_DB=-1;//表主键 ID

      public int getId_DB() {
         return id_DB;
      }
      public void setId_DB(int idDB){
         id_DB=idDB;
      }
      public String getName() {
         return name;
      }
      public void setName(String name){
         this.name=name;
      }
      public String getMoblie() {
         return moblie;
      }
      public void setMoblie(String moblie){
         this.moblie=moblie;
      }
      public String getDanwei() {
         return danwei;
      }
      public void setDanwei(String danwei){
         this.danwei=danwei;
      }
      public String getQq() {
         return qq;
      }
      public void setQq(String qq){
         this.qq=qq;
      }
      public String getAddress() {
         return address;
      }
      public void setAddress(String address){
         this.address=address;
      }
}
```

（6）手机通讯录程序添加联系人界面类：【AddContactsActivity.java】。

```
package com.demo.pr3;

import android.app.Activity;
import android.os.Bundle;
import android.view.Menu;
import android.view.MenuItem;
import android.widget.EditText;
import android.widget.Toast;
```

```
import com.demo.pr3.datax.ContactsTable;
import com.demo.pr3.datax.User;
/*******************(C)COPYRIGHT 2012*******************
*增加号码记录操作界面
*******************************************************/
public class AddContactsActivity extends Activity {

    private  EditText nameEditText;  //姓名输入框
    private  EditText mobileEditText; //手机输入框
    private  EditText qqEditText; //qq 输入框
    private  EditText danweiEditText; //单位输入框
    private  EditText addressEditText; //地址输入框
    @Override
    public void onCreate(Bundle savedInstanceState){
        super.onCreate(savedInstanceState);
        setContentView(R.layout.edit);
        setTitle("添加联系人");
        //从已设置的页面布局获得对应的控件
        nameEditText=(EditText)findViewById(R.id.name);
        mobileEditText=(EditText)findViewById(R.id.mobile);
        danweiEditText=(EditText)findViewById(R.id.danwei);
        qqEditText=(EditText)findViewById(R.id.qq);
        addressEditText=(EditText)findViewById(R.id.address);
    }
    /**
     * 创建菜单
     */
    public boolean onCreateOptionsMenu(Menu menu){

         menu.add(Menu.NONE,1,Menu.NONE,"保存");
         menu.add(Menu.NONE,2,Menu.NONE,"返回");
         return super.onCreateOptionsMenu(menu);
    }
   /**
    * 菜单事件
    */
    public boolean onOptionsItemSelected(MenuItem item){
        // TODO Auto-generated method stub
        switch(item.getItemId()){
        case 1://保存
        if(!nameEditText.getText().toString().equals(""))
            {
                User user=new User();
                user.setName(nameEditText.getText().toString());
                user.setMoblie(mobileEditText.getText().toString());
                user.setDanwei(danweiEditText.getText().toString());
                user.setQq(qqEditText.getText().toString());
                user.setAddress(addressEditText.getText().toString());
                ContactsTable ct= new ContactsTable(AddContactsActivity.this);
                if(ct.addData(user))
                {
                    Toast.makeText(AddContactsActivity.this,"添加成功！",
                    Toast.LENGTH_SHORT).show();
                    finish();
                }else
                {
                    Toast.makeText(AddContactsActivity.this,"添加失败！",
                    Toast.LENGTH_SHORT).show();

                }
            }else
            {
                Toast.makeText(AddContactsActivity.this,"请先输入数据！",
                Toast.LENGTH_SHORT).show();
            }
```

```
                break;
        case 2://返回
                finish();
            break;
        default:
            break;
        }
        return super.onOptionsItemSelected(item);
    }
}
```

（7）手机通讯录程序修改联系人界面类：【UpdateContactsActivity.java】。

```
package com.demo.pr3;
import android.app.Activity;
import android.os.Bundle;
import android.view.Menu;
import android.view.MenuItem;
import android.widget.EditText;
import android.widget.Toast;

import com.demo.pr3.datax.ContactsTable;
import com.demo.pr3.datax.User;
/********************(C)COPYRIGHT 2012********************
*修改号码记录操作界面
**************************************************************/
public class UpdateContactsActivity extends Activity {
    private  EditText nameEditText; //姓名输入框
    private  EditText mobileEditText;//手机输入框
    private  EditText qqEditText; //qq 输入框
    private  EditText danweiEditText; //单位输入框
    private  EditText addressEditText; //地址输入框
    private  User user; //修改的联系人
    public void onCreate(Bundle savedInstanceState){
        super.onCreate(savedInstanceState);
        setContentView(R.layout.edit);
        setTitle("修改联系人");
        //从已设置的页面布局获得对应的控件
        nameEditText=(EditText)findViewById(R.id.name);
        mobileEditText=(EditText)findViewById(R.id.mobile);
        danweiEditText=(EditText)findViewById(R.id.danwei);
        qqEditText=(EditText)findViewById(R.id.qq);
        addressEditText=(EditText)findViewById(R.id.address);

        //将要修改的联系人数据赋值到用户界面进行显示
        Bundle localBundle=getIntent().getExtras();
        int id=localBundle.getInt("user_ID");
        ContactsTable ct=new ContactsTable(this);
        user =ct.getUserByID(id);
        nameEditText.setText(user.getName());
        mobileEditText.setText(user.getMoblie());
        qqEditText.setText(user.getQq());
        danweiEditText.setText(user.getDanwei());
        addressEditText.setText(user.getAddress());
    }
    /**
     * 创建菜单
     */
    public boolean onCreateOptionsMenu(Menu menu){
         menu.add(Menu.NONE,1,Menu.NONE,"保存");
         menu.add(Menu.NONE,2,Menu.NONE,"返回");
         return super.onCreateOptionsMenu(menu);
    }
   /**
    * 菜单事件
    */
    public boolean onOptionsItemSelected(MenuItem item){
```

```
        // TODO Auto-generated method stub
        switch(item.getItemId()){
        case 1://保存
         if(!nameEditText.getText().toString().equals(""))
        {
                user.setName(nameEditText.getText().toString());
                user.setMoblie(mobileEditText.getText().toString());
                user.setDanwei(danweiEditText.getText().toString());
                user.setQq(qqEditText.getText().toString());
                user.setAddress(addressEditText.getText().toString());
                ContactsTable ct= new ContactsTable(UpdateContactsActivity.this);
                //修改数据库联系人信息
                if(ct.updateUser(user))
                {
Toast.makeText(UpdateContactsActivity.this,"修改成功！",Toast.LENGTH_SHORT).show();
                }else
                {
Toast.makeText(UpdateContactsActivity.this,"修改失败！", Toast.LENGTH_SHORT).show();
                }
          }else
          {
Toast.makeText(UpdateContactsActivity.this,"数据不能为空！", Toast.LENGTH_SHORT).show();
          }
           break;
        case 2://返回
            finish();
            break;
         default:
             break;
        }
        return super.onOptionsItemSelected(item);
    }
}
```

（8）手机通讯录程序显示联系人界面类：【ContactsMessageActivity.java】。

```
package com.demo.pr3;
import android.app.Activity;
import android.os.Bundle;
import android.view.Menu;
import android.view.MenuItem;
import android.widget.TextView;
import com.demo.pr3.datax.ContactsTable;
import com.demo.pr3.datax.User;
/********************(C)COPYRIGHT 2012*******************
*显示联系人界面
*********************************************************/
public class ContactsMessageActivity extends Activity {
    private  TextView nameTextView;  //姓名输入框
    private  TextView mobileTextView; //手机输入框
    private  TextView qqTextView; //qq 输入框
    private  TextView danweiTextView; //单位输入框
    private  TextView addressTextView; //地址输入框
    private  User user; //联系人
    @Override
    public void onCreate(Bundle savedInstanceState){
        super.onCreate(savedInstanceState);
        setContentView(R.layout.message);
        setTitle("联系人信息");

        //从已设置的页面布局查找对应的控件
        nameTextView=(TextView)findViewById(R.id.name);
        mobileTextView=(TextView)findViewById(R.id.mobile);
        danweiTextView=(TextView)findViewById(R.id.danwei);
        qqTextView=(TextView)findViewById(R.id.qq);
        addressTextView=(TextView)findViewById(R.id.address);

        //将要修改的联系人数据赋值到用户界面进行显示
```

```
        Bundle localBundle=getIntent().getExtras();
        int id=localBundle.getInt("user_ID");
        ContactsTable ct=new ContactsTable(this);
        user =ct.getUserByID(id);
        nameTextView.setText("姓名:"+user.getName());
        mobileTextView.setText("电话:"+user.getMoblie());
        qqTextView.setText("QQ:"+user.getQq());
        danweiTextView.setText("单位:"+user.getDanwei());
        addressTextView.setText("地址:"+user.getAddress());
    }
    /**
     * 创建菜单
     */
    public boolean onCreateOptionsMenu(Menu menu){
         menu.add(Menu.NONE,1,Menu.NONE,"返回");
         return super.onCreateOptionsMenu(menu);
    }
   /**
    * 菜单事件
    */
    public boolean onOptionsItemSelected(MenuItem item){
        // TODO Auto-generated method stub
        switch(item.getItemId()){
        case 1://返回
           finish();
           break;
         default:
             break;
        }
        return super.onOptionsItemSelected(item);
    }
}
```

（9）手机通讯录程序编辑和添加联系人布局 XML：【 /layout/edit.xml 】。

```
<?xml version="1.0" encoding="utf-8"?>
<LinearLayout
    xmlns:android="http://schemas.android.com/apk/res/android"
    android:orientation="vertical"
    android:layout_width="fill_parent"
    android:layout_height="fill_parent"
    >
  <!--姓名-->
  <LinearLayout
    android:layout_width="wrap_content"
    android:layout_height="wrap_content"
    android:orientation="horizontal"
    android:layout_marginTop="10.0dip"
    >
        <TextView android:layout_width="80.0dip"
        android:layout_height="wrap_content"
        android:text="姓名:"
        android:layout_marginLeft="5.0dip"
        />
        <EditText android:layout_width="190.0dip"
        android:id="@+id/name"
        android:layout_height="wrap_content" />
  </LinearLayout>
  <!--单位-->
  <LinearLayout
    android:layout_width="wrap_content"
    android:layout_height="wrap_content"
    android:orientation="horizontal"
    android:layout_marginTop="10.0dip"
    >
```

```
    <TextView android:layout_width="80.0dip"
    android:layout_height="wrap_content"
    android:text="单位:"
    android:layout_marginLeft="5.0dip"
    />
    <EditText android:layout_width="190.0dip"
     android:layout_height="wrap_content"
     android:id="@+id/danwei"
     android:minLines="3"
   />
</LinearLayout>
<!--手机号码-->
<LinearLayout
 android:layout_width="wrap_content"
 android:layout_height="wrap_content"
 android:orientation="horizontal"
 android:layout_marginTop="10.0dip"
>
    <TextView android:layout_width="80.0dip"
    android:layout_height="wrap_content"
    android:text="电话:"
    android:layout_marginLeft="5.0dip"
    />
    <EditText android:layout_width="190.0dip"
     android:id="@+id/mobile"
    android:layout_height="wrap_content"
    android:inputType="number" />
</LinearLayout>
 <!--qq-->
<LinearLayout
 android:layout_width="wrap_content"
 android:layout_height="wrap_content"
 android:orientation="horizontal"
 android:layout_marginTop="10.0dip"
>
   <TextView android:layout_width="80.0dip"
    android:layout_height="wrap_content"
    android:text="QQ:"
    android:layout_marginLeft="5.0dip"
    />
    <EditText android:layout_width="190.0dip"
       android:layout_height="wrap_content"
            android:id="@+id/qq"
            android:inputType="number"
   />
</LinearLayout>
<!--地址-->
<LinearLayout
 android:layout_width="wrap_content"
 android:layout_height="wrap_content"
 android:orientation="horizontal"
 android:layout_marginTop="10.0dip"
>
    <TextView android:layout_width="80.0dip"
    android:layout_height="wrap_content"
    android:text="地址:"
    android:layout_marginLeft="5.0dip"
    />
    <EditText android:layout_width="190.0dip"
    android:id="@+id/address"
    android:layout_height="wrap_content"
    android:minLines="3"
```

```
        />
    </LinearLayout>
</LinearLayout>
```

（10）手机通讯录程序查找联系人布局 XML：【 /layout/find.xml 】。

```
<?xml version="1.0" encoding="utf-8"?>
<LinearLayout
    xmlns:android="http://schemas.android.com/apk/res/android"
    android:orientation="vertical"
    android:layout_width="fill_parent"
    android:layout_height="fill_parent"
    >
   <LinearLayout
    android:layout_width="wrap_content"
    android:layout_height="wrap_content"
    android:orientation="horizontal"
   >
        <TextView android:layout_width="50.0dip"
        android:layout_height="wrap_content"
        android:text="条件:"
        android:layout_marginLeft="10.0dip"
         android:layout_marginRight="10.0dip"
        />
        <EditText android:layout_width="190.0dip"
         android:id="@+id/value"
         android:layout_height="wrap_content"  />
   </LinearLayout>
   <LinearLayout
    android:layout_width="wrap_content"
    android:layout_height="wrap_content"
    android:orientation="horizontal"
    android:layout_marginTop="10.0dip"
    android:layout_gravity="center_horizontal"
   >
    <Button
        android:layout_width="wrap_content"
        android:layout_height="wrap_content"
        android:id="@+id/find"
        android:text=" 查          询 " />
        <Button
        android:layout_width="wrap_content"
        android:layout_height="wrap_content"
        android:id="@+id/cancel"
        android:text=" 取          消 " />
   </LinearLayout>
</LinearLayout>
```

（11）手机通讯录程序主页面布局 XML：【 /layout/main.xml 】。

```
<?xml version="1.0" encoding="utf-8"?>
<LinearLayout
     xmlns:android="http://schemas.android.com/apk/res/android"
    android:orientation="vertical"
    android:layout_width="fill_parent"
    android:layout_height="fill_parent"
    >
    <ListView
        android:id="@+id/listView"
        android:layout_weight="1.0"
        android:layout_width="fill_parent"
        android:layout_height="fill_parent"
    />
</LinearLayout>
```

（12）手机通讯录程序显示联系人布局 XML：【 /layout/message.xml 】。

```
<?xml version="1.0" encoding="utf-8"?>
<LinearLayout
     xmlns:android="http://schemas.android.com/apk/res/android"
    android:orientation="vertical"
    android:layout_width="fill_parent"
    android:layout_height="fill_parent"
    >
   <!--姓名-->
   <TextView
       android:id="@+id/name"
       android:layout_width="fill_parent"
       android:layout_height="wrap_content"
       android:text="姓名:"
       android:textSize="22.0dip"
       android:layout_marginLeft="5.0dip"
   />
   <!--单位-->
   <TextView
       android:id="@+id/danwei"
       android:layout_width="fill_parent"
       android:layout_height="wrap_content"
       android:text="单位:"
       android:textSize="22.0dip"
       android:layout_marginLeft="5.0dip"
    />

   <!--手机号码-->
   <TextView
       android:id="@+id/mobile"
       android:layout_width="fill_parent"
       android:layout_height="wrap_content"
       android:text="电话:"
       android:textSize="22.0dip"
       android:layout_marginLeft="5.0dip"
   />
   <!--qq-->
    <TextView
       android:id="@+id/qq"
       android:layout_width="fill_parent"
       android:layout_height="wrap_content"
       android:text="QQ:"
       android:textSize="22.0dip"
       android:layout_marginLeft="5.0dip"
       />
    <!--地址-->
    <TextView
       android:id="@+id/address"
       android:layout_width="fill_parent"
       android:layout_height="wrap_content"
       android:text="地址:"
       android:textSize="22.0dip"
       android:layout_marginLeft="5.0dip"
    />
</LinearLayout>
```

3.10　实训项目

一、手机通讯录的改进

1. 实训目的与要求

学会利用 Intent 调用系统提供的功能。

2．实训内容

在本项目的功能基础上，增加用户拨打电话、发送短信息、保存历史通话记录等功能。

3．思考

（1）常用的通过 Intent 调用的系统功能有哪些？

（2）如何开发更具有个性化的手机通讯录功能？

二、我的移动日记

1．实训目的与要求

学会利用 SQLite 技术，并结合用户界面技术，开发涉及手机数据存储管理类的应用软件，掌握在 SQLite 中实现数据的增加、删除、修改和查询操作。

2．实训内容

要求：程序运行后打开数据显示界面，按照时间先后顺序列出已经写好的日记。

（1）数据显示界面：以“标题 + 时间”的方式列出已经写好的日记，界面上设置【退出】、【录入】、【修改】、【查询】、【删除】等菜单按钮功能。

（2）数据操作界面包括标题、时间、详细日记内容。用户可以在该界面上完成保存、退出等功能操作。

3．思考

（1）如何创建 SQLite 数据库？

（2）SQLite 与其他数据库（如 SQL Server）有何不同？

（3）SQLite 数据库文件保存在手机什么位置？

（4）SQLite 的设计与计算机上数据库的设计有何差异？

三、英语题库系统

1．实训目的与要求

学会利用 SQLite 实现题库的应用，包括随机组题、检查答题的正确性、统计分数等。

2．实训内容

（1）用户登录后，可以选择英语考试题库四级或者六级难度，系统自动组成题库。注意：可以根据需要预先向英语题库中录入一些基础数据。

（2）程序主操作界面：每道题以单选方式提供，在屏幕上显示答题者的当前得分。用户如果输入正确的答案，则显示下一道题；如果输入错误的答案，则给出答案提示。

3．思考

（1）如何根据难度来实现随机抽题、组题？

（2）如何维护、更新手机上的题库？

PART 4 项目四 开发多媒体播放器

该工作情景的目标是让学生掌握 Android 的多媒体开发技术，主要的工作任务划分为① 开发多媒体播放界面；② 播放音乐；③ 播放视频；④ 管理多媒体文件。本项目主要涉及的关键技术包括：播放控制条的控制、消息处理、声音的播放、视频图像的显示、多线程编程、服务技术的应用、多媒体文件的搜索等。

4.1 开发多媒体播放界面

一、任务分析

本任务需要实现的多媒体播放器界面效果如图 4-1 所示。

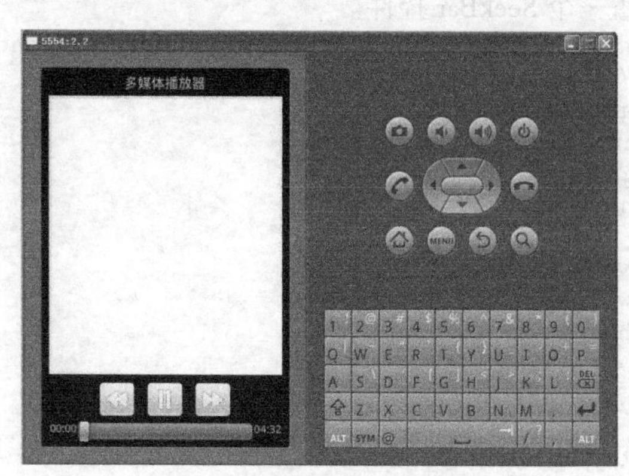

图 4-1 多媒体播放器界面

多媒体播放界面主要由标题栏、播放窗口、控制栏及进度条组成。本次任务旨在为用户提供可视的多媒体播放界面，需要思考如下问题。

（1）如何制作多媒体的播放窗口？

（2）如何将程序的窗口设置为全屏，即隐藏标题栏和状态信息栏？

二、相关知识

1. SeekBar 类

进度条可以方便地告诉用户任务执行的进度，特别是当一个程序需要花费较长时间时，如果没有进度条，用户不知道程序是否正在执行，会误以为程序假死而强制关闭程序。Android 提供 ProgressBar 控件用于表示进度条。通过设置 ProgressBar 的 XML 属性 style，就可以生成不同风格的进度条，例如设置不确定进度的小圆圈进度条：

```
style="@android:style/Widget.ProgressBar.Small"
```

系统提供的常用进度条样式还有：

- Widget.ProgressBar.Horizontal（水平进度条）；
- Widget.ProgressBar.Large（不确定进度的大圆圈进度条）。

SeekBar 控件用于表示可以拖曳的进度条，是 ProgressBar 的子类，表示完成进度的百分比，进度默认的取值范围是 0～100。下面介绍其主要方法。

（1）int getMax()：返回进度条的上限值。

（2）int getProgress()：返回进度条当前的进度。

（3）void incrementProgressBy（int diff）：增加指定数量的进度值。其中，参数 diff 为增加的进度值。

（4）void setMax（int max）：设置进度条取值范围的上限，进度条的取值范围为 0～max。

（5）void setProgress（int progress）：设置当前的进度。其中，参数 progress 表示指定的进度值。

（6）void setOnSeekBarChangeListener（SeekBar.OnSeekBar Change Listener l）：设置一个监听器，接收 SeekBar 进度改变的通知。

SeekBar 控件使用示例如下，界面效果如图 4-2 所示。

图 4-2 进度条实例

① 定义布局文件【seekbar_demo.xml】：设置两个 TextView 控件，一个用于表示当前进度条的数值，另一个用于表示是否在移动拖曳条。此外，还需设置一个 SeekBar 控件。

```
<LinearLayout xmlns:android="http://schemas.android.com/apk/res/android"
   xmlns:tools="http://schemas.android.com/tools"
   android:id="@+id/LinearLayout1"
   android:layout_width="match_parent"
   android:layout_height="match_parent"
   android:orientation="vertical" >
   <TextView
      android:id="@+id/tCurrent"
      android:layout_width="wrap_content"
      android:layout_height="wrap_content"
      android:padding="@dimen/padding_medium"
      tools:context=".SeekBarDemo" />
   <TextView
      android:id="@+id/tStatus"
      android:layout_width="wrap_content"
      android:layout_height="wrap_content" />
   <SeekBar
      android:id="@+id/seekBar1"
      android:layout_width="match_parent"
      android:layout_height="wrap_content" />
</LinearLayout>
```

② 程序代码如下。

```
public class SeekBarDemo extends Activity {
    private SeekBar seekBar;
    private TextView tCurrent;
    private TextView tStatus;

    @Override
    public void onCreate(Bundle savedInstanceState){
```

```
        super.onCreate(savedInstanceState);
        setContentView(R.layout.seekbar_demo);
        seekBar =(SeekBar)findViewById(R.id.seekBar1);
        tCurrent =(TextView)findViewById(R.id.tCurrent);
        tStatus =(TextView)findViewById(R.id.tStatus);
        seekBar.setProgress(30);
        tCurrent.setText("当前值进度: "+seekBar.getProgress());
        seekBar.setOnSeekBarChangeListener(new OnSeekBarChangeListener() {
            @Override
            public void onProgressChanged(SeekBar seekBar,int progress,
                    boolean fromUser){
                tCurrent.setText("当前值进度: "+progress);
            }

            @Override
            public void onStartTrackingTouch(SeekBar seekBar){
                tStatus.setText("正在移动拖曳条");
            }

            @Override
            public void onStopTrackingTouch(SeekBar seekBar){
                tStatus.setText("停止移动拖曳条");
            }
        });
    }
}
```


2．SurfaceView 类

SurfaceView 是 View 的子类，可以直接从内存或者 DMA 等硬件接口取得图像数据，因此是非常重要的绘图容器，可以用于游戏开发。SurfaceView 内嵌了一个专门用于绘制图像的 Surface。Surface 是 Android 图形系统中一个重要的概念和线索，这是因为 View 及其子类是画在 Surface 上的。每个 Surface 创建一个 Canvas 对象，用来管理 View 在 Surface 上的绘图操作，如画点、画线、画面等。可以把 Surface 理解为绘图的“屏幕”，在 Surface 上放着 Canvas 画布，然后在画布上绘图。在 SurfaceView 编程中，可以控制其格式、尺寸和位置。SurfaceView 的一个重要特性是能够使用后台线程绘制 Surface，数据可以被直接复制到显存中进而显示出来，有效地提高了执行效率。在进行视频或者游戏等相关项目开发时，通常会结合多线程类和 SurfaceView 类一起使用。

SurfaceView 构造函数有 3 种。

① SurfaceView（Context context）;

② SurfaceView（Context context，AttributeSet attrs）;

③ SurfaceView（Context context，AttributeSet attrs，int defStyle）。

其中，参数 context 表示上下文；参数 attrs 表示属性集；参数 defStyle 表示样式。

下面介绍 SurfaceView 类的主要方法。

（1）SurfaceHolder getHolder()：返回 SurfaceView 的 SurfaceHolder 对象。

（2）setVisibility（int visibility）：设置 SurfaceView 的可见性。其中，参数 visibility 的取值为 VISIBLE，INVISIBLE 或者 GONE，其中 INVISIBLE 和 GONE 都可以使 SurfaceView 控件不可见，区别在于前者尽管设置了不可见，但依然占用窗口布局上的控件空间；而后者则不占用窗口布局的控件空间。

（3）setLayoutParams（ViewGroup.LayoutParams params）：设置与视图有关的布局参数。其中，参数 params 表示布局参数值。LayoutParams 类的作用是说明视图如何进行布局，具体来讲就是指定视图的高度和宽度，取值为 FILL_PARENT、WRAP_CONTENT 或具体的数值。

3. SurfaceHolder

SurfaceHolder 是一个接口，用于控制 Surface，如设置 Surface 的格式、大小、像素。一般情况下还要对其进行创建、销毁、在情况改变时监控 Surface 的变化等，这就要给 SurfaceView 当前的持有者 SurfaceHolder 添加一个回调对象，相应地使用 abstract void addCallback（SurfaceHolder.Callback callback）方法。其中，参数 callback 表示接收关于 Surface 变化的信息接口，并实现该接口中的 surfaceChanged、surfaceCreated 和 surfaceDestroyed 三个抽象方法。

Android 中的 Callback 表示的是回调接口，回调指的是把一个方法传给事件源，当某一事件发生时能够调用这个方法。

SurfaceHolder 的一个重要方法是 setType(int type)，用于设置标识 Surface 实例界面的数据来源，例如，若 Surface 是与 OpenGL ES 一起使用，则 type 取值 SURFACE_TYPE_GPU；若 Surface 是被 DMA 引擎和硬件加速器访问，则 type 取值为 SURFACE_TYPE_HARDWARE。

如何理解 SurfaceView 与 SurfaceHolder 之间的关系？简单来说 SurfaceView 用于显示图像，SurfaceHolder 用于管理需显示的 SurfaceView 对象。但 SurfaceHolder 是如何管理这个对象的呢？这需要调用 SurfceHolder. addCallback()方法去添加一个 Callback 内部接口，通过实现该接口的三个抽象方法来管理或者监听 SurfaceView，这样就达到了管理 SurfaceView 的目的。

使用 SurfaceView 和 SurfaceHolder 实现图像的绘制主要有两种实现方法。

第一种方法的主要步骤介绍如下。

① 定义一个类继承 SurfaceView 并实现 SurfaceHolder.Callback 接口。除了编写构造函数之外，还需重写以下方法。

- public void surfaceChanged(SurfaceHolder holder, int format, int width，int height){}：在 Surface 的大小发生改变时触发。
- public void surfaceCreated(SurfaceHolder holder){}：在 Surface 创建时触发，一般在这里调用绘图线程。
- public void surfaceDestroyed(SurfaceHolder holder){}：在 Surface 销毁时触发，一般在这里停止、释放绘图线程。

使用 SurfaceView 有一个原则：所有的绘图工作必须在 Surface 被创建之后才能开始，也必须在 Surface 被销毁之前结束。所以 Callback 中的 surfaceCreated 和 surfaceDestroyed 就成了绘图处理代码的边界。

② 在本步骤中，Activity 类有两种实现方法：一是将在步骤①中定义的 SurfaceView 子类生成一个对象，然后将其作为参数传入 setContentView()方法，从而将 SurfaceView 对象显示在屏幕上；二是在 Activity 的布局 XML 中设置如下，其中假设在步骤①中设置的 SurfaceView 子类为 com.example.project.MySurfaceView：

```
<com.example.project.MySurfaceView
    android:id="@+id/sv"
    android:layout_width="fill_parent"
    android:layout_height="fill_parent"
/>
```

将布局文件 XML 作为 Activity 的 setContentView()方法，就可以将自定义的 SurfaceView 作为屏幕中的一部分进行显示。需要注意的是：由于默认的 XML 文件解析方法是调用 View（Context，AttributeSet）构造函数构造 View，因此在自定义的 SurfaceView 中也应该有一个参数为（Context，AttributeSet）的构造函数，并且在构造函数中执行父类的对应函数 super

(Context，AttributeSet)。

第二种方法的主要步骤介绍如下。

① 在布局文件 XML 中首先定义包括 SurfaceView 控件在内的界面布局。

② 在 Activity 程序中通过 findViewById 方法获得 SurfaceView 对象。

③ 调用 SurfaceView 对象的 getHolder()方法获得 SurfaceHolder 对象。

④ 调用 SurfaceHolder 对象的 addCallback 方法实现对 SurfaceHolder.Callback 的监听，在该步重写第一种方法提到的 surfaceChanged、surfaceCreated、surfaceDestroyed 等方法。

⑤ 调用 SurfaceHolder 对象的其他方法设置相关属性。

⑥ 将 SurfaceView 或者 SurfaceHolder 对象作为其他类相应函数的参数进行调用，如 MediaPlayer 对象的 setDisplay()方法。

4．窗体的属性设置

在开发程序时经常会遇到软件全屏显示、自定义标题（使用按钮等控件）和其他需求，本项目的媒体播放器可以将主播放页面设置为全屏。

Activity 类的 requestWindowFeature (int featureId) 是控制 Android 应用程序窗体显示的重要方法，其功能是启用窗体的扩展特性，参数 featureId 是在 Window 类中定义的常量，主要取值如下。

① DEFAULT_FEATURES：系统默认状态，一般不需要指定。

② FEATURE_CONTEXT_MENU：启用上下文菜单，相当于计算机上的右键功能，默认该项已启用，一般无需指定。

③ FEATURE_CUSTOM_TITLE：自定义标题，如将标题定义为图片加文字，需要和布局文件配合使用。

④ FEATURE_INDETERMINATE_PROGRESS：表示不确定的进度，利用圆圈状等待图标表示，需要和布局文件配合使用。

⑤ FEATURE_LEFT_ICON：在标题栏左侧显示一个图标。

⑥ FEATURE_NO_TITLE：不使用标题栏。

⑦ FEATURE_OPTIONS_PANEL：启用“选项面板”功能，默认为启用。

⑧ FEATURE_PROGRESS：在标题栏上显示加载进度。

⑨ FEATURE_RIGHT_ICON：在标题栏右侧显示一个图标。

requestWindowFeature 方法的执行必须放在 setContentView()设置布局前面，不然会报错。对于 requestWindowFeature 方法中要求设置标题图标和标记之类的操作，通常还需要调用 Window 对象的相应方法进行具体的设置。

Window 类概括了 Android 窗口的基本属性和基本功能。窗口的类型分为应用程序窗口、子窗口和系统窗口 3 种。Activity 的 getWindow()方法返回 Activity 的当前窗口对象 Window。对象具体使用时主要有以下 3 种方法，这些方法在 setContentView()设置布局后面调用。

（1）getWindow().setFeatureInt (int featureId，int value)：为窗体特征设置整型数值。其中，参数 featureId 表示希望改变的特征，取值为 Window 类中所定义的常量。

例如，设置 Window 定制标题代码为

```
getWindow().setFeatureInt(Window.FEATURE_CUSTOM_TITLE, R.layout.custom_title_1);
```

其中，R.layout.custom_title_1 为定制标题的 XML 布局文件。

（2）getWindow().setFeatureDrawable (int featureId，Drawable drawable)：为窗体特征设置

可绘制的资源。其中，参数 featureId 表示希望改变的特征；参数 drawable 表示要显示的可绘制对象。

例如，设置 Window 定制图标的代码为

```
getWindow().setFeatureDrawableResource(Window.FEATURE_LEFT_ICON    ,
android.R.drawable.ic_dialog_ alert);
```

其中，android.R.drawable.ic_dialog_alert 为 Android 系统内部提供的对话框警告图标。

（3）getWindow().setFlags（int flags，int mask）：设置窗体的标记。其中，参数 flags 表示窗体的标记，取值来自 WindowManager.LayoutParams 类中定义的常量，WindowManager.LayoutParams 的作用是向 WindowManager 描述 Window 的管理策略，例如锁屏、亮度设置、焦点获得、触摸屏屏蔽、窗口类型设置、软键盘显示等；参数 mask 表示需要更改的窗口标记位。示例如下。

① 设置 Windows 全屏幕的代码为

```
getWindow().setFlags(WindowManager.LayoutParams.FLAG_FULLSCREEN,
        WindowManager.LayoutParams.FLAG_FULLSCREEN);
```

② 设置窗体始终点亮的代码为

```
getWindow().setFlags(WindowManager.LayoutParams.FLAG_KEEP_SCREEN_ON,
        WindowManager.LayoutParams.FLAG_KEEP_SCREEN_ON);
```

③ 设置窗体背景模糊的代码为

```
getWindow().setFlags(WindowManager.LayoutParams.FLAG_BLUR_BEHIND,
        WindowManager.LayoutParams.FLAG_BLUR_BEHIND);
```

综合使用 requestWindowFeature 方法和 Window 对象方法的示例如下。

① 隐藏窗体标题，并全屏显示窗体。

```
requestWindowFeature(Window.FEATURE_NO_TITLE); //隐藏标题栏
getWindow().setFlags(WindowManager.LayoutParams.FLAG_FULLSCREEN,WindowManager.LayoutP
arams.FLAG_FULLSCREEN);//隐藏状态信息栏（例如电池电量、手机网络信号等）
setContentView(R.layout.main);//此处根据实际情况，指定需要显示的布局文件
```

② 在标题栏左侧显示图标，效果如图 4-3 所示。

```
requestWindowFeature(Window.FEATURE_LEFT_ICON);
setContentView(R.layout.main);
getWindow().setFeatureDrawableResource(Window.FEATURE_LEFT_ICON,R.drawable.icon);
```

图 4-3 标题栏图标

三、任务实施

多媒体播放器需要向用户提供控制操作，因此需在用户界面设计中提供进度控制、开始、快退、快进、暂停等控件。

（1）创建项目：创建一个 Android Project，项目名称为【MediaPlayer】，包名为【com.demo.pr4】，选择 Android SDK 2.2 版本，并创建一个 Activity 名为【MediaPlayerActivity】。

（2）创建用户界面布局：根据任务分析大致可以用 4 个步骤来设置 XML 的布局规划。

① 设置项目主题区域。将程序的主题“多媒体播放器”放在第一行，要求字体居中，大小为 18 像素点，字体样式为 bold，距离顶部 5 像素点。下面选择 LinearLayout 和 TextView 两种 XML 元素进行布局。

```
<LinearLayout
   android:orientation="horizontal"
   android:layout_width="wrap_content"
   android:layout_height="wrap_content"
   android:layout_marginTop="5.0dip"
>
```

```
    <TextView
        android:id="@+id/title"
        android:text="多媒体播放器"
        android:layout_width="fill_parent"
        android:layout_height="wrap_content"
        android:textSize="18.0dip"
        android:textStyle="bold"
        android:gravity="center"
    ></TextView>
</LinearLayout>
```

② 设置媒体播放区域。媒体播放区域用于播放媒体视频，需要使用 SurfaceView 控件。SurfaceView 控件以流的方式来显示视频。设计要求：媒体窗口要占满除控制栏、进度条、标题栏外的所有区域，因此需设置 layout_weight 属性。layout_weight 属性值越大，其权重就越大，可以随屏幕的增大而相应增大。下面是媒体播放区域的设置。

```
<LinearLayout
android:orientation="horizontal"
android:layout_weight="1"          //权重设置
android:layout_height="fill_parent"
android:layout_width="fill_parent"
>
   <SurfaceView android:layout_gravity="center"
     android:layout_weight="1"
     android:layout_height="fill_parent"
     android:layout_width="fill_parent"
     android:id="@+id/surfaceview">
   </SurfaceView>
</LinearLayout>
```

③ 设置控制区域。控制区域主要由快退、暂停、开始和快进组成。要求：控制区域排布居中，每个控件间隙 5 像素点；距离媒体播放区域 5 像素点；【开始】和【暂停】按钮不能同时出现，必须隐藏其中的一个。下面选择 LinearLayout 和 ImageButton 两种 XML 元素进行布局。

```
<LinearLayout
    android:orientation="horizontal"
    android:layout_width="wrap_content"
    android:layout_height="wrap_content"
    android:layout_gravity="center_horizontal"
    android:layout_marginTop="5.0dip"
    >
     <!--快退-->
    <ImageButton
    android:id="@+id/rew"
    android:layout_width="wrap_content"
    android:layout_height="wrap_content"
     android:src="@drawable/ic_media_rew"
    ></ImageButton>
    <!--暂停-->
     <ImageButton
     android:layout_marginLeft="10.0dip"
    android:id="@+id/pause"
    android:layout_width="wrap_content"
    android:layout_height="wrap_content"
    android:src="@drawable/ic_media_pause"
    ></ImageButton>
     <!--开始-->
    <ImageButton
     android:layout_marginLeft="10.0dip"
    android:id="@+id/start"
    android:layout_width="wrap_content"
    android:layout_height="wrap_content"
    android:src="@drawable/ic_media_play"
```

```
    android:visibility="gone"
    ></ImageButton>
     <!--快进-->
     <ImageButton
     android:id="@+id/ff"
     android:layout_marginLeft="10.0dip"
     android:layout_width="wrap_content"
     android:layout_height="wrap_content"
     android:src="@drawable/ic_media_ff"
    ></ImageButton>
</LinearLayout>
```

④ 设置进度区域。进度区域要求显示媒体文件总时长、已播放时长、进度，选择 LinearLayout、SeekBar 和 TextView 等 3 种 XML 元素进行布局。

```
<LinearLayout
        android:layout_width="fill_parent"
        android:layout_height="wrap_content"
        android:layout_gravity="center"
        android:orientation="horizontal">
    <!-- 已经播放时长-->
     <TextView
          android:id="@+id/play_time"
          android:text="00:00"
            android:layout_width="wrap_content"
            android:layout_marginLeft="5.0dip"
            android:textSize="15.0dip"
            android:layout_height="wrap_content">
    </TextView>
    <!-- 进度-->
    <SeekBar android:id="@+id/seekbar"
            android:layout_width="fill_parent"
            android:layout_height="wrap_content"
            android:layout_weight="1.0"
            android:layout_marginLeft="2.0dip"
            android:layout_marginRight="2.0dip"
    />
    <!-- 总时长-->
    <TextView
         android:id="@+id/all_time"
         android:text="04:32"
         android:layout_width="wrap_content"
         android:layout_marginRight="5.0dip"
         android:textSize="15.0dip"
         android:layout_height="wrap_content">
         </TextView>
</LinearLayout>
```

（3）隐藏 Android 系统标题栏：多媒体应用程序要求设置为全屏。在 onCreate（Bundle savedInstanceState）方法中添加如下两行代码即可实现相应效果。

```
requestWindowFeature(Window.FEATURE_NO_TITLE); //隐藏标题栏
getWindow().setFlags(WindowManager.LayoutParams.FLAG_FULLSCREEN,WindowManager.LayoutP
arams.FLAG_FULLSCREEN);//隐藏状态信息栏（例如电池电量、手机网络信号等）
```

4.2 播放音乐

一、任务分析

本任务需要实现的效果如图 4-4 所示。

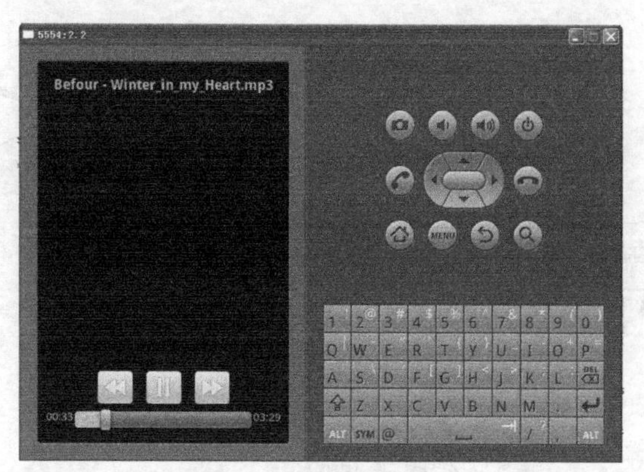

图 4-4 音频播放示例

本次任务要求实现音乐文件的播放，在播放界面上显示播放音乐的文件名、已播放的时间、音乐的总时间、音乐播放控制等。需要思考如下几个问题

（1）如何控制播放音乐？

（2）如何查找手机上的音频文件？

二、相关知识

1．MediaPlayer 的生命周期

音乐播放是 Android 中常见的应用，市场上已经出现了各种各样的手机音乐播放器。MediaPlayer 类用于管理媒体播放的操作，不仅可以播放音频文件还可以结合 SurfaceView 来实现视频文件的播放，是一种比较灵活的多媒体播放组合。本次任务将重点讲述如何在 MediaPlayer 中实现音频播放功能。

下面先介绍 MediaPlayer 的生命周期，如图 4-5 所示。

① Idle 状态：当使用 new()方法创建一个 MediaPlayer 对象或者调用了其 reset()方法时，该 MediaPlayer 对象处于 Idle 状态。这两种方法的一个重要差别是：通过 reset()方法进入 Idle 状态，在这个状态下再调用 getDuration()等方法，将会触发 OnErrorListener.onError()，MediaPlayer 会进入 Error 状态；但如果是用 new()方法创建的 MediaPlayer 对象，则并不会触发 onError()事件，也不会进入 Error 状态。

② End 状态：通过 release()方法可以进入 End 状态，只要 MediaPlayer 对象不再被使用，就应当尽快将其通过 release()方法进行释放，以便释放相关的软硬件资源。如果 MediaPlayer 对象处于 End 状态，则不会再进入任何其他状态。

③ Initialized 状态：这个状态比较简单，MediaPlayer 只要调用 setDataSource()方法就可以进入 Initialized 状态，表示此时要播放的文件已经设置好。

④ Prepared 状态：初始化完成之后还需要调用 prepare()或 prepareAsync()方法。这两个方法一个是同步，一个是异步，只有进入 Prepared 状态，才表明 MediaPlayer 到目前为止都没有错误，可以进行文件播放。

⑤ Preparing 状态：主要和 prepareAsync()相配合。在异步准备完成后，会触发 OnPreparedListener.onPrepared()，进而进入 Prepared 状态。

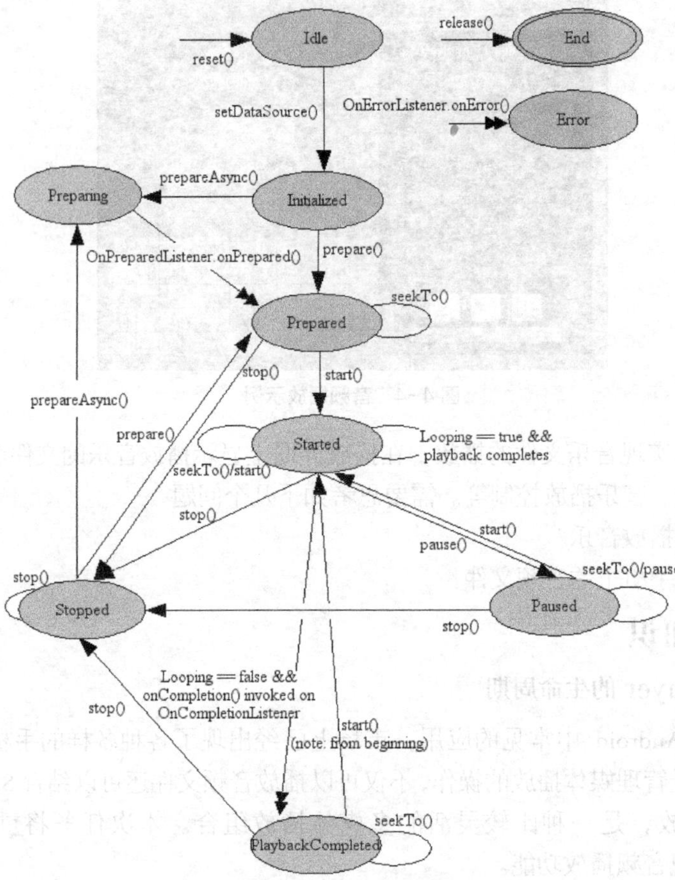

图 4-5 MediaPlayer 的生命周期

⑥ Started 状态：MediaPlayer 一旦准备好，就可以调用 start()方法，这样 MediaPlayer 就处于 Started 状态，表明 MediaPlayer 正在播放音乐文件。可以使用 isPlaying()检测 MediaPlayer 是否处于 Started 状态。如果音乐播放完毕，但因设置了循环播放，MediaPlayer 仍会处于 Started 状态。类似地，如果在该状态下 MediaPlayer 调用了 seekTo()或者 start()方法均可以让 MediaPlayer 停留在 Started 状态。

⑦ Paused 状态：在 Started 状态下 MediaPlayer 调用 pause()方法可以实现暂停，从而进入 Paused 状态。MediaPlayer 暂停后再次调用 start()则可以继续播放，使其处于 Started 状态。在暂停状态可以调用 seekTo()方法，但不会改变其状态。

⑧ Stopped 状态：Started 或者 Paused 状态下均可调用 stop()方法停止 MediaPlayer，而处于 Stopped 状态的 MediaPlayer 要想重新播放，需要通过 prepareAsync()或 prepare()回到先前的 Prepared 状态才可以重新开始。

⑨ PlaybackCompleted 状态：当文件正常播放完毕，而又没有设置循环播放时就进入该状态，并会触发 OnCompletionListener 的 onCompletion()方法。此时可以调用 start()方法重新从头播放文件，也可以调用 stop()方法停止 MediaPlayer，或者使用 seekTo()方法来重新定位播放位置。

⑩ Error 状态：如果由于某种原因 MediaPlayer 出现了错误，会触发 OnErrorListener.onError()事件，此时 MediaPlayer 进入 Error 状态。及时捕捉并妥善处理这些错误较为重要，有助于及

时释放相关的软硬件资源，也可以改善用户体验。可以通过 setOnErrorListener (android.media. MediaPlayer.OnErrorListener) 方法设置对错误的监听器。一旦 MediaPlayer 进入了 Error 状态，就可以通过调用 reset()方法来恢复，使得 MediaPlayer 能够重新返回到 Idle 状态。

2. MediaPlayer 类控制多媒体

Android 多媒体框架支持播放不同位置存放的多媒体资源，包括：

①本地资源，如来自于应用程序的 RAW 资源或者文件系统；

②内部 URI，如来自于 Content Resolver；

③外部 URL，以流的方式提供。

MediaPlayer 类的构造函数较为简单，只有一个无参的构造函数：MediaPlayer()。下面介绍 MediaPlayer 类的主要方法：

(1) MediaPlayer create (Context context，Uri uri，SurfaceHolder holder);

(2) MediaPlayer create (Context context，int resid);

(3) MediaPlayer create (Context context，Uri uri)。

上面三个方法都是静态方法，用于创建 MediaPlayer 对象。其中，参数 context 表示上下文；参数 uri 表示多媒体文件的位置；参数 holder 表示显示视频的 SurfaceHolder 对象；参数 resid 表示存放在本地的多媒体资源 ID，以 R.raw.XXX 形式描述。

(4) int getCurrentPosition()：获取多媒体当前的播放位置，时间单位为毫秒 (ms)。可以将该方法的运行结果作为参数传入 Time 类的构造函数中，生成一个 Time 对象，然后再使用 Time 对象的 toString()将时间转换为 HHMMSS 格式。

(5) int getDuration ()：获取多媒体总的时间，时间单位为毫秒 (ms)。

(6) int getVideoHeight ()：返回视频的高度，若没有视频，则返回 0。

(7) int getVideoWidth ()：返回视频的宽度，若没有视频，则返回 0。

(8)boolean isLooping ()：判断媒体播放器是否循环播放，如果正在循环播放，则返回 true。

(9) boolean isPlaying ()：判断媒体播放器是否正在播放，如果正在播放，则返回 true。

(10) void pause ()：暂停播放。在调用 start()方法后恢复播放。

(11) void prepare ()：准备播放器进行同步播放。

(12) void repareAsync ()：准备播放器进行异步播放。

在设置好播放数据源和显示的 Surface 后，需要调用 prepare()或者 prepareAsync()方法。若对象为文件，可以调用 prepare()方法；若对象为流媒体，应调用 prepareAsync()方法，这样可以避免产生阻塞。

(13) void release ()：释放与 MediaPlayer 对象有关的资源。在不需要使用 MediaPlayer 时，应调用该方法。

(14) void reset ()：将 MediaPlayer 重置为非初始化状态。调用该方法之后，需要重新设置数据源和调用 prepare()进行初始化。

(15) void seekTo (int msec)：移到指定时间位置的媒体内容。其中，参数 msec 是偏于开始位置的时间，单位为毫秒 (ms)。

(16) void setAudioStreamType (int streamtype)：设置声音流的类型。其中，参数 streamtype 表示音频流的类型，已在 AudioManager 类中定义。

- STREAM_ALARM：用于警告的音频流。

● STREAM_DTMF：用于双音多频的音频流。

● STREAM_MUSIC：用于音乐播放的音频流。多媒体播放器常选择该类型。

● STREAM_NOTIFICATION：用于通知的音频流。

● STREAM_RING：用于铃声的音频流。

● STREAM_SYSTEM：用于系统声音的音频流。

● STREAM_VOICE_CALL：用于电话呼叫的音频流。

（17）void setDataSource（String path）。

（18）void setDataSource（Context context，Uri uri）。

（19）void setDataSource（FileDescriptor fd，long offset，long length）。

（20）void setDataSource（FileDescriptor fd）。

以上 4 个方法都用来设置多媒体资源所在的位置，以便于进行播放。通过设置不同的参数来打开不同数据源的多媒体。其中，参数 path 表示文件的路径或者流媒体的 URL（使用 http 或者 rtsp）；参数 context 表示上下文；参数 uri 表示播放数据的 URI 位置；参数 fd 表示对多媒体文件的描述，FileDescriptor 类用于包装一个 Unix 文件的描述；参数 offset 表示开始播放多媒体文件的位置；参数 length 表示播放多媒体数据的长度。

（21）void setDisplay（SurfaceHolder sh）：将 SurfaceHolder 设置为展示多媒体的视频。在播放一个视频时，如果没有调用该方法或者 setSurface（Surface）方法，则只会播放声音。

（22）void setLooping（boolean looping）：将播放器设置成循环或非循环播放。参数 looping 取值为 true 时，播放器循环播放；取值为 false 时，播放器不进行循环播放。

（23）void setOnBufferingUpdateListener（MediaPlayer.OnBufferingUpdateListener listener）：注册一个回调事件，当网络流的缓冲状态被改变时调用。

（24）void setOnCompletionListener（MediaPlayer.OnCompletionListener listener）：注册一个回调事件，当已经播放完媒体资源之后调用。

（25）void setOnErrorListener（MediaPlayer.OnErrorListener listener）：注册一个回调事件，在进行异步操作期间发生错误时调用。

（26）void setOnInfoListener（MediaPlayer.OnInfoListener listener）：注册一个回调事件，当有信息或者警告时调用。

（27）void setOnPreparedListener（MediaPlayer.OnPreparedListener listener）：注册一个回调事件，当多媒体源已经准备好播放时调用。

（28）void setOnSeekCompleteListener（MediaPlayer.OnSeekCompleteListener listener）：注册一个回调事件，在一个 Seek 搜寻操作完成后调用。

（29）void setOnVideoSizeChangedListener（MediaPlayer.OnVideoSizeChangedListener listener）：注册一个回调事件，当视频的大小被更新时调用。

（30）void setVolume（float leftVolume，float rightVolume）：设置播放器的音量。其中，参数 leftVolume 表示左声道音量；参数 rightVolume 表示右声道音量。两个参数的取值范围都是从 0.0～1.0。

（31）void start ()：开始或者恢复播放。当还没有播放或者已经停止播放时，执行该方法则开始播放；当播放处于暂停状态时，执行该方法则从暂停的地方继续播放。

（32）void stop ()：停止播放。

下面举例说明如何播放一个存放在应用程序 res/raw/目录下的名为【music_file】的音

频文件。

```
MediaPlayer mediaPlayer=MediaPlayer.create(context,R.raw.music_file);
mediaPlayer.start();
```

MediaPlayer 主要用于播放音频，并不提供图像的输出界面。要结合使用 SurfaceView 控件，才能够实现视频的输出。可通过 MediaPlayer.setDisplay（SurfaceHolder sh）来将视频画面输出到 SurfaceView 上。

从 MediaPlayer 的生命周期图来看，使用 MediaPlayer 播放媒体文件并不复杂，下面举例说明如何使用 MeidaPlayer 播放音频。

（1）在【AndroidManifest.xml】中使用 mimeType 属性定义音频播放器可以处理的音频类型，即<data android:mimeType="audio/*" />。这样在选择播放音频文件时，系统才会去调用该程序。

```
<?xml version="1.0" encoding="utf-8"?>
<manifest xmlns:android="http://schemas.android.com/apk/res/android"
     package="com.demo.simpleMediaPlayer"
     android:versionCode="1"
     android:versionName="1.0">
   <uses-sdk android:minSdkVersion="8" />
   <application android:icon="@drawable/icon"
              android:label="@string/app_name">
      <activity android:name=".SimpleMediaPlayerActivity"
              android:label="@string/app_name">
          <intent-filter>
             <action android:name="android.intent.action.MAIN" />
             <category android:name="android.intent.category.LAUNCHER" />
          </intent-filter>
             <intent-filter>
            <action android:name="android.intent.action.VIEW" />
            <category android:name="android.intent.category.DEFAULT" />
             <data android:mimeType="audio/*" />
         </intent-filter>
      </activity>
   </application>
</manifest>
```

（2）程序代码：【SimpleMediaPlayerActivity.java】。

```
package com.demo.simpleMediaPlayer;
public class SimpleMediaPlayerActivity extends Activity {
   private static final String TAG="SimpleMediaPlayer2";
   public static final int STOP_MENU_ID  =Menu.FIRST;
   public static final int START_MENU_ID =Menu.FIRST+1;
   private MediaPlayer mMediaPlayer;
   private String mPath;
   @Override
   public void onCreate(Bundle savedInstanceState){
      super.onCreate(savedInstanceState);
      Intent intent=getIntent();
      Uri uri=intent.getData();
      if (uri != null) {
            mPath = uri.getPath(); // 获得媒体路径
            setTitle(mPath);// 设置标题
            setContentView(android.R.layout.simple_list_item_1);
            // 创建 MediePlayer
            mMediaPlayer = new MediaPlayer();
            // 设置媒体路径
            try {
                mMediaPlayer.setDataSource(mPath);
                mMediaPlayer.prepare();
            } catch (Exception ex) {
                ex.printStackTrace();
            }
```

```
        }

}
@Override
public boolean onCreateOptionsMenu(Menu menu){
        super.onCreateOptionsMenu(menu);
        menu.add(0,STOP_MENU_ID,0,"暂停");
        menu.add(0,START_MENU_ID,0,"开始");
        return true;
    }
  @Override
  public boolean onOptionsItemSelected(MenuItem item){
        switch(item.getItemId()){
        case STOP_MENU_ID:     //暂停
            if(mMediaPlayer==null || ! mMediaPlayer.isPlaying()){
                    Toast.makeText(this,
                                "没有音乐文件，无需暂停", Toast.LENGTH_LONG).show();
                }else{
                    mMediaPlayer.pause();
                }
            break;
        case START_MENU_ID:
            if(mMediaPlayer ==null){
                    Toast.makeText(this,
                "没有选中音乐文件，请到文件夹中点击音乐文件后再播放", Toast.LENGTH_LONG).show();
                }else{
                    mMediaPlayer.start(); // 开始
                }
            break;
        }
        return super.onOptionsItemSelected(item);
    }
@Override
protected void onDestroy() {
      super.onDestroy();
      mMediaPlayer.stop(); //停止
    }
}
```

注意：测试本程序，需要单击手机文件系统上的音频文件。Android 系统会提示可供选择的播放器，其中的 SimpleMediaPlayer 就是用上述代码实现的一个音频播放器，如图 4-6 所示。

（3）选择 SimpleMediaPlayer，可体验自行编写的音频播放器的播放效果，如图 4-7 所示。

图 4-6　提示可供选择的播放器

图 4-7　播放音频文件示例

3. Intent Filter

应用程序的组件为了告诉Android自己能响应和处理哪些隐式Intent请求，可以在Android的主配置文件【AndroidManifest.xml】中声明一个或者多个 Intent Filter，用来指明 Activity、Service、BroadcastReceiver 组件可以响应的隐式 Intents。给组件添加 Intent Filters 的方法是在声明组件元素的子元素中添加<intent-filter>元素。例如，具有编写新邮件 Activity 的 E-mail 应用程序可以在其 AndroidManifest.xml 中声明一个 Intent Filter 以便于响应“发送”Intent（注：用于发送邮件），在应用程序的 Activity 中就可以创建一个“发送”动作（ACTION_SEND）的 Intent，通过 startActivity()调用 Intent。当用户希望发送邮件时，Android 系统可以匹配 E-mail 应用程序中的“发送”Activity，并启动它。

Intent Filter 由 action（动作）、data（数据）和 category（类别）3 部分构成，用于描述该组件所能响应 Intent 请求的能力。如，对于邮件请求发送，邮件发送程序的 Intent Filter 就应该声明其所希望接收的 Intent Action 是 ACTION_SEND，以及与之相关的请求数据，包括邮件地址、邮件标题、邮件内容等。

一个隐式 Intent 请求要能够被传递给目标组件，必须通过上述 3 部分的检查。如果任何一部分不匹配，Android 都不会将该隐式 Intent 传递给目标组件。若两个组件有相同的<intent-filter>，则系统将会弹出一个选择组件的窗口供用户选择。对不同组件定义的 Intent Filter，其 action、data、catergory 的赋值可能部分相同，例如，Intent.ACTION_VIEW 用于显示用户的数据。下面的代码使用 Intent.ACTION_VIEW 来触发不同的组件，显示用户不同类型的数据。

① 使用浏览器打开浏览网址。

```
Uri uri=Uri.parse("http://www.baidu.com");
Intent it =new Intent(Intent.ACTION_VIEW,uri);
startActivity(it);
```

② 使用媒体播放器播放视频。

```
Uri uri=Uri.parse("file:///sdcard/media.mp4");
Intent intent=new Intent(Intent.ACTION_VIEW);
intent.setDataAndType(uri,"video/*");
startActivity(intent);
```

可以把 Intent Filter 理解为应用程序对组件的过滤器。了解 Intent Filter 的规则很重要，这样一方面便于开发者为自己的组件定义 Intent Filter，另一方面也可以在程序中更好地调用 Android 自带的系统功能。下面对 Intent Filter 的定义规则进行介绍。

（1）动作检查。

<intent-filter>元素中可以包括子元素<action>，比如：

```
<intent-filter>
     <action android:name="android.intent.action.MAIN" />
     <action android:name="android.intent.action.VIEW" />
     <action android:name="android.intent.action.EDIT" />
  <action android:name="com.example.project.SAVE_DATA" />
</intent-filter>
```

每个 action、category、data 元素赋值都是一行，如果有多个就写多行。Intent 的 action 属性名字是一个普通的字符串，表示该 Intent 所要完成的一个抽象“动作”。字符串的取值既可以由 Android 定义，也可以由程序员定义。如果想将某个 Intent 定义为程序入口点，则需要对 action 的属性赋值为“android.intent.action.MAIN”，并将 category 的属性赋值为“android.intent.category.LAUNCHER”。例如，对于音乐播放器应用程序，当用户单击相应图标启动程序后，应用程序启动后显示的第一个界面往往就是程序的入口点。

一条<intent-filter>元素至少应该包含一个<action>，否则任何 Intent 请求都不能和该<intent-filter>匹配。如果 Intent 请求的 Action 和<intent-filter>中的某一条<action>匹配，那么该 Intent 就可以通过这条<intent-filter>的动作检查。如果 Intent 请求没有说明具体的 Action 类型，那么只要<intent-filter>中包含有 Action 类型，Intent 请求就能顺利地通过<intent-filter>的动作检查。

（2）类别检查。

<intent-filter>元素可以包含<category>子元素，比如：

```
<intent-filter . . . >
    <category android:name="android.intent.category.LAUNCHER" />
    <category android:name="android.intent.category.DEFAULT" />
    <category android:name="android.intent.category.ALTERNATIVE" />
</intent-filter>
```

Intent 的 category 属性是一个普通的字符串，为 action 增加附加类别信息，通常二者结合使用。"android.intent.category.DEFAULT" 用来指明组件是否可以接收到隐式 Intents，所以说除了指明程序入口点的<intent filter>不用包含 DEFAULT 类别之外，其余的<intent filter>应包含。

只有当 Intent 请求中所有的 category 与组件中某一个 Intent Filter 的<category>完全匹配时，才会允许 Intent 请求通过检查，Intent Filter 中多余的<category>声明并不会导致匹配失败。一个没有指定任何类别要求的 Intent Filter 只会匹配没有设置类别的 Intent 请求。

（3）数据检查。

数据在<intent-filter>中的描述如下：

```
<intent-filter . . . >
    <data android:mimeType="video/mpeg" >
    <data android:mimeType="audio/mpeg"
    android:content="com.example.project:300/folder/subfolder/aduio"/>
</intent-filter>
```

<data>元素指定了希望接收的 Intent 请求的数据类型（由 mimeType 属性指定）和数据 URI（使用 content 属性指定）。每个 URI 包括四个属性参数(scheme、host、port、path)，形如 scheme://host:port/path。如 content://com.example.project:300/folder/subfolder/aduio 中的"scheme"是 content，"host"是 com.example.project，"port"是 300，"path"是 folder/subfolder/audio。数据的数据类型和 URI 无需同时指定。例如，<data android:mimeType="video/mpeg" >只是指定了数据类型，那么包含相同的数据类型但是没有指定 URI 的 Intent 请求可以通过检查。

三、任务实施

（1）定义控件类，在 onCreate 方法中获取 XML 布局文件中的控件。

```
private Display currDisplay;
private SurfaceView surfaceView; //显示视频的 SurfaceView
private SurfaceHolder holder;
private MediaPlayer player;
private int vWidth,vHeight;//视频宽高
private Timer    timer;        //计时器
private ImageButton  rew; //快退
private ImageButton  pause;//暂停
private ImageButton  start;//开始
private ImageButton  ff;//快进
private TextView   play_time;//已播放时间
private TextView   all_time;//总播放时间
private TextView   title;//播放文件名称
private SeekBar    seekbar;//进度条
public void onCreate(Bundle savedInstanceState){
    super.onCreate(savedInstanceState);
```

```
        ……//隐藏系统标题栏,在 4.1 节中已经实现
        setContentView(R.layout.main);
        //获得当前布局文件的控件对象
        title=(TextView)findViewById(R.id.title);
        surfaceView =(SurfaceView)findViewById(R.id.surfaceview);
        rew=(ImageButton)findViewById(R.id.rew);
        pause=(ImageButton)findViewById(R.id.pause);
        start=(ImageButton)findViewById(R.id.start);
        ff=(ImageButton)findViewById(R.id.ff);
        play_time=(TextView)findViewById(R.id.play_time);
        all_time=(TextView)findViewById(R.id.all_time);
        seekbar=(SeekBar)findViewById(R.id.seekbar);
        ……
}
```

（2）创建 MediaPlayer：通过 Intent 获得播放媒体的位置，然后初始化 MediaPlayer，设置媒体准备好的监听器和媒体播放完成的监听器，并播放音乐。

```
public void onCreate(Bundle savedInstanceState){
        ……//略,接第(1)步
        //获取传递过来的媒体路径
        Intent intent=getIntent();
        Uri uri=intent.getData();
        String  mPath;
        if(uri!=null)
        {
           mPath=uri.getPath(); //获得媒体路径
        }else
        {
          //默认加载一个音乐播放文件，测试时应在 sdcard 目录下放一个名为 music.mp3 的音乐
          mPath="/mnt/sdcard/music.mp3";
        }
        //下面开始实例化 MediaPlayer 对象
        player=new MediaPlayer();
        //设置 prepare 完成监听器
        player.setOnPreparedListener(new OnPreparedListener() {
          @Override
          public void onPrepared(MediaPlayer mp){
               // 当 prepare 完成后,该方法触发,播放音频
                player.start();
                //启动时间更新及进度条更新任务,每 0.5s 更新一次
               timer=new Timer();
               timer.schedule(new MyTask(),50,500);
          }
        });
        //设置播放完成监听器
        player.setOnCompletionListener(new OnCompletionListener() {
          @Override
          public void onCompletion(MediaPlayer mp){
               // 当 MediaPlayer 播放完成后触发
               if(timer!=null)
               {
                    timer.cancel();
                    timer=null;
               }
          }
        });

        player.setAudioStreamType(AudioManager.STREAM_MUSIC);
        try {
            //从多媒体文件路径截取,设置播放文件名称
            title.setText(mPath.substring(mPath.lastIndexOf("/")+1));
            //指定需要播放文件的路径,初始化 MediaPlayer
            player.setDataSource(mPath);
        } catch(Exception e){
            e.printStackTrace();
        }
```

```
    ......
}
```

（3）按键事件处理：按键包括【开始】、【暂停】、【快进】、【快退】。【开始】和【暂停】按钮不能同时出现。当开始播放时，启动进度条更新任务线程。快进和快退都需要在播放音乐的情况下执行，改变时间的幅度都设置为 10 s。

```
public void onCreate(Bundle savedInstanceState){
    ......//略,接第(2)步
    //播放操作
    start.setOnClickListener(new OnClickListener() {
        @Override
        public void onClick(View v){
            //按下开始操作后,暂停按钮设为可见,开始按钮设为隐藏
            start.setVisibility(View.GONE);
            pause.setVisibility(View.VISIBLE);
            player.start();//启动播放
            if(timer!=null)
            {
                timer.cancel();
                timer=null;
            }
            //启动时间更新及进度条更新任务,每 0.5s 更新一次
            timer=new Timer();
            timer.schedule(new MyTask(),50,500);
        }
    });
    //暂停操作
    pause.setOnClickListener(new OnClickListener() {
        @Override
        public void onClick(View v){
            //按下暂停操作后,暂停按钮设为隐藏,开始按钮设为可见
            pause.setVisibility(View.GONE);
            start.setVisibility(View.VISIBLE);
            player.pause();
            //按下暂停后，需要停止进度条控制线程
            if(timer!=null)
            {
                timer.cancel();
                timer=null;
            }
        }
    });

    //快退操作,每次快退 10s
    rew.setOnClickListener(new OnClickListener() {
        @Override
        public void onClick(View v){
            //判断是否正在播放
            if(player.isPlaying())
            {
                int currentPosition=player.getCurrentPosition();
                //将 10 秒转换成毫秒，即 10000
                if(currentPosition-10000>0)
                {
                    player.seekTo(currentPosition-10000);
                }
            }
        }
    });
    //快进操作,每次快进 10s
    ff.setOnClickListener(new OnClickListener() {

        @Override
        public void onClick(View v){
            //判断是否正在播放
            if(player.isPlaying())
```

```
            {
                int currentPosition=player.getCurrentPosition();
                if(currentPosition+10000<player.getDuration())
                {
                    player.seekTo(currentPosition+10000);
                }
            }
        }
    });
}
```


4.3 播放视频

一、任务分析

本任务需要实现的效果如图 4-8 所示。

图 4-8 视频播放界面

本次任务要求实现视频文件的播放，在播放界面上显示视频文件的名称、视频图像、播放的时间、视频的总时间、视频播放控制等。需要思考如下几个问题。

（1）如何打开视频？

（2）如何显示视频图像？

（3）如何对播放的视频进行管理？

二、相关知识

除了可以使用 MediaPlayer 播放多媒体文件之外，还可以使用另一种相对简单的方式来播放多媒体文件，那就是 VideoView。

VideoView 是 SurfaceView 的子类，用于显示一个视频文件。VideoView 类能够加载来自不同数据源的图像，并可以设置视频的控制栏，能够对视频进行计算测量，使其可以应用在各种布局管理器中。

VideoView 构造函数有 3 种：

① VideoView（Context context）;

② VideoView（Context context，AttributeSet attrs）;

③ VideoView（Context context，AttributeSet attrs，int defStyle）。

其中，参数 context 表示上下文；参数 attrs 表示属性集；参数 defStyle 表示样式。

VideoView 类的很多方法和功能与 MeidaPlayer 类相似，主要方法介绍如下。

（1）void setVideoPath（String path）：设置视频资源的所在位置。

（2）void setVideoURI（Uri uri）：设置视频资源的 URI。

（3）void setMediaController（MediaController controller）：设置播放控制进度条。

（4）void start ()：开始播放视频。

（5）void pause ()：暂停播放视频。

（6）void resume ()：继续播放视频。

① 在【AndroidManifest.xml】中使用 mimeType 属性定义视频播放器可以处理的视频类型，即：<data android:mimeType="video/*" />，这样在单击播放视频文件时，系统才能够调用该程序。

```
<activity android:name="SimpleMediaPlayer1" android:label="MediaPlayer">
        <intent-filter>
            <action android:name="android.intent.action.VIEW" />
            <category android:name="android.intent.category.DEFAULT" />
            <data android:mimeType="video/*" />
         </intent-filter>
<activity>
```

② 使用 VideoView 进行视频播放。

```
public class SimpleMediaPlayer1 extends Activity {
    private static final String TAG="SimpleMediaPlayer1";
    public VideoView mVideoView;
    @Override
    public void onCreate(Bundle savedInstanceState){
       super.onCreate(savedInstanceState);
       mVideoView=new VideoView(this); // 建立一个 VideoView
          Intent intent=getIntent();   // 获取 Intent
       Uri uri=intent.getData();    // 从 Intent 中获取数据
       if (uri!=null){
              String path=uri.getPath();  // 获取路径
              setTitle(path);          // 设置标题为媒体的路径
               mVideoView.setVideoURI(uri);       // 设置 URI
           mVideoView.setMediaController(new MediaController(this));// 设置媒体控制条
               mVideoView.start();               // 开始播放
               setContentView(mVideoView);       // 将 VideoView 设置为屏幕界面
          } else{
             Toast.makeText(this, "没有选中多媒体文件，请到文件夹中点击多媒体文件后再播放",
Toast.LENGTH_LONG).show();
          }

    }
}
```

注意：测试本程序，需要点击手机文件系统上的视频文件。当单击视频文件时，Android 系统会提示可供选择的播放器，而其中的 SimpleMediaPlayer1 就是上面代码实现的一个视频播放器，如图 4-9 所示。

③ 选择 SimpleMediaPlayer1，可体验自行编写的视频播放器的播放效果，如图 4-10 所示。

图 4-9　媒体播放器选择

图 4-10　播放视频

三、任务实施

（1）设置视频显示 SurfaceView 的行为。接 4.2 节，使 MediaPlayer 能够同时播放音频文件和视频文件。

```
public void onCreate(Bundle savedInstanceState){
        ...... //略,接 4.2 节的内容
    //给 SurfaceView 添加 CallBack 监听
        holder=surfaceView.getHolder();
        holder.addCallback(new SurfaceHolder.Callback() {
            @Override
            public void surfaceDestroyed(SurfaceHolder holder){
                // TODO Auto-generated method stub
            }
        // 当 SurfaceView 中的 Surface 被创建时调用
            @Override
            public void surfaceCreated(SurfaceHolder holder){
                //设置 MediaPlayer 在指定的 Surface 中进行播放
                player.setDisplay(holder);
                //在指定了 MediaPlayer 播放的容器后,用 prepareAsync 准备播放
                player.prepareAsync();
            }

            @Override
        public void surfaceChanged(SurfaceHolder holder,int format,int width,int height){
                // TODO Auto-generated method stub
            }
        });
        //为了可以播放视频或者使用 Camera 预览,需要指定其 Buffer 类型
        holder.setType(SurfaceHolder.SURFACE_TYPE_PUSH_BUFFERS);
    //获得程序窗口默认的 Display 对象,通过 Display 可以获得屏幕的大小
    currDisplay=this.getWindowManager().getDefaultDisplay();
}
```

（2）修改 OnPreparedListener 监听器，使其同时播放视频。

```
player.setOnPreparedListener(new OnPreparedListener() {
    @Override
        public void onPrepared(MediaPlayer mp){
            // 当 prepare 完成后,该方法触发。在该方法内编写播放视频代码
            //首先取得视频的宽和高
            vWidth=player.getVideoWidth();
            vHeight=player.getVideoHeight();
        if(vWidth > currDisplay.getWidth() || vHeight > currDisplay.getHeight()){
            //如果视频的宽或者高超出了当前屏幕的大小,则需要进行缩放
                float wRatio =(float)vWidth/(float)currDisplay.getWidth();
                 float hRatio=(float)vHeight/(float)currDisplay.getHeight();
                //选择大的比例进行缩放
                float ratio=Math.max(wRatio,hRatio);
                vWidth =(int)Math.ceil((float)vWidth/ratio);
                vHeight =(int)Math.ceil((float)vHeight/ratio);
                //重新设置 surfaceView 的布局参数
            surfaceView.setLayoutParams(new
                        LinearLayout.LayoutParams(vWidth,vHeight));
                //然后开始播放视频
                player.start();
            }else
            {
                player.start();
            }
            if(timer!=null)
            {
                timer.cancel();
                timer=null;
            }
            //启动时间和进度条更新任务,每 0.5s 更新一次
            timer=new Timer();
```

```
            timer.schedule(new MyTask(),50,500);
        }
});
```

4.4 管理多媒体文件

一、任务分析

本任务需要实现的效果如图 4-11 所示。

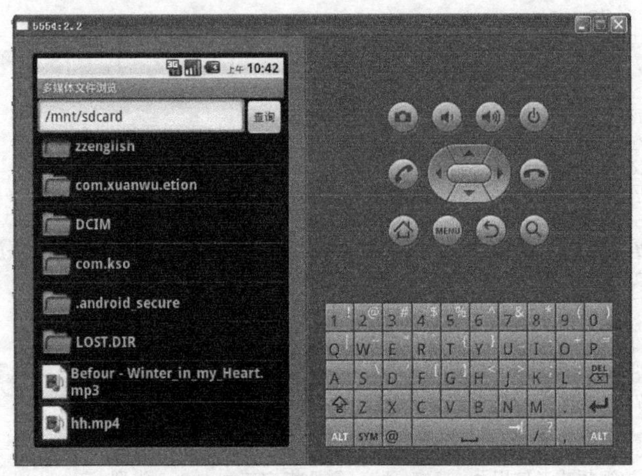

图 4-11　多媒体文件浏览界面

本次任务是浏览手机目录下的音频和视频文件，以便可以在多媒体播放器上选择播放。需要思考如下几个问题。

（1）如何搜索手机上的文件？

（2）如何在界面上列出媒体文件？

二、相关知识

1. Android 的文件结构

Android 的核心是基于 Linux 操作系统，没有所谓的盘符，只有目录，而且根目录并不像 Windows 系统那样分为 C 盘、D 盘、E 盘等。Android 的文件系统结构采用树形管理，其入口从“/”目录（也称为根节点），如图 4-12 所示。在根节点下包含很多的目录，比如 sdcard、data、mnt、sbin 等，当然在这些目录中，还有很多子目录或文件。

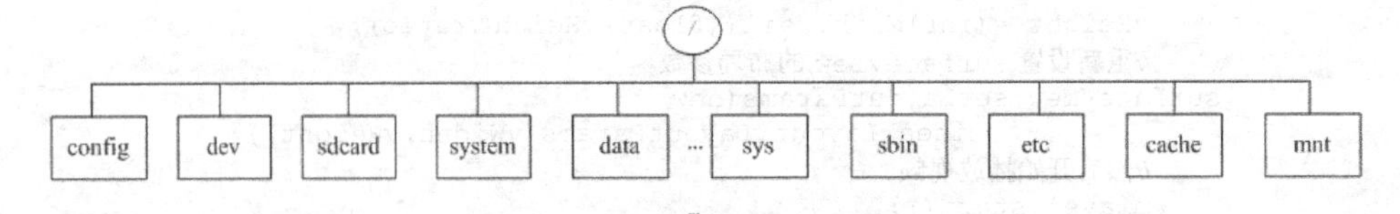

图 4-12　Android 文件系统结构

除了可以使用 DDMS 可视化的方式查看 Android 的文件系统之外，还可以通过命令方式登录到 Android 文件系统中，启动的命令是 adb shell，用法如下。

① 在 Windows 中，开启一个 DOS 命令视窗（或执行 cmd.exe 程序）。

② 输入命令：adb shell。当出现#提示符时，表示已经进入模拟器的 Linux 系统。接下来，就可以使用 Linux 的指令，例如 cd、ls、pwd、cat、rm 等来对文件系统进行管理。需要注意的

是，执行上述命令需要先打开 Android 的模拟器，否则会提示找不到设备的错误信息。另外，若提示没有 adb 命令，则需要在 Android 的 sdk\platform-tools 目录下执行 adb shell 命令。

图 4-12 列出的只是 Android 文件系统中的部分目录，下面介绍其中几个比较重要的目录。

（1）/system 目录：存放操作系统的大部分文件。

① 在/system/app 目录下存放预载入的应用程序（*.apk），该文件夹下的程序为系统默认的组件，如闹钟程序 AlarmClock.apk、浏览器程序 Browser.apk、计算器程序 Calculator.apk 等，如图 4-13 所示。用户自己安装的软件不会出现在这里，而是存放在/data/文件夹中。

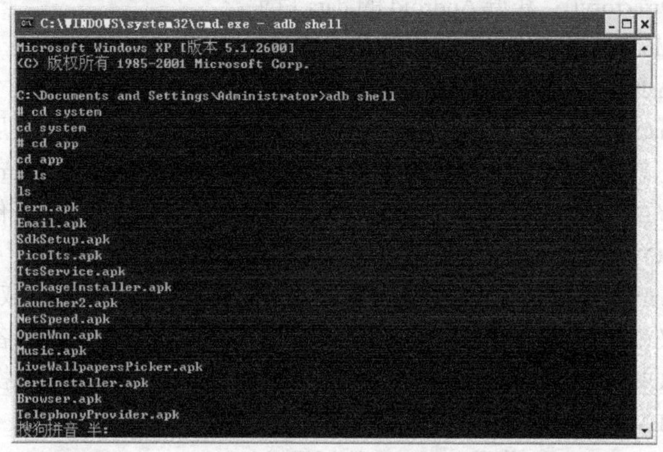

图 4-13　通过命令查看 system/app 目录下的 apk 程序

② 在/system/framework 目录下存放着 Android 系统的核心程序库，用于应用程序与硬件的交互，如 core.jar、framework-res.apk、ext.jar 等。

③ 在/system/media/audio/目录下存放着系统的声音文件，例如，闹铃声、来电铃声等，这些声音文件大多为 OGG 格式。

④ /system/etc 目录用于存放系统的配置文件，比如 APN 接入点设置等核心配置。

⑤ /system/bin 目录用于存放系统的二进制可执行程序，主要是 Linux 自带的组件，如 Ping 程序。

⑥ /system/fonts 目录用于存放字体文件，包括标准字体和粗体、斜体，还可能包括中文字库。

⑦ /system/lib 目录主要用于存放系统底层库。

（2）/data 目录：存放用户安装的软件以及各种数据。

① 在/data/app 目录下存放着用户安装的软件。

② 在/data/data/目录下存放着每个应用程序以包命名的文件夹，即<app-package-name>。由 Context.openFileOutput()方法建立的文件都存放在该目录下的 files 子目录内，这在项目三中已经进行了介绍；而用 Context.getShared Preferences()建立的 preferences（*.xml）文件，则存放在 shared_pref 子目录中。

③ 在/data/anr/目录下保存着 ANR（Application is Not Responding）错误信息，可以通过 cat 命令查看【traces.txt】文件，以便追踪出现错误的原因。

（3）/mnt 目录：外部设备都会挂载到这个目录下，如将 SD 卡挂载上去后，会生成一个/mnt/sdcard 目录。

（4）/sdcard 目录：表示 SD 卡的路径，是 mnt/sdcard 的虚拟或快捷方式。mnt 是 Android Lunix 系统下挂载外部设备的专用目录，mnt/sdcard 是实际文件所在的路径。

（5）/dev 目录：用于存放设备节点文件。

（6）/sbin 目录：存放一个用于调试的 adbd 程序。

Environment 类一方面提供了静态的变量，用于表示外部存储设备的状态和存放文件的标准目录，另外一方面还提供静态的方法，用于获得与文件路径有关的环境变量。这样可以帮助程序员更方便地去操作 Android 系统中的文件。下面介绍 Environment 类的主要方法。

（1）getDataDirectory()：获得 Android 的 data 目录。

（2）getDownloadCacheDirectory()：获得 Android 的下载缓冲内容目录。

（3）getExternalStorageDirectory()：获得 Android 的外部存储目录。

（4）getExternalStoragePublicDirectory（String type）：获得外部存储的顶层目录，用于放置特定类型的文件。其中，参数 type 表示存放文件的类型，已经在 Environment 的常量中进行了定义。例如，DIRECTORY_DOWNLOADS 表示放置用户下载文件的标准目录；DIRECTORY_MOVIES 表示放置用户使用的电影文件的标准目录。

（5）getExternalStorageState()：获得外部存储设备的当前状态。状态的取值在 Environment 的常量中已经定义，例如，MEDIA_MOUNTED 表示外部存储设备已经准备好，可以进行读写；MEDIA_MOUNTED_READ_ONLY 则表示外部存储设备只能进行读操作，不能进行写操作；MEDIA_REMOVED 则表示外部存储设备不存在。

（6）getRootDirectory ()：获得 Android 的根目录。

此外，Activity 类也提供了一个名为 getFilesDir 的方法，用于获得由当前应用程序调用 openFileOutput()方法创建的文件所在目录。

2. File 类

File 类提供了查找、获取和设置 Android 文件系统中文件的操作。注意：这里所指的文件包含目录在内，即目录也被当作一个文件。

File 的构造函数有四种。

① File（File dir，String name）：根据给出的目录和文件名构造一个新的文件。

② File（String path）：根据给出的路径构造一个新的文件。

③ File（String dirPath，String name）：根据给出的目录路径字符串和文件名构造一个新的文件。

④ File（URI uri）：根据给出的 URI 路径构造一个新的文件。

下面介绍 File 类的主要方法。

（1）boolean createNewFile()：创建一个新的空文件。

（2）boolean delete()：删除本文件。

（3）void deleteOnExit()：当虚拟器正常终止后，自动删除本文件。

（4）boolean exists()：判断文件（夹）是否存在，返回 true 表示存在，返回 false 表示不存在。

（5）File getAbsolutePath()：返回文件的绝对路径。

（6）String getName()：返回文件的名字。

（7）String getParent()：返回文件上一级目录的路径名。

（8）String getPath()：返回文件的相对路径。

（9）boolean isDirectory()：判断是否为文件夹，返回 true 表示是文件夹，返回 false 表示不是。

（10）boolean isFile ()：判断是否为文件，返回 true 表示是文件，返回 false 表示不是。

（11）boolean isHidden ()：判断文件是否隐藏。

（12）long lastModified ()：返回文件夹最新的修改时间。

（13）long length ()：返回文件的字节数。

（14）String[] list ()：返回当前目录下的所有文件名，并用一个字符串数组表示。若调用方法的对象不是目录，则调用该方法会返回 null。

（15）File[] listFiles()方法：列出文件夹下的所有文件和文件夹，用一个 File 数组表示。

（16）File[] listRoots()：获取根目录。

（17）boolean mkdir()：根据目录名，创建最后一层的目录。

（18）boolean mkdirs()：根据目录名，创建最后一层的目录，若前一层的目录不存在，也需同时创建。

（19）boolean renameTo（File newPath）：将文件更改为新的路径。

（20）boolean setLastModified（long time）：设置文件最后更改的时间。其中参数 time 的单位为毫秒（ms）。

下面列出了常用的文件操作实现方法。

（1）获取根目录。

```
File[] fileList=File.listRoots();
```

说明

实际上，Android 或者其他的 Unix 系统，只有唯一的根目录："/"，所以上述操作虽然得到一个数组，但数组中的元素只有 1 个。

（2）列出根目录下的文件。

```
File file=new File("/");
File[] fileList=file.listFiles();
```

（3）获取系统存储根目录下的文件。

说明

这里说的系统仅仅是指/system，不包括外部存储。

```
File file=Environment.getRootDirectory();
File[] fileList=file.listFiles();
```

（4）获取 SD 卡存储根目录下的文件。

```
//要获取 SD 卡存储根目录首先要确认 SD 卡是否装载,如果返回 true,则已装载,否则未装载
boolean is=Environment.getExternalStorageState().equals(Environment.MEDIA_MOUNTED);
 if(is){
File file=Environment.getExternalStorageDirectory();//与 File file=new File("/sdcard");等效
 File[] fileList=file.listFiles();
}
```

（5）获取 data 根目录下的文件。

```
File file=Environment.getDataDirectory();//与 File file=new File("/data");等效
File[] fileList=file.listFiles();
```

说明

由于 data 文件夹是 Android 系统中一个非常重要的文件夹，一般权限无法获取该目录下的文件，即 fileList.length 会返回为 0。

（6）获取私有文件路径。

```
Context context=this;//首先，在 Activity 中获取 Context
File file=context.getFilesDir();
String path=file.getAbsolutePath();
```

说明

此处返回的路径为/data/data/包/files，这里的包是指主 Activity 所在的包，其路径也在 data 文件夹下。程序本身可以对自己的私有文件进行操作，但很多私有的数据将会写入到私有文件路径下，这也是 Android 为什么对 data 目录中的数据采取保护的原因之一。

（7）获取文件（夹）绝对路径、相对路径、文件（夹）名、父目录。

```
File file=……
String relativePath=file.getPath();//相对路径
String absolutePath=file.getAbsolutePath();//绝对路径
String fileName=file.getName();//文件(夹)名
String parentPath=file.getParent();//父目录
```

（8）新建文件夹。

```
void createnewFolder(String path){
    try {
        // 实例化一个 File 对象
        File myFile=new File(path);
        // 判断是否存在，不存在则创建
        if(!myFile.exists()){
            // mkdir()只创建最后一层的目录，而 mkdirs()会创建必须但不存在的父目录
            myFile.mkdir();/创建文件夹
        }
    } catch(Exception e){
        // 创建失败，捕捉异常
        Log.v("directory exception",e.getMessage());
    }
}
```

（9）新建文件。

```
void createnewFile(String path,String content){
    try {
        File myFile=new File(path);
        if(!myFile.exists()){
            // 创建一个空的文件
            myFile.createNewFile();
        }
        //向文件中输入内容
        FileWriter fw=new FileWriter(myFile);
        fw.write(content);
        if(fw != null)
            fw.close();
    } catch(IOException e){
        Log.v("file exception",e.getMessage());
    }
}
```

（10）重命名文件（夹）。

```
File file=……
String parentPath=file.getParent();
String newName="newname";//重命名后的文件或者文件夹名
File filex=new File(parentPath,newName);
file.renameTo(filex);
```

（11）删除文件（夹）。

```
File file=……
file.delete();//立即删除
```

下面列举完整示例说明 Environment 和 File 的使用方法，通过不同的按钮来显示 Android 的文件路径，效果如图 4-14 所示。

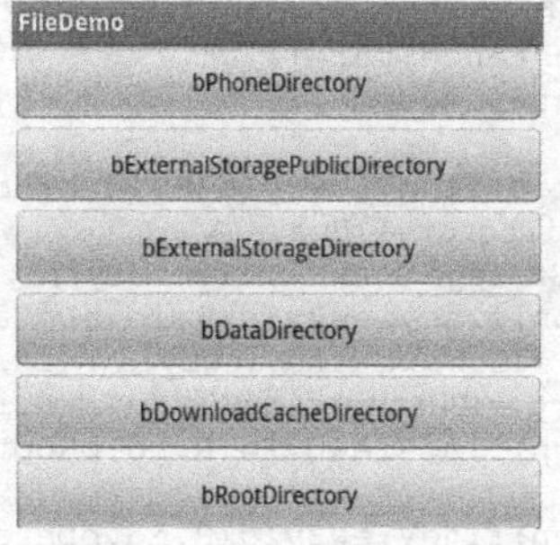

图 4-14　文件管理示例效果

（1）布局文件【filedemo.xml】的定义如下。

```
<?xml version="1.0" encoding="utf-8"?>
<LinearLayout xmlns:android="http://schemas.android.com/apk/res/android"
    android:orientation="vertical" android:layout_width="fill_parent"
    android:layout_height="fill_parent">


    <Button android:id="@+id/bPhoneDirectory" android:layout_width="fill_parent"
        android:layout_height="wrap_content" android:text="bPhoneDirectory" />

    <Button
       android:id="@+id/bExternalStoragePublicDirectory"
       android:layout_width="fill_parent"
       android:layout_height="wrap_content"
       android:text="bExternalStoragePublicDirectory" />

    <Button android:id="@+id/bExternalStorageDirectory"
        android:layout_width="fill_parent" android:layout_height="wrap_content"
        android:text="bExternalStorageDirectory" />

    <Button android:id="@ + id/bDataDirectory" android:layout_width="fill_parent"
        android:layout_height="wrap_content" android:text="bDataDirectory" />
    <Button android:id="@+id/bDownloadCacheDirectory"
        android:layout_width="fill_parent" android:layout_height="wrap_content"
        android:text="bDownloadCacheDirectory" />
    <Button android:id="@+id/bRootDirectory" android:layout_width="fill_parent"
        android:layout_height="wrap_content" android:text="bRootDirectory" />
</LinearLayout>
```

（2）程序实现代码如下。

```
public class FileDemo extends Activity {
    private int iFileNum=0;
    private Button bPhoneDirectory;
    private Button bExternalStoragePublicDirectory;
    private Button bDataDirectory;
    private Button bDownloadCacheDirectory;
    private Button bExternalStorageDirectory;
    private Button bRootDirectory;


    private File fPhoneDirectory;
    private File fExternalStoragePublicDirectory;
    private File fDataDirectory;
    private File fDownloadCacheDirectory;
    private File fExternalStorageDirectory;
    private File fRootDirectory;

    private String name;

    /** Called when the activity is first created. */
    @Override
```

```
    public void onCreate(Bundle savedInstanceState){
        super.onCreate(savedInstanceState);
        setContentView(R.layout.filedemo);

        bPhoneDirectory =(Button)findViewById(R.id.bPhoneDirectory);
        bExternalStoragePublicDirectory=(Button)
    findViewById(R.id.bExternalStorage PublicDirectory);
        bRootDirectory =(Button)findViewById(R.id.bDataDirectory);
        bDownloadCacheDirectory =(Button)findViewById(R.id.bDownloadCacheDirectory);
        bExternalStorageDirectory =(Button)findViewById(R.id.bExternalStorageDirectory);
        bRootDirectory =(Button)findViewById(R.id.bRootDirectory);

        bDataDirectory =(Button)findViewById(R.id.bDataDirectory);

        fPhoneDirectory=this.getFilesDir();
        fExternalStoragePublicDirectory=

    Environment.getExternalStoragePublicDirectory(Environment.DIRECTORY_MUSIC);
        fDataDirectory=Environment.getDataDirectory();
        fDownloadCacheDirectory=Environment.getDownloadCacheDirectory();
        fExternalStorageDirectory=Environment.getExternalStorageDirectory();
        fRootDirectory=Environment.getRootDirectory();

        /*
         * 没有存储卡的时候按键无效
         */
        if(Environment.getExternalStorageState().equals(
                Environment.MEDIA_REMOVED)){
            bExternalStorageDirectory.setEnabled(false);
        }

        /*
         * 访问应用程序手机内存
         */
        bPhoneDirectory.setOnClickListener(new Button.OnClickListener() {
            public void onClick(View v){
                String path=fPhoneDirectory.getPath();
                try{
                    FileOutputStream     outStream=FileDemo.this.openFileOutput("test.txt",
Activity.MODE_PRIVATE);
                    outStream.write("hello".getBytes());
                  outStream.close();
                }catch(Exception ex){
                    return;
                }
                name=path+"\n";
                ListFiles(path);
                Toast.makeText(FileDemo.this,name,Toast.LENGTH_LONG).show();
            }
        });
        bExternalStoragePublicDirectory.setOnClickListener(new Button.OnClickListener() {
            public void onClick(View v){
                String path=fExternalStoragePublicDirectory.getAbsolutePath();
                Toast.makeText(FileDemo.this,path,Toast.LENGTH_LONG)
                        .show();
            }
        });
        /*
         * 访问存储卡
         */
        bExternalStorageDirectory
                .setOnClickListener(new Button.OnClickListener() {
                    public void onClick(View v){
                        String path=fExternalStorageDirectory.getPath();
                        name=path+"\n";
                        ListFiles(path);

                        Toast.makeText(FileDemo.this,name,Toast.LENGTH_LONG)
```

```
                                .show();
                    }
                });

        bDownloadCacheDirectory
                .setOnClickListener(new Button.OnClickListener() {
                    public void onClick(View v){
                        String path=fDownloadCacheDirectory.getAbsolutePath();
                        Toast.makeText(FileDemo.this,path,Toast.LENGTH_LONG)
                                .show();
                    }
                });

        bRootDirectory.setOnClickListener(new Button.OnClickListener() {
            public void onClick(View v){
                String path=fRootDirectory.getAbsolutePath();
                Toast.makeText(FileDemo.this,path,Toast.LENGTH_LONG).show();
            }
        });

        bDataDirectory.setOnClickListener(new Button.OnClickListener() {
            public void onClick(View v){
                String path=fDataDirectory.getAbsolutePath();
                Toast.makeText(FileDemo.this,path,Toast.LENGTH_LONG).show();
            }
        });

    }

    private boolean ListFiles(String path){
        File file=new File(path);

        for(File f : file.listFiles()){
            path=f.getAbsolutePath();
            name=name+f.getName()+"\n";
        }

        return true;
    }
}
```

3．多媒体文件扫描

多媒体文件扫描实质就是查找出手机上的多媒体文件，以方便用户有选择地播放。实现该功能需要定义多媒体文件类型的筛选标准。下面的代码将需要支持的多媒体文件的后缀放入到一个数组中。

```
private final String[] FILE_MapTable ={
                        ".3gp",".mov",".avi",".rmvb",".wmv",".mp3", ".mp4" };
```

接下来在指定的文件夹下扫描媒体文件，主要的原理是：先使用 listFiles()方法查找出当前目录下的文件，然后将这些文件的后缀和上一步定义的数组进行比较，以检查它们是否属于多媒体文件。

```
File f=new File("/mnt/sdcard");
File[] files=f.listFiles();
for(int i=0; i < files.length; i++){
  if(files[i].isFile()){
    String fileName=files[i].getName();
    int index=fileName.lastIndexOf(".");
    if(index>0){
     String endName=fileName.substring(index,fileName.length()).toLowerCase();
     boolean bFind =false;
     for(int x=0; x<FILE_MapTable.length; x++){
         //符合预先定义的多媒体格式的文件才会在界面中显示
         if(endName.equals(FILE_MapTable[x])){
               bFind=true;
               break;
         }
      } else{
```

```
            //下面省略将目录信息保存下来的操作代码
        }
        if(bFind){
            System.out.println("该文件为媒体文件");
            //下面省略将多媒体文件信息保存下来的操作代码
        }
    }
```

最后，定义一个 Adapter 的子类，通过 ListView 控件将筛选出来的媒体文件显示给用户。

此外，采用 MediaStore 类是另一种更为便捷的获得多媒体文件信息的方式。MediaStore 用于存放多媒体信息，此类包含三个内部类，分别如下。

① MediaStore.Audio：存放音频信息。

② MediaStore.Image：存放图片信息。

③ MediaStore.Vedio：存放视频信息。

上述每个类又有相应的字段表示多媒体信息。当手机开机或者有 SD 卡插拔等事件发生时，系统将会自动扫描 SD 卡和手机内存上的多媒体文件，如声音、视频、图片等，将相应的信息放入到定义好的数据库表格中。在程序中，并不需要关心如何去扫描手机系统中的文件，只要通过 ContentProvider 方式提供的查询接口，我们便可以得到各种手机上的多媒体信息。例如，是想要查询手机上所有在外部存储卡上的音乐文件的各种信息，其代码如下。

```
Cursor cursor =getContentResolver().query(
            MediaStore.Audio.Media.EXTERNAL_CONTENT_URI,
            new String[] { MediaStore.Audio.Media.TITLE,
                    MediaStore.Audio.Media._ID,
                    MediaStore.Audio.Media.DATA,
                    MediaStore.Audio.Media.DURATION }, null, null, null);
```

其中，MediaStore.Audio.Media.TITLE 表示歌曲的名称；MediaStore.Audio.Media._ID 表示歌曲 ID；MediaStore.Audio.Media.DATA 表示歌曲文件的路径；MediaStore.Audio.Media.DURATION 表示歌曲的总播放时长。

三、任务实施

（1）定义列出多媒体文件的界面布局文件【myfile.xml】，使用一个 EditText 表示查询目录，使用一个 Button 表示查询按键，使用一个 ListView 来显示文件。

```
    <?xml version="1.0" encoding="utf-8"?>
<LinearLayout
  xmlns:android="http://schemas.android.com/apk/res/android"
  android:orientation="vertical"
  android:layout_width="fill_parent"
  android:layout_height="fill_parent">
  <!--查询栏-->
  <LinearLayout
  android:orientation="horizontal"
  android:layout_width="fill_parent"
  android:layout_height="wrap_content">
  <!--文件夹路径-->
  <EditText android:id="@+id/path_edit"
  android:layout_width="fill_parent"
  android:layout_height="wrap_content"
  android:layout_weight="1.0"
  />
  <!--查询按键-->
  <Button android:id="@+id/qry_button"
  android:layout_width="wrap_content"
  android:layout_height="wrap_content"
  android:text="查询"
  />
```

```
  </LinearLayout>
  <!--文件列表-->
  <ListView android:id="@+id/file_listview"
    android:layout_width="fill_parent"
    android:layout_height="fill_parent"
    android:layout_weight="1"
  />
</LinearLayout>
```

（2）设置一个 ListView 列表子项布局文件【file_item.xml】，用于显示每条文件记录信息。文件列表图标分为文件夹或多媒体图标，分别用一个 ImageView 元素表示，并用一个 TextView 元素表示文件名称。

```
<?xml version="1.0" encoding="utf-8"?>
<LinearLayout
    xmlns:android="http://schemas.android.com/apk/res/android"
    android:layout_width="fill_parent"
    android:layout_height="wrap_content"
    >
    <!-- 文件夹图标 -->
    <ImageView
    android:id="@+id/folder"
    android:layout_width="wrap_content"
    android:layout_height="wrap_content"
    android:src="@drawable/folder"
  />
  <!—多媒体文件图标 -->
    <ImageView
    android:id="@+id/music"
    android:layout_width="wrap_content"
    android:layout_height="wrap_content"
    android:src="@drawable/music"
    android:visibility="gone"
  />
  <!-- 文件名称-->
    <TextView
    android:id="@+id/name"
    android:layout_width="fill_parent"
    android:layout_height="fill_parent"
    android:text="fdfsfsf"
    android:textSize="18.0dip"
    android:layout_weight="1.0"
    android:textStyle="bold"
    android:gravity="center_vertical"
    ></TextView>
</LinearLayout>
```

（3）在 onCreate 方法中加载 XML 布局文件，获得界面上的控件对象，编写按钮单击事件和文件被选中事件的处理代码。

```
    // 支持的媒体格式
    private final String[]FILE_MapTable={
            ".3gp",".mov",".avi",".rmvb",".wmv",".mp3", ".mp4" };
    private Vector<String> items=null; // items: 存放显示的名称
    private Vector<String> paths=null; // paths: 存放文件路径
    private Vector<String> sizes=null; // sizes: 存放文件大小
    private String   rootPath="/mnt/sdcard"; //起始文件夹
    private EditText pathEditText;  // 目录路径
    private Button   queryButton;  //查询按钮
    private ListView  fileListView;//文件列表
    @Override
    protected void onCreate(Bundle icicle){
        super.onCreate(icicle);
        this.setTitle("多媒体文件浏览");
```

```
        setContentView(R.layout.myfile);
        //从布局文件【myfile.xml】中找到对应的控件对象
        pathEditText =(EditText)findViewById(R.id.path_edit);
        queryButton =(Button)findViewById(R.id.qry_button);
        fileListView=(ListView)findViewById(R.id.file_listview);
        //单击查询按钮事件
        queryButton.setOnClickListener(new Button.OnClickListener() {
            public void onClick(View arg0){
                File file=new File(pathEditText.getText().toString());
                if(file.exists()){
                    if(file.isFile()){
                        //如果是媒体文件则直接打开播放
                        openFile(pathEditText.getText().toString());
                    } else {
                        //如果是目录则打开目录下的文件
                        getFileDir(pathEditText.getText().toString());
                    }
                } else {
                    Toast.makeText(MyFileActivity.this,
             "找不到该位置,请确定位置是否正确!",Toast.LENGTH_SHORT).show();
                }
            }
        });
        //设置 ListItem 中的文件被单击时要做的动作
        fileListView.setOnItemClickListener(new OnItemClickListener() {
            @Override
            public void onItemClick(AdapterView<?> arg0,View arg1,
              int position,long arg3){
                fileOrDir(paths.get(position));
            }
        });
        //打开默认文件夹
        getFileDir(rootPath);
    }
```

（4）打开指定路径的文件夹，获得多媒体文件和文件夹信息，将其保存在集合中，并传入到 ListView 中进行显示。

```
private void getFileDir(String filePath){
        /* 设置目前所在路径 */
        pathEditText.setText(filePath);
        items=new Vector<String>();
        paths=new Vector<String>();
        sizes=new Vector<String>();
        File f=new File(filePath);
        File[] files=f.listFiles();
        if(files != null){
            /* 将所有文件添加 ArrayList 中 */
            for(int i=0; i < files.length; i++){
                if(files[i].isDirectory()){
                    items.add(files[i].getName());
                    paths.add(files[i].getPath());
                    sizes.add("");//文件夹不记录空间大小
                }
            }

            for(int i=0; i < files.length; i++){
                if(files[i].isFile()){
                    String fileName=files[i].getName();
                    int index=fileName.lastIndexOf(".");
                        if(index > 0){
                        String endName=fileName.substring(index,
                                fileName.length()).toLowerCase();
                        String type=null;
                        for(int x=0; x < FILE_MapTable.length; x++){
                            //符合预先定义的多媒体格式的文件才会在界面中显示
                            if(endName.equals(FILE_MapTable[x])){
```

```
                            type=FILE_MapTable[x];
                            break;
                        }
                    }
                    if(type != null){
                        items.add(files[i].getName());
                        paths.add(files[i].getPath());
                        sizes.add(files[i].length()+"");
                    }
                }
            }
        }
    }
    /* 使用自定义的 FileListAdapter 将数据传入 ListView 中 */
    fileListView.setAdapter(new FileListAdapter(this,items));
}
```

（5）定义一个 BaseAdapter 的子类作为 ListView 的文件列表适配器，需要实现 getCount()、getItem ()、getItemId()、getView()方法。

```
class  FileListAdapter  extends  BaseAdapter
{
 private Vector<String> items=null; // items 用于存放显示的名称
 private MyFileActivity myFile;
   public FileListAdapter(MyFileActivity myFile,Vector<String> items){
          this.items=items;
          this.myFile=myFile;
      }
      @Override
      public int getCount() {
          return items.size();
      }
      @Override
      public Object getItem(int position){
          return items.elementAt(position);
      }
      @Override
      public long getItemId(int position){
          return items.size();
      }
      @Override
      public View getView(int position,View convertView, ViewGroup parent){
     if(convertView==null)
     {
           //加载列表项布局文件【file_item.xml】
      convertView=myFile.getLayoutInflater().inflate(R.layout.file_item,null);
     }
     //文件名称
TextView name =(TextView)convertView.findViewById(R.id.name);
     //媒体文件类型
ImageView   music=(ImageView)convertView.findViewById(R.id.music);
     //文件夹类型
ImageView folder=(ImageView)convertView.findViewById(R.id.folder);
   name.setText(items.elementAt(position));
      if(sizes.elementAt(position).equals(""))
      {
          //隐藏媒体文件图标,显示文件夹图标
          music.setVisibility(View.GONE);
          folder.setVisibility(View.VISIBLE);
      }else
      {
          //隐藏文件夹图标,显示媒体文件图标
          folder.setVisibility(View.GONE);
          music.setVisibility(View.VISIBLE);
      }
     return convertView;
      }
}
```

（6）打开文件或者文件夹，如果是文件，则将文件信息通过 Intent 传给播放多媒体的 Activity 类 MediaPlayerActivity 进行播放；如果是文件夹，则列出该文件夹下的文件。

```
//处理文件或者目录的方法
private void fileOrDir(String path){
        File file=new File(path);
        if(file.isDirectory()){
            getFileDir(file.getPath());
        } else {
            openFile(path);
        }
}
//打开媒体文件
private void openFile(String  path){
          //打开媒体播放器，进行播放
     Intent intent=new Intent(MyFileActivity.this,MediaPlayerActivity.class);
     intent.putExtra("path",path);
     startActivity(intent);
     finish();
}
```

（7）修改 MediaPlayerActivity 类。

① 使 MediaPlayerActivity 的 onCreate 方法可以接收到 MyFileActivity 传递的媒体路径。

```
public void onCreate(Bundle savedInstanceState){
       super.onCreate(savedInstanceState);
       ...... //略
       //获取传递过来的媒体路径
       Intent intent=getIntent();
       //获取外部单击多媒体文件传递过来的路径
       Uri uri=intent.getData();
       String mPath="/mnt/sdcard/music.mp3";
       if(uri!=null)
       {
         mPath=uri.getPath(); //获得媒体路径
       }else
       {
         //从多媒体文件预览界面传来的点播文件路径
         Bundle localBundle=getIntent().getExtras();
         if(localBundle!=null)
         {
              String t_path=localBundle.getString("path");
              if(t_path!=null&&!"".equals(t_path))
              {
                 mPath=t_path;
              }
         }
       }
   ...... //略
}
```

② 添加从 MediaPlayerActivity 跳转到 MyFileActivity 的方法，通过单击菜单的方式实现。

```
@Override
public boolean onCreateOptionsMenu(Menu menu){
    menu.add(0,1,0,"文件夹");
    return super.onCreateOptionsMenu(menu);
}
@Override
public boolean onOptionsItemSelected(MenuItem item){
    if(item.getItemId()==1)
    {    //跳转到媒体文件浏览器
        Intent intent=new
            Intent(MediaPlayerActivity.this,MyFileActivity.class);
        startActivity(intent);
        finish();
    }
```

```
        return super.onOptionsItemSelected(item);
    }
```

（8）定义返回键的操作：当按下【返回】按钮时，如果不是在根目录则返回上一级文件夹，如果是根目录则直接返回事件，结束当前的 Activity。

```
public boolean onKeyDown(int keyCode, KeyEvent event){
    //判断触发按键是否为back键
    if(keyCode == KeyEvent.KEYCODE_BACK){
        pathEditText =(EditText)findViewById(R.id.path_edit);
        File file=new File(pathEditText.getText().toString());
        //如果是默认路径就直接返回back事件，其中rootPath取值为"/mnt/sdcard"
        if(rootPath.equals(
            pathEditText.getText().toString().trim())){
            return super.onKeyDown(keyCode,event);
        } else {
            getFileDir(file.getParent());
            return true;
        }
    //如果不是back键则正常响应
    } else {
        return super.onKeyDown(keyCode,event);
    }
}
```

4.5 多线程开发

一、任务分析

当我们需要在 android 程序中运行比较耗时的操作时，如网络请求、下载等，为了不阻塞主线程（UI 线程），我们需要使用多线程。

然而，Android 中只允许在主线程中完成 UI 更新操作，所以在多线程中需要同步 UI 操作时，必须借助 Handler 或者 AsyncTask 来完成异步处理。

本次的任务是在多线程中实现更新 UI 操作。

二、相关知识

1. 多线程技术

现代操作系统是一个多任务的操作系统，即一次可以运行或提交多个作业。多线程技术正是实现多任务的基础，其意义在于一个应用程序中，有多个部分可以同时执行，从而可以获得更高的处理效率。每个程序至少有一个进程，一个进程至少有一个线程。程序、进程、线程三个概念既有联系又有区别，三者的关系如图 4-15 所示。

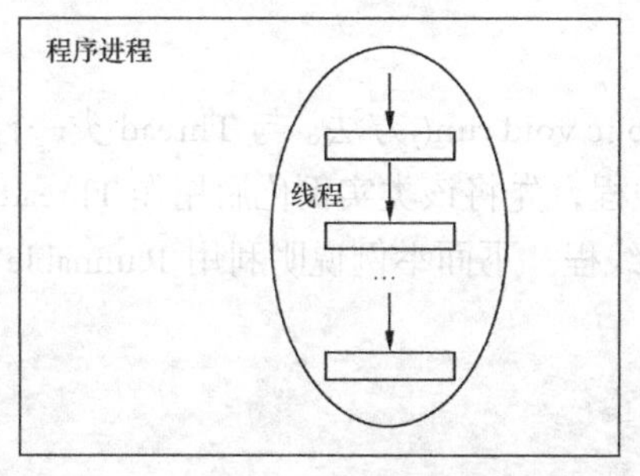

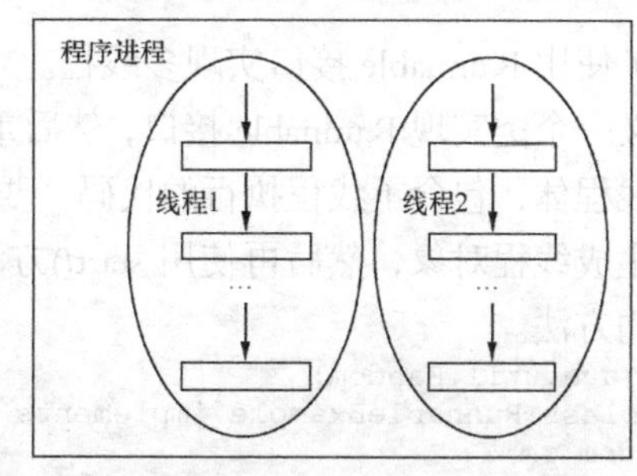

图 4-15 程序、进程和线程的联系与区别

① 程序是一组指令的有序集合，本身没有任何运行的含义，只是一个静态的实体。

② 进程是在内存中运行的应用程序，反映了程序在一定的数据集上运行的全部动态过

程，即一个程序如果没有被执行，就不会产生进程。每个进程都有自己独立的一块内存空间，具有自己的生命周期。即进程通过创建而产生，通过系统调度而运行，当等待资源或事件时处于等待状态，在完成任务后被撤销。

③ 线程是进程的一个实体，是 CPU 调度和分派的基本单位。一个进程可以启动多个线程。线程不能够独立运行，总是属于某个进程，进程中的多个线程共享进程的内存。一个线程可以创建和撤销另一个线程；同一个进程中的多个线程之间可以并发执行。使用线程的优点在于线程创建、销毁和切换的负荷远小于进程。

与 Java 一样，Android 可以使用两种方式来实现多线程操作，这两种方式依次是：①提供 Thread 类或者 Runnable 接口编写代码来定义、实例化和启动新线程；②使用 java.util 包中的 Timer 和 TimerTask 类。

（1）使用 Thread 类实现多线程。

定义一个线程类继承 Thread 类，然后重写 public void run()方法。其中 run()方法称为线程体，包含了线程执行的代码。当 run()方法执行完后，线程结束。要运行线程，先用线程类定义一个对象，再调用对象的 start()方法即可。下面举例说明利用 Thread 类实现线程的方法。

```
import java.util.Random;
public class ThreadExample extends Thread {
    Random rm;
    //创建以 tName 为名称的线程
    public ThreadExample(String tName){
        super(tName);
        rm=new Random();
    }
    public void run() {
        for(int i=1; i <=10; i++){
            System.out.println(i+" "+getName());
            try { //随机休眠一段时间,以打乱线程的执行顺序
                sleep(rm.nextInt(1000));
            } catch(InterruptedException e){
                e.printStackTrace();
            }
        }
        System.out.println(getName()+"完成! ");
    }
    public static void main(String[] args){
        //定义两个线程
        ThreadExample thread1=new ThreadExample("线程一");
        thread1.start();
        ThreadExample thread2=new ThreadExample("线程二");
        thread2.start();
    }
}
```

（2）使用 Runnable 接口实现多线程。

定义一个类实现 Runnable 接口，然后重写 public void run()方法。与 Thread 类一样，run()方法为线程体，包含了线程执行的代码。要运行线程，先将该类实例化后用作 Thread 对象的参数，生成线程对象，然后再使用 start()方法开始线程。下面举例说明利用 Runnable 接口实现线程的方法。

```
import java.util.Random;
public class RunnableExample implements Runnable{
    Random rm;
    String name;
    //创建以 tName 为名字的线程
    public RunnableExample(String tName){
        this.name=tName;
        rm=new Random();
```

```
    }

    public void run() {
        for(int i=1; i <=10; i++){
            System.out.println(i+" "+name);
            try {//随机休眠一段时间,以打乱线程的执行顺序
                Thread.sleep(rm.nextInt(1000));
            } catch(InterruptedException e){
                e.printStackTrace();
            }
        }
        System.out.println(name+"完成! ");
    }
    public static void main(String[] args){
    //创建线程对象的方法是将RunnableExample实例化后作为参数传到Thread类中
        Thread thread1=new Thread(new RunnableExample("线程一"));
        thread1.start();
        Thread thread2=new Thread(new RunnableExample("线程二"));
        thread2.start();
    }
}
```

（3）使用 Timer 和 TimerTask 类实现多线程。

Timer 类负责创建、管理和执行线程。Timer 类的主要方法是 schedule 方法，可以在方法当中指定线程运行的任务、任务执行的开始时间、任务执行的间隔时间等参数。也可以通过调用 cancel()方法停止一个 Timer 并终止后台线程。

TimerTask 类表示一个任务，在 TimerTask 子类中需实现 run()方法。要使用 Timer 线程，首先要定义一个 TimerTask 的子类，然后在主程序中分别定义一个 Timer 对象和 TimerTask 对象，把 TimerTask 对象作为 Timer 对象的 schedule()方法的参数进行任务调度。下面举例说明利用 Timer 和 TimerTask 类实现多线程的方法。

```
import java.util.Random;
import java.util.Timer;
import java.util.TimerTask;

public class TimerExample  {
        public static void main(String[] args){
            Timer  timer1=new Timer();
            Timer  timer2=new Timer();
            MyThread  thread1= new MyThread("线程一");
            MyThread  thread2= new MyThread("线程二");
            timer1.schedule(thread1,0);
            timer2.schedule(thread2,0);
        }
    }
class MyThread extends TimerTask {
    Random rm;
    String name;
        public MyThread(String tName){
            this.name=tName;
            rm=new Random();
        }
        public void run(){
            for(int i=1; i <=10; i++){
                System.out.println(i+" "+name);
                try {  //随机休眠一段时间,以打乱线程的执行顺序
                Thread.sleep(rm.nextInt(1000));
            } catch(InterruptedException e){
                e.printStackTrace();
            }
        }
        System.out.println(name+"完成! ");
    }
}
```

2. Handler 消息处理器

多线程能够较好地提高 Android 应用程序的效率。当需要进行一项耗时操作时，如联网读取数据或者读取本地较大的文件，不能把这些操作放在主线程中，否则界面可能会出现假死现象。如果在 5 秒内未能完成 UI 事件的响应，可能会收到 Android 系统给出的错误提示"强制关闭"。因此，建议把耗时的操作放在子线程中运行。

Android 应用程序的消息处理机制是服务于线程的，每个线程都可以有自己的消息队列和消息循环。当应用程序启动时，Android 会首先开启一个主线程，主线程负责进行事件分发。Activity 是一个 UI 线程，运行于主线程中。在一个 Activity 中可以创建多个组件或者子线程。如果这些线程或者组件把自身的消息放入 Activity 的主线程消息队列，那么这些消息就会在主线程中被统一处理，而主线程一般负责界面的更新操作，因此这种方式可以较好地实现 Android 界面的更新。

线程怎样把消息放入主线程的消息队列呢？答案是通过 Android 中的消息发送器 Handler。实际上 Android 可以利用 Handler 来实现 UI 线程的更新。Handler 的主要作用是接收子线程发送的数据，并利用这些数据来配合主线程更新 UI。Handle 对象属于创建它的线程，如在 Activity 中创建的 Handler 就属于该 Activity。

对于一个每隔 3 秒更新其标题的程序，未使用 Handler 的程序代码如下。

```
public class ThreadNoHandler extends Activity {
    int num=0;
    @Override
    public void onCreate(Bundle savedInstanceState){
        super.onCreate(savedInstanceState);
        setContentView(android.R.layout.simple list item);
        Timer timer=new Timer();
        timer.schedule(new MyTask(),1,3000);
    }

    class MyTask extends TimerTask {
     public void run() {
            setTitle("当前的刷新次数： "+num);
             num++;
     }
    }
}
```

上述代码的作用是利用 Timer 定时调用多线程来更改窗口标题。但执行该代码并没有得到预期的效果，甚至还报错。原因就在于 Android 出于安全考虑，不允许子线程修改主线程的显示。因此需要采用 Handler 类来解决该问题。

Handler 包含了两个队列，其中一个是线程队列，另外一个是消息队列。使用 post（Runnabler）方法会将线程对象放到该 Handler 的线程队列中，使用 sendMessage（Message message）方法则可将消息放到消息队列中。从消息队列读取消息时会自动执行 Handler 中的 public void handleMessage（Message msg）方法，因此在创建 Handler 时需要使用匿名内部类重写该方法，并在方法中编写实现消息逻辑处理的有关代码。

Message 类用于定义发送给 Handler 的消息，可以包含任意的数据类型。Message 对象可以通过直接使用无参的构造函数来生成，但更好的方法是调用 Message.obtain()方法或者 Handler.obtainMessage()方法，这样可以从可回收对象池中获取 Message 对象。此外，Message 类还有一些重要的方法，例如，getData()获得消息的数据；getTarget()获得处理该消息的 Handler 对象；setTarget(Handler target)设置处理的 Handler 对象；setData(Bundle data)设置消息的数据。

Message 类有几个 public 类型的字段，可以用于设置 Message 携带的信息。

① public int arg1。

② public int arg2。

如果只需要存储几个整型数据，建议直接对 arg1 和 arg2 进行赋值，而不是调用 setData() 方法。

③ public Object obj：发送给接收器的对象。

④ public Messenger replyTo：指明响应此 message 的 Messenger 对象。

⑤ public int what：用户自定义的消息代码，可以理解为消息的 ID。每个 Handler 对消息代码有自己的命名空间，所以不用担心自定义的消息跟其他 Handlers 发生冲突。

有了上面的基础知识之后，下面使用 Handler 对上述代码进行改进。

```
public class ThreadWithHandler extends Activity {
     int num=0;
     Handler myHandler=new Handler(){
    @Override
    public void handleMessage(Message msg){
         super.handleMessage(msg);
         switch(msg.what){
         case 1:
              updateTitle();
         }
    }
   };
public void onCreate(Bundle savedInstanceState){
   super.onCreate(savedInstanceState);
   setContentView(android.R.layout.simple_list_item);
   Timer timer=new Timer();
   timer.schedule(new MyTask(),1,3000);
}
class MyTask extends TimerTask {
   public void run() {
     Message msg=new Message();
          msg.what=1;
     myHandler.sendMessage(msg);
   }
}
public void updateTitle(){
     setTitle("当前的刷新次数：  "+num);
      num++;
}
  }
```

程序运行效果如图 4-16 所示。

从上述代码可以看出，更新标题的代码操作放在主线程中。Handler 对象在主线程中进行创建。子线程再通过 Handler 对象利用 Message 对象作为信息载体发送信息通知，并把该信息放入主线程的消息队列中，从而配合了主线程更新 UI 界面的操作。

图 4-16　TimerTask 实例效果

除此之外，Handler 还可以通过使用 post（Runnable r）方法来实现 UI 控件的更新。Post() 方法将线程 r 放入消息队列，r 自会在创建 Handler 的线程中执行。下面通过动态更新进度条的程序进行说明，实现效果如图 4-17 所示。

图 4-17　ProgressBar 实例效果

（1）定义一个布局文件【progressbar_handler_demo.xml】，在布局上建立一个进度条控件。

```
<LinearLayout xmlns:android="http://schemas.android.com/apk/res/android"
```

```
    xmlns:tools="http://schemas.android.com/tools"
    android:id="@+id/LinearLayout1"
    android:layout_width="match_parent"
    android:layout_height="match_parent"
    android:orientation="vertical" >
    <ProgressBar
        android:id="@+id/progressBar"
        style="@android:style/Widget.ProgressBar.Horizontal"
        android:layout_width="match_parent"
        android:layout_height="wrap_content" />
</LinearLayout>
```

（2）创建 Activity 主程序。

```
public class ProgressBarHandlerDemo extends Activity {
    private ProgressBar mProgress;
    private int mProgressStatus=0;
    private Handler mHandler=new Handler();
    private int i=0;
    protected void onCreate(Bundle icicle){
        super.onCreate(icicle);
        setContentView(R.layout.progressbar_handler_demo);
        mProgress =(ProgressBar)findViewById(R.id.progressBar);
        new Thread(new Runnable() {
            public void run() {
                while(mProgressStatus < 100){
                    mProgressStatus=doWork();
                    mHandler.post(new Runnable() {
                        public void run() {
                            //更新进度条
                            mProgress.setProgress(mProgressStatus);
                        }
                    });
                }
            }
        }).start();
    }
    //进度条数值变化的规则：每隔 1 秒增加 1
    public int doWork(){
     i++;
     try{
          Thread.sleep(1000);
        }catch(Exception ex){
            ex.printStackTrace();
        }
     return i;
    }
}
```

3．AsyncTask 实现异步操作

AsyncTask 是 Android 提供的轻量级异步任务类，能够较容易地实现 UI 线程的更新。异步任务在后台线程中运行，它提供的接口既可以传递当前异步执行的进度信息，实现 UI 界面的更新，又能够在最后将执行的结果反馈给 UI 主线程。

一个异步任务是由 3 个泛型类型 Params，Progress 和 Result 和 4 个方法 PreExecute, doInBackground, onProgressUpdate 及 onPostExecute 构成的。其中 Params 参数表示启动任务执行时需要输入的参数；Progress 表示后台任务执行期间得到的数值；Result 表示后台执行任务最终返回的结果。

使用 AsyncTask 进行多线程编程时，需要继承 AsyncTask，必须重写 doInBackground (Params...)方法。该方法是后台执行方法，比较耗时的操作都可以放在这里。任务的最后执行结果由该方法返回。在执行过程中也可以通过调用 publicProgress(Progress…)方法来更新任务

的进度，onProgressUpdate(Progress...)方法将能够接收这些数值。

另外，根据功能的需要，选择重写以下三个方法。

① onPreExecute() ：在任务开始执行前调用该方法，执行 UI 线程操作。例如，可以在这里显示进度对话框。

② onProgressUpdate(Progress…) ：若在 doInBackground(Params...)方法内调用 publishProgress(Progress...)方法，该方法将会被调用，用于在界面上显示任务执行的进度。

③ onPostExecute(Result) ： 后台 doInBackground(Params...) 方法执行完后，会执行该方法。该方法的参数 Result 实际上就是 doInBackground 返回的结果。

为了正确使用 AsyncTask 类，以下是几条应该遵守的线程准则：

- AsyncTask 类的实例必须在 UI 线程中创建，即在主线程中创建；
- execute(Params...)方法必须在 UI 线程中调用；
- 不要手动调用 onPreExecute()，onPostExecute(Result)，doInBackground(Params...)，onProgressUpdate(Progress...)这 4 个方法；
- AsyncTask 任务只能被执行一次，若多次调用将会抛出异常。

下面使用 AsyncTask 类来实现每隔 3 秒更新标题信息：

```
public class AsyncTaskDemo extends Activity {
    @Override
   protected void onCreate(Bundle savedInstanceState) {
        super.onCreate(savedInstanceState);
        new UpdateTitle(this).execute(1);
       }
}

class UpdateTitle extends AsyncTask<Integer, Integer, Void> {

    Activity context;
    public UpdateTitle(Activity context) {
       this.context = context;
    }
    @Override
    protected void onProgressUpdate(Integer... values) {
       String sTitle = String.valueOf(values[0]);
       context.setTitle(sTitle);
       super.onProgressUpdate(values);
    }
    @Override
    protected Void doInBackground(Integer... params) {
       //后台执行方法，此方法在非 UI 线程中运行
       int num = params[0].intValue();
       while (true) {
           num++;
           publishProgress(num);
           try {
               Thread.sleep(3000);
           } catch (Exception ex) {
               ex.printStackTrace();
           }
       }
   }
 }
```

三、任务实施

(1)进度时间栏更新：主要通过 TimerTask 任务线程发送 UI 消息来实现时间进度栏的更新。

```
// 进度栏任务
public  class  MyTask extends  TimerTask
```

```
{
    public void run() {
     Message message=new Message();
     message.what=1;
     //发送消息更新进度栏和时间显示
     handler.sendMessage(message);
   }
}
//处理进度栏和时间显示
private final Handler handler=new Handler(){
  public void handleMessage(Message msg){
       switch(msg.what){
            case 1:
                 Time progress=new Time(player.getCurrentPosition());
                 Time allTime=new Time(player.getDuration());
                 String  timeStr=progress.toString();
                 String  timeStr2=allTime.toString();
                 //已播放时间
                 play_time.setText(timeStr.substring(timeStr.indexOf(":")+1));
                  //总时间
                  all_time.setText(timeStr2.substring(timeStr.indexOf(":")+1));
                  int progressValue=0;
                  //计算当前的播放进度
                  if(player.getDuration()>0){
                       progressValue=seekbar.getMax()*
                       player.getCurrentPosition()/player.getDuration();
                 }
                 //设置进度栏进度
                 seekbar.setProgress(progressValue);
                 break;
            }
            super.handleMessage(msg);
       }
   };
```

4.6 后台服务 Service

一、任务分析

为了在播放音乐的同时，让用户也可以在移动终端上做其他事情，需要让播放程序在后台运行，这时需要使用到后台服务 Service。

本次任务需要考虑以下几个问题：

（1）Service 和 Activity 之间的异同以及两者之间是如何交互的；

（2）Service 和多线程之间的异同。

二、相关知识

1．Service 概念与生命周期

Service 是 Android 系统中的四大组件之一（Activity、Service、BroadcastReceiver、ContentProvider）。与 Activity 类似，它的运行也具有生命周期，都是 Context 的子类，但它没有用户界面且只能在后台运行。即使用户切换到其他的应用程序，Service 还可以继续在后台运行。Service 可以应用在很多场合，例如，在播放多媒体时，用户若启动了其他的 Activity，还可以在后台继续播放多媒体；当需要检测 SD 卡上文件的变化，或者在后台记录客户地理信息位置的改变，或者通过网络上传和下载文件时，都可以通过执行 Service 来实现。Service 还可以提供给其他应用复用，比如一个天气预报服务可以被其他应用调用。

按服务启动的方式，Service 一般分为 Started 服务和 Bound 服务两种。

（1）Started 服务。

Started 服务是指组件通过调用 startService()来启动。即使调用 Stated 服务的组件被销毁，该服务依然可能继续运行。一般来说 Stated 服务执行单一的操作，不返回结果给调用的组件。Service 服务可以调用 stopSelf() 或者组件使用 stopService()来结束。无论调用了多少次 startService()，只需要调用一次 stopService()就可以停止 Service。

（2）Bound 服务。

Bound 服务是指组件通过调用 bindService()来绑定服务。Bound 服务允许组件与服务进行交互，例如发送请求和获取结果。多个组件可以绑定至同一个服务。只要还有其他的组件与 Bound 服务绑定，那么该服务就能够一直在运行。反之，当没有组件和服务绑定时，该服务将会被销毁。某个组件可以调用 unbindService()来解除与服务的绑定。

要创建一个服务，需要定义 Service 的子类，在类中重写一些回退方法（callback methods），用于处理服务的生命周期，以及提供某种机制来为组件绑定服务。下面讲解当中比较重要的方法。

① onStartCommand()：当组件通过 startService()方法请求服务启动时，该方法将被系统调用。在该方法内，会接收到组件传过来的 Intent，里面包含有关服务的数据。一旦该方法开始执行，服务就能够在后台运行。对于 Bound 服务，不需要实现该方法。

② onBind()：这是一个必须实现的方法。当一个组件想通过 bindService()绑定服务，该方法将会被系统调用。该方法通过返回 IBinder 来提供组件与服务的通信接口。如果用户不允许绑定，那么只需要返回 null。

③ onCreate()：当服务第一次被创建时，该方法将会被系统调用。该方法在 onStartCommand()或 onBind()方法之前执行。

④ onDestroy()：当服务不再使用或正在被销毁时，该方法将被系统调用。在该方法中，实现对线程、监听器等资源的清理或者释放。

定义 Started 服务类的操作代码如下：

```
public class StartedService extends Service{
   private static final String TAG = "LocalService";
   MediaPlayer mediaPlayer=null;
      @Override
      public IBinder onBind(Intent intent) {
         return null;
      }
      @Override
      public void onCreate() {
         super.onCreate();
         Log.i(TAG, "调用 onCreate");
     //test 为放在 raw 目录下一个 mp3 文件
         mediaPlayer=MediaPlayer.create(this,R.raw.test);
         mediaPlayer.setLooping(true);
      }
      @Override
      public int onStartCommand(Intent intent, int flags, int startId) {
          Log.i(TAG, "调用 onStartCommand");
           //这里可以启动媒体播放器,播放音乐
          mediaPlayer.start();
          return START_STICKY;
      }

      @Override
      public void onDestroy() {
          Log.i(TAG, "调用 onDestroy");
          mediaPlayer.stop();
```

```
        mediaPlayer.release();
        super.onDestroy();
    }
}
```

定义调用及关闭 Started 服务的代码如下。

```
public class ServiceMusic extends Activity {

    @Override
    protected void onCreate(Bundle savedInstanceState) {
        super.onCreate(savedInstanceState);
        this.setTitle("请点击菜单执行命令");
    }

    @Override
    public boolean onCreateOptionsMenu(Menu menu) {
        menu.add(1, 1, 1, "以服务方式播放音乐");
        menu.add(1, 2, 2, "停止播放音乐");
        return true;
    }

    @Override
    public boolean onOptionsItemSelected(MenuItem item) {
        int id = item.getItemId();
    //建立打开服务的 Intent
        Intent intent=new Intent(ServiceMusic.this, StartedService.class);
        switch(id){
          case 1:
            startService(intent);  //启动服务
            break;
          case 2:
            stopService(intent);
           break;
        }
        return super.onOptionsItemSelected(item);
    }
}
```

此外，必须在 Manifest.xml 文档中注册上面定义的 Service 才可以正常使用

```
        <service android:name=".StartedService" />
```

（3）服务的生命周期。

图 4-18 是两种类型 Service 的生命周期，从图中可以看出，这两种服务的生命周期都是在 onCreate()方法和 onDestroy()之间。其中 onCreate()方法用于服务的初始化设置，onDestroy()方法用于释放所有与服务有关的资源。服务的活跃期是从 onStartCommand()或者 onBind()开始。

2. Service 与多线程

Service 与线程是既有区别又有联系的。线程是程序执行的最小单元，它是分配 CPU 的基本单位。可以用线程来执行一些异步操作。服务 Service 是 Android 的一种机制，虽然说它在后台运行没有显示出界面，但它是运行在应用程序的主线程上的。所以它不能在本地服务上运行比较耗时的操作（如网络请求，下载等）时运行，否则将阻塞主线程（UI 线程），在这种情况下，就需要使用多线程。

线程既可以在 Activity 中被创建，也可以在 Service 中被创建。在 Activity 类中创建的线程生命周期不应该超出整个应用程序的生命周期，也就是在整个应用程序退出之后，线程都应该完全退出，这样才不会出现内存泄漏或者僵尸线程。如果开发者希望整个应用程序都退出之后依然能运行该线程，那么就应该把线程放到 Service 中去创建和启动了。其中在 Service 中创建的线程，适合长期执行一些独立于应用程序的后台任务，比较常见的是在 Service 中保持与服务器端的长连接。

总的来说，在 Android 中，当应用程序被关闭后，服务仍然可以在后台运行，不容易被系统销毁。而线程则是可以提高系统效率，因此可以将两者结合起来一起使用。

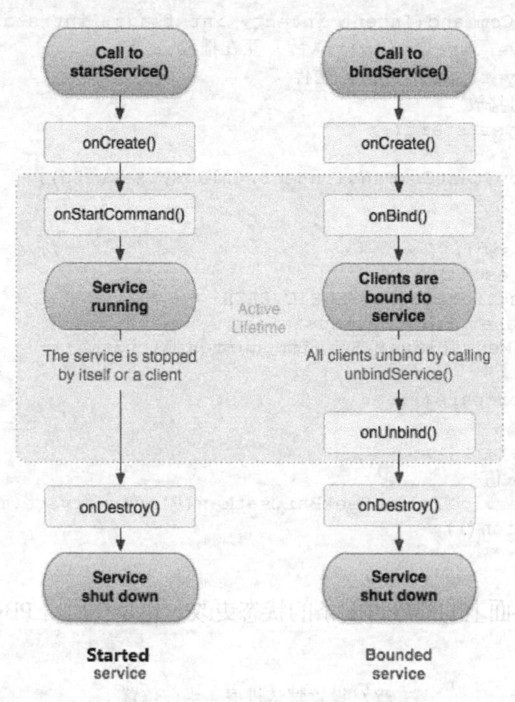

图 4-18　Service 生命周期图

三、任务实施

本教材提供了一个使用 Started 服务与多线程配合的播放音频功能的实例项目，详细的完整代码请查看 com.demo.pr4_2 包。其中主运行 A ctivity 是 MediaPlayer_Service 类，实现了包括播放列表的生成，【上一首】、【下一首】、【播放】、【暂停】按钮的布局以及事件监听，播放进度条、音量控制进度条的布局以及事件监听等。下面是项目中与服务有关的主要代码。

```
        // 播放
        playButton.setOnClickListener(new OnClickListener() {
            //@Override
            public void onClick(View v) {
                playMusic(PlayerMag.PAUSE);
            }
        });
        public void playMusic(int action) {
            Intent intent = new Intent();
            intent.putExtra("MSG", action);
            intent.setClass(MediaPlayer_Service.this, PlayerService.class);
            //启动播放服务
            startService(intent);  //通过 Service 播放音频
        }
```

后台服务 PlayerService 类实现了播放功能，其主要代码如下。

```
//播放 MediaPlayer 后台服务定义
public class PlayerService extends Service implements Runnable,
        MediaPlayer.OnCompletionListener {

    public static MediaPlayer mMediaPlayer = null;
    @Override
    public void onCreate() {
        super.onCreate();
        mMediaPlayer = new MediaPlayer();
```

```
        mMediaPlayer.setOnCompletionListener(this);
    }
//启动 service 时执行的方法
    @Override
    public int onStartCommand(Intent intent, int flags, int startId) {
        if (mMediaPlayer.isPlaying()) {// 正在播放
            mMediaPlayer.pause();// 暂停
        } else {// 没有播放
            mMediaPlayer.start();
        }
        return super.onStartCommand(intent, flags, startId);
    }

    public void playMusic() {
        mMediaPlayer.reset();
        mMediaPlayer.setDataSource(MUSIC_PATH
            + MediaPlayer_Service.mMusicList
                .get(MediaPlayer_Service.currentListItme));
        //准备播放
        mMediaPlayer.prepare();
        //开始播放
        mMediaPlayer.start();
        // 设置进度条最大值
        MediaPlayer_Service.audioSeekBar.setMax(PlayerService.mMediaPlayer
            .getDuration());
    }
}
```

同时，对于播放器界面上的播放进度条的状态更改，也是在后台 PlayerService 类中实现的。

```
    // 刷新进度条
    public void run() {
        int CurrentPosition = 0;// 设置默认进度条当前位置
        int total = mMediaPlayer.getDuration()
        while (mMediaPlayer != null && CurrentPosition < total) {
            try {
                Thread.sleep(1000);
                if (mMediaPlayer != null) {
                    CurrentPosition = mMediaPlayer.getCurrentPosition();
                }
            } catch (InterruptedException e) {
                e.printStackTrace();
            }
            MediaPlayer_Service.audioSeekBar.setProgress(CurrentPosition);
        }
    }
```

使用服务实现播放音频的效果如图 4-19。

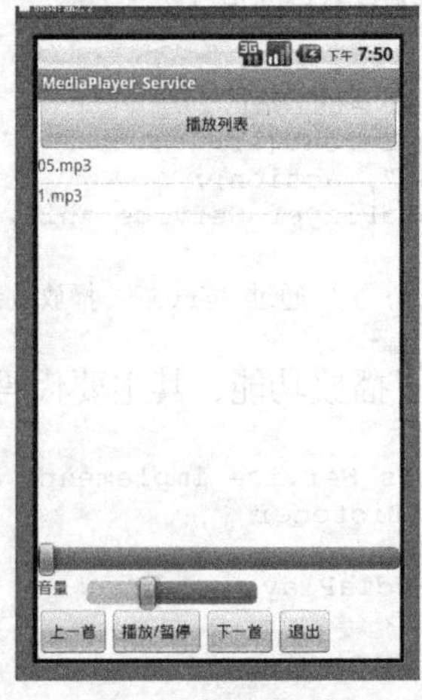

图 4-19　后台播放音频任务界面

4.7 完整项目实施

手机多媒体播放器由多个类和XML布局文件组成，下面分别对每个类的实现作详细介绍。

（1）手机多媒体程序运行的AndroidManifest描述【AndroidManifest.xml】，在此文件中声明程序启动的Activity、其他Activity，以及支持的多媒体类型。

```
<?xml version="1.0" encoding="utf-8"?>
<manifest xmlns:android="http://schemas.android.com/apk/res/android"
     package="com.demo.pr4"
     android:versionCode="1"
     android:versionName="1.0">
   <uses-sdk android:minSdkVersion="8" />
   <application android:icon="@drawable/icon"
                android:label="@string/app_name">
       <activity android:name=".MediaPlayerActivity"
                android:label="@string/app_name"
                android:screenOrientation="portrait">
           <intent-filter>
               <action android:name="android.intent.action.MAIN" />
           <category android:name="android.intent.category.LAUNCHER" />
           </intent-filter>
       </activity>
       <!--媒体文件 filter-->
       <activity android:name=".MediaPlayerActivity"
               android:label="@string/app_name">
           <intent-filter>
            <action android:name="android.intent.action.VIEW" />
            <category android:name="android.intent.category.DEFAULT" />
               <data android:mimeType="video/*" />
               <data android:mimeType="audio/*" />
           </intent-filter>
       </activity>
       <activity android:name=".MyFileActivity"></activity>
   </application>
</manifest>
```

（2）手机多媒体程序播放界面XML布局：【main.xml】。

```
<LinearLayout
   xmlns:android="http://schemas.android.com/apk/res/android"
   android:layout_width="fill_parent"
   android:layout_height="fill_parent"
   android:orientation="vertical">
 <!-- 媒体名称-->
  <LinearLayout
  android:orientation="horizontal"
  android:layout_width="fill_parent"
  android:layout_height="wrap_content"
  android:layout_marginTop="10.0dip"
  >
  <TextView
      android:id="@+id/title"
      android:text="多媒体播放器"
      android:layout_width="fill_parent"
      android:layout_height="wrap_content"
      android:textSize="18.0dip"
      android:textStyle="bold"
      android:gravity="center"
  ></TextView>
 </LinearLayout>
  <!-- 视频播放窗口 -->
  <LinearLayout
  android:orientation="horizontal"
  android:layout_weight="1"
```

```
 android:layout_height="fill_parent"
 android:layout_width="fill_parent"
 >
  <SurfaceView android:layout_gravity="center"
       android:layout_weight="1"
       android:layout_height="fill_parent"
       android:layout_width="fill_parent"
       android:id="@+id/surfaceview">
  </SurfaceView>
 </LinearLayout>
<!-- 播放控制 -->
<LinearLayout
   android:orientation="horizontal"
   android:layout_width="wrap_content"
   android:layout_height="wrap_content"
   android:layout_gravity="center_horizontal"
   android:layout_marginTop="5.0dip"
   >
    <!--快退-->
   <ImageButton
   android:id="@+id/rew"
   android:layout_width="wrap_content"
   android:layout_height="wrap_content"
    android:src="@drawable/ic_media_rew"
   ></ImageButton>
   <!--暂停-->
    <ImageButton
    android:layout_marginLeft="10.0dip"
   android:id="@+id/pause"
   android:layout_width="wrap_content"
   android:layout_height="wrap_content"
   android:src="@drawable/ic_media_pause"
   ></ImageButton>
    <!--开始-->
   <ImageButton
    android:layout_marginLeft="10.0dip"
   android:id="@+id/start"
   android:layout_width="wrap_content"
   android:layout_height="wrap_content"
   android:src="@drawable/ic_media_play"
   android:visibility="gone"
   ></ImageButton>
    <!--快进-->
    <ImageButton
    android:id="@+id/ff"
    android:layout_marginLeft="10.0dip"
    android:layout_width="wrap_content"
    android:layout_height="wrap_content"
    android:src="@drawable/ic_media_ff"
   ></ImageButton>
</LinearLayout>

<!-- 进度条-->
<LinearLayout
        android:layout_width="fill_parent"
        android:layout_height="wrap_content"
        android:layout_gravity="center"
        android:layout_marginBottom="10.0dip"
        android:orientation="horizontal">
     <!-- 已经播放的时间-->
      <TextView
          android:id="@+id/play_time"
```

```
            android:text="00:00"
              android:layout_width="wrap_content"
              android:layout_marginLeft="5.0dip"
              android:textSize="15.0dip"
              android:layout_height="wrap_content">
     </TextView>
     <!-- 进度-->
     <SeekBar   android:id="@+id/seekbar"
              android:layout_width="fill_parent"
              android:layout_height="wrap_content"
              android:layout_weight="1.0"
              android:layout_marginLeft="2.0dip"
              android:layout_marginRight="2.0dip"
     />
     <!-- 总共要播放的时间-->
     <TextView
           android:id="@+id/all_time"
           android:text="04:32"
              android:layout_width="wrap_content"
              android:layout_marginRight="5.0dip"
              android:textSize="15.0dip"
              android:layout_height="wrap_content">
         </TextView>
</LinearLayout>
</LinearLayout>
```

（3）手机多媒体程序媒体文件浏览界面 XML 布局：【myfile.xml】。

```
<?xml version="1.0" encoding="utf-8"?>
<LinearLayout
  xmlns:android="http://schemas.android.com/apk/res/android"
  android:orientation="vertical"
  android:layout_width="fill_parent"
  android:layout_height="fill_parent">
  <!--查询栏-->
  <LinearLayout
  android:orientation="horizontal"
  android:layout_width="fill_parent"
  android:layout_height="wrap_content">
  <!--文件夹路径-->
  <EditText android:id="@+id/path_edit"
  android:layout_width="fill_parent"
  android:layout_height="wrap_content"
  android:layout_weight="1.0"
  />
  <!--查询按键-->
  <Button android:id="@+id/qry_button"
  android:layout_width="wrap_content"
  android:layout_height="wrap_content"
  android:text="查询"
  />
  </LinearLayout>
  <!--文件列表-->
  <ListView android:id="@+id/file_listview"
    android:layout_width="fill_parent"
    android:layout_height="fill_parent"
    android:layout_weight="1"
  />
</LinearLayout>
```

（4）手机多媒体程序媒体文件浏览界面列表子项 XML 布局：【file_item.xml】。

```
<?xml version="1.0" encoding="utf-8"?>
<LinearLayout
   xmlns:android="http://schemas.android.com/apk/res/android"
```

```
    android:layout_width="fill_parent"
    android:layout_height="wrap_content"
    >
    <!-- 文件夹图标 -->
    <ImageView
    android:id="@+id/folder"
    android:layout_width="wrap_content"
    android:layout_height="wrap_content"
    android:src="@drawable/folder"
   />
   <!-- 媒体文件图标 -->
    <ImageView
    android:id="@+id/music"
    android:layout_width="wrap_content"
    android:layout_height="wrap_content"
    android:src="@drawable/music"
    android:visibility="gone"
   />
   <!-- 文件名称-->
    <TextView
    android:id="@+id/name"
    android:layout_width="fill_parent"
    android:layout_height="fill_parent"
    android:text="fdfsfsf"
    android:textSize="18.0dip"
    android:layout_weight="1.0"
    android:textStyle="bold"
    android:gravity="center_vertical"
    ></TextView>
</LinearLayout>
```

（5）手机多媒体程序播放处理类：【MediaPlayerActivity.java】。

```
package com.demo.pr4;
import java.sql.Time;
import java.util.Timer;
import java.util.TimerTask;
import android.app.Activity;
import android.content.Intent;
import android.media.AudioManager;
import android.media.MediaPlayer;
import android.media.MediaPlayer.OnCompletionListener;
import android.media.MediaPlayer.OnPreparedListener;
import android.net.Uri;
import android.os.Bundle;
import android.os.Handler;
import android.os.Message;
import android.view.Display;
import android.view.Menu;
import android.view.MenuItem;
import android.view.SurfaceHolder;
import android.view.SurfaceView;
import android.view.View;
import android.view.Window;
import android.view.WindowManager;
import android.view.View.OnClickListener;
import android.widget.ImageButton;
import android.widget.LinearLayout;
import android.widget.SeekBar;
import android.widget.TextView;
public class MediaPlayerActivity extends Activity{
    private Display currDisplay;
    private SurfaceView surfaceView;
    private SurfaceHolder holder;
    private MediaPlayer player;
    private int vWidth,vHeight;
    private Timer  timer;
```

```
private ImageButton  rew; //快退
private ImageButton  pause;//暂停
private ImageButton  start;//开始
private ImageButton  ff;//快进
private TextView   play_time;//已播放时间
private TextView   all_time;//总播放时间
private TextView   title;//播放文件名称
private SeekBar    seekbar;//进度条

public void onCreate(Bundle savedInstanceState){
    super.onCreate(savedInstanceState);
    requestWindowFeature(Window.FEATURE_NO_TITLE); // 隐藏标题
    getWindow().setFlags(
    WindowManager.LayoutParams.FLAG_FULLSCREEN,
    WindowManager.LayoutParams.FLAG_FULLSCREEN);
    setContentView(R.layout.main);
    //获取外部单击多媒体文件传递过来的路径
    Intent intent=getIntent();
    Uri uri=intent.getData();
    String  mPath="/mnt/sdcard/music.mp3";
    if(uri!=null)
    {
         mPath=uri.getPath(); //获得媒体路径
    }else
    {
        //从多媒体文件预览界面传来的点播文件路径
        Bundle localBundle=getIntent().getExtras();
        if(localBundle!=null)
        {
             String t_path=localBundle.getString("path");
             if(t_path!=null&&!"".equals(t_path))
             {
                mPath=t_path;
             }
        }
    }
    //获取当前布局文件的控件对象
    title=(TextView)findViewById(R.id.title);
    surfaceView =(SurfaceView)findViewById(R.id.surfaceview);
    rew=(ImageButton)findViewById(R.id.rew);
    pause=(ImageButton)findViewById(R.id.pause);
    start=(ImageButton)findViewById(R.id.start);
    ff=(ImageButton)findViewById(R.id.ff);
    play_time=(TextView)findViewById(R.id.play_time);
    all_time=(TextView)findViewById(R.id.all_time);
    seekbar=(SeekBar)findViewById(R.id.seekbar);
    //给 SurfaceView 添加 CallBack 监听器
    holder=surfaceView.getHolder();
    holder.addCallback(new SurfaceHolder.Callback() {

        @Override
        public void surfaceDestroyed(SurfaceHolder holder){}
        //当 SurfaceView 中的 Surface 被创建时调用
        @Override
        public void surfaceCreated(SurfaceHolder holder){
            //设置 MediaPlayer 在指定的 Surface 中进行播放
            player.setDisplay(holder);
            //在指定了 MediaPlayer 播放的容器后,使用 prepareAsync 准备播放
            player.prepareAsync();
        }
        @Override
        public void surfaceChanged(SurfaceHolder holder,int format,
            int width,int height)
            {
            }
```

```
    });
    //为了可以播放视频或者使用Camera预览,需要指定其Buffer类型
    holder.setType(SurfaceHolder.SURFACE_TYPE_PUSH_BUFFERS);
    //下面开始实例化MediaPlayer对象
    player=new MediaPlayer();
    //设置播放完成监听器
    player.setOnCompletionListener(new OnCompletionListener() {
        @Override
        public void onCompletion(MediaPlayer mp){
            // 当MediaPlayer播放完成后触发
            if(timer!=null)
            {
                timer.cancel();
                timer=null;
            }
        }
    });
    //设置prepare完成监听器
    player.setOnPreparedListener(new OnPreparedListener() {
        @Override
        public void onPrepared(MediaPlayer mp){
            // 当prepare完成后,该方法触发用于播放视频
            //首先取得视频的宽和高
            vWidth=player.getVideoWidth();
            vHeight=player.getVideoHeight();
        if(vWidth > currDisplay.getWidth() || vHeight > currDisplay.getHeight()){
                //如果视频的宽或者高超出了当前屏幕的大小,则要进行缩放
                float wRatio =(float)vWidth/(float)currDisplay.getWidth();
                float hRatio =(float)vHeight/(float)currDisplay.getHeight();
                //选择大的比例进行缩放
                float ratio=Math.max(wRatio,hRatio);
                vWidth =(int)Math.ceil((float)vWidth/ratio);
                vHeight =(int)Math.ceil((float)vHeight/ratio);
                //重新设置surfaceView的布局参数
  surfaceView.setLayoutParams(new LinearLayout.LayoutParams(vWidth,vHeight));
                //然后开始播放视频
                player.start();
            }else
            {
                player.start();
            }
            if(timer!=null)
            {
                timer.cancel();
                timer=null;
            }
            //启动时间更新及进度条更新任务,每0.5s更新一次
            timer=new Timer();
            timer.schedule(new MyTask(),50,500);
        }
    });
    player.setAudioStreamType(AudioManager.STREAM_MUSIC);
    try {
        //指定需要播放文件的路径,初始化MediaPlayer
        title.setText(mPath.substring(mPath.lastIndexOf("/")+1));
        player.setDataSource(mPath);
    } catch(Exception e){
        e.printStackTrace();
    }
    //暂停操作
    pause.setOnClickListener(new OnClickListener() {
        @Override
        public void onClick(View v){
            //按下暂停操作后,pause按钮设为隐藏,start按钮设为可见
            pause.setVisibility(View.GONE);
            start.setVisibility(View.VISIBLE);
```

```
            player.pause();
            if(timer!=null)
            {
                timer.cancel();
                timer=null;
            }
        }
    });
    //播放操作
    start.setOnClickListener(new OnClickListener() {
        @Override
        public void onClick(View v){
            //按下开始操作后,pause 按钮设为可见,start 按钮设为隐藏
            start.setVisibility(View.GONE);
            pause.setVisibility(View.VISIBLE);
            //启动播放
            player.start();
            if(timer!=null)
            {
                timer.cancel();
                timer=null;
            }
            //启动时间更新及进度条更新任务,每 0.5s 更新一次
            timer=new Timer();
            timer.schedule(new MyTask(),50,500);
        }
    });

    //快退操作,每次快退 10s
    rew.setOnClickListener(new OnClickListener() {

        @Override
        public void onClick(View v){
            //判断是否正在播放
            if(player.isPlaying())
            {
                int currentPosition=player.getCurrentPosition();
                if(currentPosition-10000>0)
                {
                    player.seekTo(currentPosition-10000);
                }
            }
        }
    });
    //快进操作,每次快进 10s
    ff.setOnClickListener(new OnClickListener() {
        @Override
        public void onClick(View v){
            //判断是否正在播放
            if(player.isPlaying())
            {
                int currentPosition=player.getCurrentPosition();
                if(currentPosition+10000<player.getDuration())
                {
                    player.seekTo(currentPosition+10000);
                }
            }
        }
    });
    //取得当前 Display 对象
    currDisplay=this.getWindowManager().getDefaultDisplay();
}
@Override
public boolean onCreateOptionsMenu(Menu menu){
    menu.add(0,1,0,"文件夹");
    return super.onCreateOptionsMenu(menu);
}
```

```
    @Override
    public boolean onOptionsItemSelected(MenuItem item){
        if(item.getItemId()==1)
        {
            Intent intent=new
            Intent(MediaPlayerActivity.this,MyFileActivity.class);
            startActivity(intent);
            finish();
        }
        return super.onOptionsItemSelected(item);
    }
    // 进度栏任务类
    public  class  MyTask extends  TimerTask
    {
        public void run() {
             Message message=new Message();
             message.what=1;
             //发生消息更新进度栏和时间显示
             handler.sendMessage(message);
         }
    }
    // 处理进度栏和时间显示
    private final Handler handler=new Handler(){
      public void handleMessage(Message msg){
         switch(msg.what){
             case 1:
                 Time progress=new Time(player.getCurrentPosition());
                 Time allTime=new Time(player.getDuration());
                 String  timeStr=progress.toString();
                 String  timeStr2=allTime.toString();
                 //已播放时间
                 play_time.setText(timeStr.substring(timeStr.indexOf(":")+1));
                    //总时间
                   all_time.setText(timeStr2.substring(timeStr.indexOf(":")+1));
                 int progressValue=0;
                 if(player.getDuration()>0){
                        progressValue=seekbar.getMax()*
                        player.getCurrentPosition()/player.getDuration();
                 }
                 //进度栏进度
                 seekbar.setProgress(progressValue);
                 break;
             }
             super.handleMessage(msg);
         }
    };
}
```

（6）手机多媒体程序媒体文件浏览处理类：【MyFileActivity.java】。

```
package com.demo.pr4;
import java.io.File;
import java.util.Vector;
import android.app.Activity;
import android.content.Intent;
import android.os.Bundle;
import android.view.KeyEvent;
import android.view.View;
import android.view.ViewGroup;
import android.widget.AdapterView;
import android.widget.BaseAdapter;
import android.widget.Button;
import android.widget.EditText;
import android.widget.ImageView;
import android.widget.ListView;
import android.widget.TextView;
import android.widget.Toast;
import android.widget.AdapterView.OnItemClickListener;
```

```
public class MyFileActivity extends Activity  {
   // 支持的媒体格式
   private final String[]FILE_MapTable={
        ".3gp",".mov",".avi",".rmvb",".wmv",".mp3", ".mp4" };
   private Vector<String> items=null; // items: 存放显示的名称
   private Vector<String> paths=null; // paths: 存放文件路径
   private Vector<String> sizes=null; // sizes: 存放文件大小
   private String   rootPath="/mnt/sdcard"; //起始文件夹
   private EditText pathEditText;  // 路径
   private Button   queryButton;  //查询按钮
   private ListView  fileListView;//文件列表
   @Override
   protected void onCreate(Bundle icicle){
      super.onCreate(icicle);
      this.setTitle("多媒体文件浏览");
      setContentView(R.layout.myfile);
      //从【myfile.xml】中找到对应的控件对象
      pathEditText =(EditText)findViewById(R.id.path_edit);
      queryButton =(Button)findViewById(R.id.qry_button);
      fileListView=(ListView)findViewById(R.id.file_listview);
      //单击查询按钮事件
      queryButton.setOnClickListener(new Button.OnClickListener() {
         public void onClick(View arg0){
            File file=new File(pathEditText.getText().toString());
            if(file.exists()){
               if(file.isFile()){
                  //如果是媒体文件则直接打开播放
                  openFile(pathEditText.getText().toString());
               } else {
                  //如果是目录则打开目录下的文件
                  getFileDir(pathEditText.getText().toString());
               }
            } else {
               Toast.makeText(MyFileActivity.this,
            "找不到该位置,请确定位置是否正确!",Toast.LENGTH_SHORT).show();
            }
         }
      });
      //设置ListItem中的文件被单击时要做的动作
      fileListView.setOnItemClickListener(new OnItemClickListener() {
         @Override
         public void onItemClick(AdapterView<?> arg0,View arg1,
         int position,long arg3){
            fileOrDir(paths.get(position));
         }
      });

      //打开默认文件夹
      getFileDir(rootPath);
   }
   /**
    * 重写返回键功能:返回上一级文件夹
    */
   @Override
   public boolean onKeyDown(int keyCode,KeyEvent event){
      //判断触发按键是否为back键
      if(keyCode == KeyEvent.KEYCODE_BACK){
         pathEditText =(EditText)findViewById(R.id.path_edit);
         File file=new File(pathEditText.getText().toString());
         if(rootPath.equals(
            pathEditText.getText().toString().trim())){
            return super.onKeyDown(keyCode,event);
         } else {
            getFileDir(file.getParent());
            return true;
```

```
        }
        //如果不是 back 键则正常响应
    } else {
        return super.onKeyDown(keyCode,event);
    }
}
/**
 * 处理文件或者目录的方法
 */
private void fileOrDir(String path){
    File file=new File(path);
    if(file.isDirectory()){
        getFileDir(file.getPath());
    } else {
        openFile(path);
    }
}
/**
 * 取得文件结构的方法
 */
private void getFileDir(String filePath){
    /* 设置目前所在路径 */
    pathEditText.setText(filePath);
    items=new Vector<String>();
    paths=new Vector<String>();
    sizes=new Vector<String>();
    File f=new File(filePath);
    File[] files=f.listFiles();
    if(files != null){
        /* 将所有文件添加到 ArrayList 中 */
        for(int i=0; i < files.length; i++){
            if(files[i].isDirectory()){
                items.add(files[i].getName());
                paths.add(files[i].getPath());
                sizes.add("");
            }
        }
        for(int i=0; i < files.length; i++){
            if(files[i].isFile()){
                String fileName=files[i].getName();
                int index=fileName.lastIndexOf(".");
                if(index > 0){
                    String endName=fileName.substring(index,
                        fileName.length()).toLowerCase();
                    String type=null;
                    for(int x=0; x < FILE_MapTable.length; x++){
                        //符合预先定义的多媒体格式的文件才会在界面中显示
                        if(endName.equals(FILE_MapTable[x])){
                            type=FILE_MapTable[x];
                            break;
                        }
                    }
                    if(type != null){
                        items.add(files[i].getName());
                        paths.add(files[i].getPath());
                        sizes.add(files[i].length()+"");
                    }
                }
            }
        }
    }
    /* 使用自定义的 FileListAdapter 将数据传入 ListView 中*/
    fileListView.setAdapter(new FileListAdapter(this,items));
}
/**
 * 打开媒体文件
 */
```

```
    private void openFile(String  path){
        //打开媒体播放器，进行播放
    Intent intent=new Intent(MyFileActivity.this,
        MediaPlayerActivity.class);
        intent.putExtra("path",path);
        startActivity(intent);
        finish();
    }
    /**
     *ListView 列表适配器
     */
    class  FileListAdapter  extends  BaseAdapter
    {
        private Vector<String> items=null; // items 用于存放显示的名称
        private MyFileActivity myFile;
        public FileListAdapter(MyFileActivity myFile, Vector<String> items)
        {
            this.items=items;
            this.myFile=myFile;
        }
        @Override
        public int getCount() {
            return items.size();
        }
        @Override
        public Object getItem(int position){
            return items.elementAt(position);
        }
        @Override
        public long getItemId(int position){
            return items.size();
        }
        @Override
        public View getView(int position,View convertView,
        ViewGroup parent){
            if(convertView==null)
            {
               //加载列表项布局文件【file_item.xml】
               convertView=myFile.getLayoutInflater().inflate(R.layout.file_item,null);
            }
             //文件名称
            TextView   name =(TextView)convertView.findViewById(R.id.name);
             //媒体文件类型
             ImageView   music=(ImageView)convertView.findViewById(R.id.music);
             //文件夹类型
            ImageView folder=(ImageView)convertView.findViewById(R.id.folder);
              name.setText(items.elementAt(position));
                if(sizes.elementAt(position).equals(""))
                {
                 //隐藏媒体文件图标,显示文件夹图标
                 music.setVisibility(View.GONE);
                 folder.setVisibility(View.VISIBLE);
                }else
                {
                 //隐藏文件夹图标,显示媒体文件图标
                 folder.setVisibility(View.GONE);
                 music.setVisibility(View.VISIBLE);
               }
            return convertView;
         }
    }
}
```

4.8 实训项目

一. 开发多媒体播放器

1. 实训目的与要求

综合利用 MediaPlayer、SurfaceView、数据存储技术，开发一款较为完整的多媒体播放器。

2. 实训内容

在完成本项目的多个任务之后，实现如下功能以进一步完善多媒体播放器。

（1）在播放界面中显示播放音乐文件列表，可以将音乐文件添加到列表上。

（2）在播放界面上增加播放模式功能：单曲循环、顺序播放、列表循环、随机播放。

（3）在播放界面上增加音量调节功能。

（4）在播放界面中增加播放历史功能。

3. 思考

（1）如何保存播放音乐文件列表？

（2）如何较好地实现随机播放？

（3）如何采用数据结构的层次遍历算法，遍历出手机上所有的音频或视频文件？

（4）如何使用 VideoView 类实现多媒体播放？

PART 5 项目五 开发手机相机

本项目工作情景的目标是让学生掌握利用 Android 系统的相机和传感器技术，开发与手机硬件有关的应用。主要的工作任务划分为① 制作相机打开界面；② 拍照控制；③ 相片保存和预览；④ 照片浏览。本项目主要涉及的关键技术包括：相机参数的设置、相机拍照、图片的显示、图片的切换、手机晃动方向传感检测等。本项目的程序涉及硬件的操作，建议将程序部署到真实的手机上进行测试。

5.1 相机打开界面

一、任务分析

本任务需要实现的效果如图 5-1 所示。

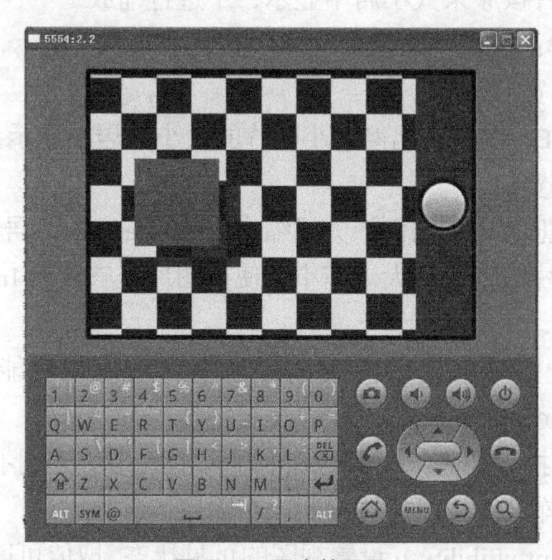

图 5-1 程序效果图

手机相机程序主要功能包括拍照、浏览照片等，是智能手机都具备的基本功能。手机相机程序界面一般包括预览拍照窗口和显示照片窗口。要完成本次任务，需要思考以下两个问题。

（1）如何使用相机程序进行照片预览？

（2）如何调用系统的照相功能？

二、相关知识

1．ImageView 类

ImageView 类用于显示各种图像，例如，图标、图片。ImageView 可以加载来自不同数据源的图像，例如，资源、ContentProvider。

下面介绍 ImageView 的主要方法。

（1）void setAdjustViewBounds（boolean adjustViewBounds）：取值为真时，表示调整 ImageView 的边界，使其显示的图像能保持横纵比。需与 setMaxHeight、setMaxWidth 方法一起使用，才有效果。

（2）void setImageBitmap（Bitmap bm）：显示 Bitmap 图像。

（3）void setImageDrawable（Drawable drawable）：显示 Drawable 图像。

（4）void setImageResource（int resId）：显示 Drawable 图像。其中，参数 resId 表示 drawable 的标识符。

（5）void setImageURI（Uri uri）：显示指定 uri 的图像。

（6）void setMaxHeight（int maxHeight）：设置 ImageView 的最大高度。

（7）void setMaxWidth（int maxWidth）：设置 ImageView 的最大宽度。

说明

setMaxHeight 和 setMaxWidth 方法要发挥作用，需要先调用 setAdjustViewBounds(true) 方法，并将 ImageView 控件的宽度和高度设置为 WRAP_CONTENT。

（8）void setScaleType(ImageView.ScaleType scaleType)：控制图像如何调整大小或者移动，以适应 ImageView 的大小。其中，参数 scaleType 的取值在 ImageView.ScaleType 中定义如下。

① CENTER：将图片按原来大小居中显示，不进行缩放。

② CENTER_CROP：按统一比例扩大图片的大小并居中显示，使得图片的长和宽等于或大于 ImageView 的长和宽。

③ CENTER_INSIDE：按统一比例缩小图片的大小并居中显示，使得图片的长和宽等于或小于 ImageView 的长和宽。

④ FIT_CENTER：把图片按比例扩大或缩小到 ImageView，居中显示。

⑤ FIT_END：把图片按比例扩大或缩小，使得图片能够放入 ImageView，与 ImageView 的底部边缘对齐进行显示。

⑥ FIT_START：把图片按比例扩大或者缩小，使得图片能够放入 ImageView，与 ImageView 的顶部边缘对齐进行显示。

⑦ FIT_XY：把图片扩大或者缩小到 ImageView 的大小（比例不固定）。

⑧ MATRIX：用矩阵来缩放图片的绘制。在运用该参数之前，应先创建一个矩阵。

（9）void setVisibility(int visibility)：设置视图的可见状态。取值可以是 VISIBLE、INVISIBLE 或者 GONE。

下面举例说明 ImageView 的使用方法，使用本例需要在本项目的 drawable 目录中存放一个名为 image1.jpg 图片文件。

（1）定义一个布局文件，只有一个 ImageView 控件。

```
<LinearLayout xmlns:android="http://schemas.android.com/apk/res/android"
    xmlns:tools="http://schemas.android.com/tools"
```

```
    android:id="@+id/LinearLayout1"
    android:layout_width="match_parent"
    android:layout_height="match_parent"
    android:orientation="vertical" >
    <ImageView
        android:id="@+id/imageView1"
        android:layout_width="wrap_content"
        android:layout_height="wrap_content" />
</LinearLayout>
```

（2）定义 Activity 类：程序默认效果如图 5-2 所示；如果去除 iv.setScaleType（ImageView. Scale Type.FIT_XY）代码的注释，则效果如图 5-3 所示。

图 5-2 等比例缩小

图 5-3 非固定比例扩大显示

```
public class ImageViewDemo extends Activity {
    @Override
    public void onCreate(Bundle savedInstanceState){
        super.onCreate(savedInstanceState);
        setContentView(R.layout.activity_image_view_demo);
        ImageView iv=(ImageView)(this.findViewById(R.id.imageView1));
        iv.setImageResource(R.drawable.image1);
        //iv.setScaleType(ImageView.ScaleType.FIT_XY);
    }
}
```

2．调用系统的拍照功能

利用 Android 系统强大的组件特性，使应用开发者只需通过 Intent 就可以用很少的代码方便地调用系统自带的相机程序，获取所拍摄的照片。

MediaStore 类是媒体提供者，包含所有内部和外部存储设备可用的媒体源数据。其中，属性 ACTION_IMAGE_CAPTURE 表示标准的启动相机程序捕捉照片的 Intent 动作；属性 ACTION_VIDEO_CAPTURE 则表示标准的启动相机程序捕捉视频的 Intent 动作。

下面来看一个调用系统相机的简单案例。

（1）创建布局文件【camera_system_demo.xml】，在布局中定义一个 ImageView 组件。

```
<LinearLayout xmlns:android="http://schemas.android.com/apk/res/android"
    xmlns:tools="http://schemas.android.com/tools"
    android:id="@+id/LinearLayout1"
    android:layout_width="match_parent"
    android:layout_height="match_parent"
    android:orientation="vertical" >
    <ImageView
        android:id="@+id/imageView1"
```

```
        android:layout_width="fill_parent"
        android:layout_height="fill_parent"
        android:src="@drawable/ic_action_search" />
</LinearLayout>
```

（2）创建 Activity 程序【CameraSystemDemo】。

```
public class CameraSystemDemo extends Activity {
    // 定义照片保存的路径
    private String imgPath="/sdcard/";
    private ImageView imageView;
    private File file;

    @Override
    public void onCreate(Bundle savedInstanceState){
        super.onCreate(savedInstanceState);
        setContentView(R.layout.camera_system_demo);
        imageView =(ImageView)this.findViewById(R.id.imageView1);
       //以当前时间作为照片名字,其中 yyyy 表示年,MM 表示月,dd 表示日
        //hh 表示时,mm 表示分,ss 表示秒
        SimpleDateFormat formatter=new SimpleDateFormat("yyyyMMddhhmmss");
        Date curDate=new Date(System.currentTimeMillis());
        String str=formatter.format(curDate);
        imgPath=imgPath+str+".jpeg";
        // 创建文件
        file=new File(imgPath);
        if(!file.exists()){
            // 创建目录
            File vDirPath=file.getParentFile();
            boolean result=vDirPath.mkdirs();
            if(!result){
                Log.v("create directory","创建文件夹不成功");
            } else {
                Log.v("create directory","创建文件夹成功");
            }
        }

        Uri uri=Uri.fromFile(file);
        Intent intent=new Intent(MediaStore.ACTION_IMAGE_CAPTURE);
        intent.putExtra(MediaStore.EXTRA_OUTPUT,uri);
        // 打开系统相机
        startActivityForResult(intent,10);
    }

    @Override
    protected void onActivityResult(int requestCode,int resultCode,Intent data){
        // TODO Auto-generated method stub
        if(file.exists()){
            imageView.setImageURI(Uri.fromFile(file));
        }

    }
}
```

上述代码使用了 startActivityForResult 方法，表示希望在调用执行新的 Activity 之后，能够获得其返回的结果，所以应在 Activity 程序中重载 void onActivityResult（int requestCode，int resultCode，Intent data）方法，对结果做进一步的处理。根据本程序的要求，在这里只需取出新照相片的 Uri，将其显示到 ImageView 中即可，因此没有用到 onActivityResult 方法传递过来的参数值，但这些参数值往往会在其他应用中被用到。

此外，如果只是调用系统相机进行拍照，不对照片文件进行特殊命名，也不对拍照结果做进一步处理，则可以简单地使用如下代码实现系统相机程序（见图 5-4）的调用。

```
Intent intent=new Intent(); //调用相机
```

```
intent.setAction("android.media.action.STILL_IMAGE_CAMERA");
startActivity(intent);
```

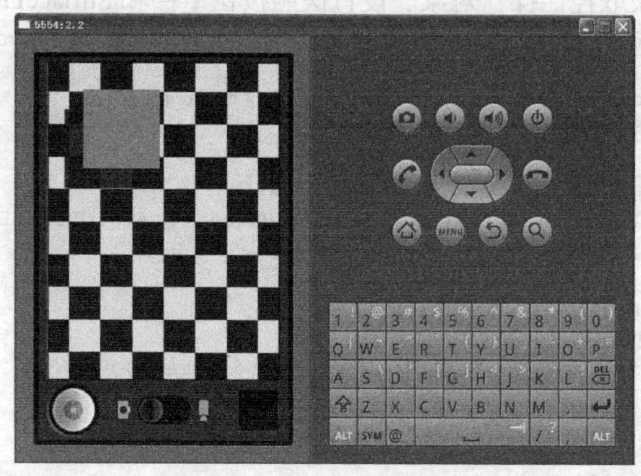

图 5-4　系统相机程序

三、任务实施

（1）相机界面布局文件【cameralayout.xml】的定义：在界面上设置一个 SurfaceView 控件用于相机图像的预览，设置一个 ImageView 用于照片的预览，再设置一个 ImageView 用于设置拍照按钮。

```
<?xml version="1.0" encoding="utf-8"?>
<LinearLayout xmlns:android="http://schemas.android.com/apk/res/android"
   android:orientation="horizontal"
   android:layout_width="fill_parent"
   android:layout_height="fill_parent"

   >
   <!--相机拍照预览窗口-->
   <SurfaceView android:id="@+id/camera"
      android:layout_width="fill_parent"
      android:layout_height="fill_parent"
      android:layout_weight="1.0"
      android:layout_marginLeft="-20.0dip"
   />
   <!--照片-->
   <ImageView android:id="@+id/image"
      android:layout_width="fill_parent"
      android:layout_weight="1.0"
      android:layout_height="fill_parent"/>
   <!--快照-->
   <LinearLayout
   android:orientation="horizontal"
   android:layout_width="72.0dip"
   android:layout_height="fill_parent"
   android:layout_gravity="center"
   android:background="@drawable/bg_camera_pattern"
   >
   <ImageView android:id="@+id/shutter"
      android:layout_width="wrap_content"
      android:layout_gravity="center"
      android:src="@drawable/btn_shutter"
      android:clickable="true"
      android:layout_height="wrap_content"/>
  </LinearLayout>
</LinearLayout>
```

（2）定义一个 Activity 类 CameraActivity，用于实现照相功能。在 onCreate 方法中隐藏标题、设置布局文件、获取控件操作对象、设置快照按钮事件和 SurfaceHolder 回调事件。

```
public class CameraActivity extends Activity implements OnClickListener, Callback {
  ...../略
  protected void onCreate(Bundle savedInstanceState){
      super.onCreate(savedInstanceState);
      //隐藏标题
      requestWindowFeature(Window.FEATURE_NO_TITLE);
      getWindow().setFlags(WindowManager.LayoutParams.FLAG_FULLSCREEN,
      WindowManager.LayoutParams.FLAG_FULLSCREEN);
      //设置布局文件
      setContentView(R.layout.cameralayout);
      //获取控件操作对象
      mSurfaceView =(SurfaceView)findViewById(R.id.camera);
      mImageView =(ImageView)findViewById(R.id.image);
      shutter=(ImageView)findViewById(R.id.shutter);
      //设置快照按钮事件
      shutter.setOnClickListener(this);
      mImageView.setVisibility(View.GONE);
      mSurfaceHolder=mSurfaceView.getHolder();
      //设置 SurfaceHolder 回调事件
      mSurfaceHolder.addCallback(this);
      //SURFACE_TYPE_PUSH_BUFFERS 标识数据来源于其他对象，比如照相机
      mSurfaceHolder.setType(SurfaceHolder.SURFACE_TYPE_PUSH_BUFFERS);
  }
  ...//略
}
```

5.2 相机拍照控制

一、任务分析

实现相机功能有两种方法：一种是上一节介绍的利用 Intent 直接调用系统的相机程序；另外一种是利用 Android 提供的 API 来定制一个自己的相机处理程序。程序使用手机进行拍照时，需要先获得访问手机相机硬件的接口。Android 系统提供了调用系统硬件的接口，本任务是要调用手机 API 来设置相机参数。要完成本次任务，需要思考以下两个问题。

（1）如何打开手机上的相机?

（2）如何设置相机的硬件参数，如自动对焦、开启闪光灯等?

二、相关知识

1. 使用 Camera 类

虽然可以直接使用系统的拍照功能，但对于个人定制的相机程序来说该功能仍显得不够方便。利用 Android 提供的 Camera API 可以定制相机程序。Camera 是 Android 系统实现拍照的重要类，能够管理相机硬件，设置相机的图像捕捉参数，打开相机并抓拍照片，也可以开始或者停止预览图像。要在程序中使用相机设备，应在 Android 的 Manifest 文件中使用<uses-permission>声明应用程序使用相机的权限，并使用<uses-feature>声明应用程序用到的相机特性，例如使用相机的自动对焦功能，在 Manifest 中应声明如下：

```
<uses-permission android:name="android.permission.CAMERA" />
 <uses-feature android:name="android.hardware.camera" />
 <uses-feature android:name="android.hardware.camera.autofocus" />
```

下面介绍 Camera 类的主要方法。

（1）void autoFocus（Camera.AutoFocusCallback cb）：让相机进行自动对焦，当相机自动

聚焦后，执行回调对象 cb 中的函数。在回调函数中编写与自动对焦有关的处理方法，如聚焦成功后调用 MediaActionSound 类播放相应的声音提醒用户，执行该方法要求相机处于预览状态，即该方法应放在 startPreview()和 stopPreview()方法之间运行。需要注意的是，有些设备可能不支持自动聚焦，可以通过 getFocusMode 方法进行查询。

（2）void cancelAutoFocus()：取消自动对焦。

（3）Camera.Parameters getParameters()：获得相机当前的参数。更改 Parameters 对象的参数后，必须将其作为 setParameters（Camera.Parameters）的参数进行传递并执行，才能使参数生效。

（4）void lock()：对相机进行锁定，避免其他进程调用。相机对象在默认情况下处于锁定状态，除非之前调用过 unlock()方法。在一般情况下，该方法不常用。

（5）Camera open()：创建一个新的相机对象，可以访问相机设备。该方法为静态方法，可以直接调用。

（6）void reconnect()：当其他进程使用了相机服务之后，重新连接相机服务，此操作会请求对相机加锁。若在程序中调用 unlock()，其他的进程可能会使用相机，若不再使用，则程序必须重新调用 reconnect()方法连接相机。

（7）void release()：断开连接，释放相机资源。

（8）void setDisplayOrientation（int degrees）：设置相机预览显示旋转的角度。

（9）void setErrorCallback（Camera.ErrorCallback cb）：设置相机在出错时进行回调处理。

（10）void setParameters（Camera.Parameters params）：设置相机服务的参数。

（11）void setPreviewCallback（Camera.PreviewCallback cb）：设置相机的预览回调处理。

（12）void setPreviewDisplay（SurfaceHolder holder）：设置用于相机预览显示的 SurfaceHolder。

（13）void startPreview()：开始在屏幕上捕捉和画出预览帧。在执行该方法之前，应先执行 setPreviewDisplay（SurfaceHolder）方法。

（14）void startSmoothZoom（int value）：开始平滑变焦。其中，参数 value 表示变焦值。

（15）void stopPreview()：停止在 Surface 上捕捉和画出预览帧。

（16）void stopSmoothZoom ()：停止平滑变焦。

（17）void takePicture（Camera.ShutterCallback shutter，Camera.PictureCallback raw，Camera.Picture- Callback postview，Camera.PictureCallback jpeg）：以异步方式进行图像捕捉，即拍照。相机服务将在图像捕捉过程中开始一系列的回调操作。参数 shutter 所表示的回调发生在图像被捕捉之后，可以设置播放“咔嚓”声之类的操作，以便让用户知道已经拍摄了一张照片。后面有 3 个 PictureCallback 接口参数，用于捕捉照片图像数据的回调，分别对应 3 种形式的图像数据。图像数据可以在 PictureCallback 接口的 void onPictureTaken(byte[] data，Camera camera）中获得。这 3 种数据对应的回调函数正好是按照参数放置的先后顺序进行调用的。参数 raw 表示的回调发生在原始图像（未被压缩）可用的时候；参数 postview 所表示的回调发生在已经处理过的图像可用时；参数 jpeg 表示的回调发生在有压缩图像时。如果应用程序不需要特定的回调操作，可以将其参数设置为 null。最简单的形式是将所有的参数都设置为 null，这样尽管能够捕获照片，但是不能对照片数据进行处理。通常只关心 JPEG 图像的数据，此时可将前面两个 PictureCallback 接口的参数直接设置为 null。

（18）void takePicture（Camera.ShutterCallback shutter, Camera.PictureCallback raw, Camera.Picture- Callback jpeg）：与 takePicture（shutter, raw, null, jpeg）等价。

（19）void unlock ()：对相机解锁，以便其他的进程可以使用。当其他进程使用完毕后，应调用 reconnect()方法，以声明对相机的使用。

上面的方法中，有很多用到回调接口作为参数。下面对这些回调接口进行介绍。

① Camera.AutoFocusCallback 是一个用于通知相机自动聚焦完成的回调接口，实现方法为 onAutoFocus（boolean success，Camera camera）。该方法在相机自动聚焦完成时被调用，可以将拍照片的代码放在该方法中，例如，takePicture 方法。其中参数 success 为 true 时表示聚焦成功，若为 false 则表示聚焦失败；参数 camera 表示相机对象。

② Camera.ErrorCallback 是一个用于相机错误通知的回调接口，实现方法为 onError（int error，Camera camera）。该方法在相机出错时被调用。其中，参数 error 表示错误代码，取值有 CAMERA_ERROR_ UNKNOWN 和 CAMERA_ERROR_SERVER_DIED；参数 camera 表示相机对象。

③ Camera.PictureCallback 是一个用于从照片捕捉中获得图像数据的回调接口，实现方法为 onPictureTaken（byte[] data，Camera camera）。该方法在捕捉照片后图像数据可用时被自动调用。其中，参数 data 是图像数据字节数组，数据的格式依赖于回调的上下文 context 以及 Camera.Parameters 参数的设置；参数 camera 表示相机对象。

④ Camera.PreviewCallback 用于预览帧被显示时，传送预览帧的副本，实现方法为 onPreviewFrame（byte[] data，Camera camera）。该方法在预览帧显示时被自动调用。其中，参数 data 是预览帧的内容，以字节数组形式表示。其格式由 ImageFormat 定义，并可以通过 getPreviewFormat()方法查询获得。如果没有调用过 setPreviewFormat（int）方法进行格式设置，那么默认是 YCbCr_420_SP（NV21）。参数 camera 表示相机对象。

⑤ Camera.ShutterCallback 是一个用在实际图像捕捉时刻的回调接口，实现方法为 onShutter()。该方法在从传感器中捕捉照片的一刻即被自动调用。可以在该方法中编写按下快门按键发出的声音或者其他的反馈操作。

2．使用 Camera.Parameters 类

Camera.Parameters 类用于对相机服务进行设置。不同手机有不同的相机性能，应用程序应该在设置参数之前查询相应信息。例如，应用程序在调用 setColorEffect（String）方法之前，先调用 getSupportedColorEffects()查询相机是否支持颜色效果。

相机的主要参数包括颜色效果（color effect）、抗条带（antibanding）、闪光灯模式（flash mode）、聚焦模式（focus mode）、情景模式（scene mode）、白平衡（white balance）、预览大小等。下面介绍 Camera.Parameters 类主要的设置方法。

（1）setAntibanding（String antibanding）：设置抗条带或防条带功能。主要用于防止照片场景在渐变时出现严重的一条一条的带状纹理。其中，参数 antibanding 表示新的抗条带数值，取值可以为 ANTIBANDING_60HZ、ANTIBANDING_50HZ、ANTIBANDING_AUTO 等。如果想关闭该功能，则可以将参数设置为 ANTIBANDING_OFF。

（2）setColorEffect（String value）：设置颜色效果。其中，参数 value 表示新的颜色效果，取值可以为 EFFECT_NONE、EFFECT_MONO、EFFECT_NEGATIVE、EFFECT_SOLARIZE、EFFECT_SEPIA、EFFECT_POSTERIZE、EFFECT_WHITEBOARD、EFFECT_AQUA 等。

（3）setExposureCompensation（int value）：设置曝光补偿的索引值。其中，参数 value 表示曝光补偿的索引值，取值范围介于 getMinExposureCompensation()和 getMaxExposure Compensation()之间。若取值为 0，则表示不进行曝光。

（4）setFlashMode（String value）：设置闪光模式。其中，参数 value 表示闪光模式，取值可以为 FLASH_MODE_OFF、FLASH_MODE_AUTO、FLASH_MODE_ON、FLASH_MODE_RED_EYE 或 FLASH_MODE_TORCH。

（5）setFocusMode（String value）：设置聚焦模式。其中，参数 value 表示聚焦模式，取值可以为 FOCUS_MODE_AUTO、FOCUS_MODE_INFINITY、FOCUS_MODE_MACRO、FOCUS_MODE_ FIXED、FOCUS_MODE_EDOF 或 FOCUS_MODE_CONTINUOUS_VIDEO。

（6）setGpsAltitude（double altitude）：设置 GPS 的高度值，以便于将该值存放在 JPEG 文件的 Exif 头部。Exif（Exchangeable image file format）用于表示图像的信息，例如，拍照时的解析度、时间、光圈、相机厂商、相机型号等，但这些信息不直接显示在图像上。

（7）setGpsLatitude（double latitude）：设置 GPS 的纬度，以便于将该值存放在 JPEG 文件的 Exif 头部。

（8）setGpsLongitude（double longitude）：设置 GPS 的经度，以便于将该值存放在 JPEG 文件的 Exif 头部。

（9）setGpsProcessingMethod（String processing_method）：设置 GPS 的处理方法，以便于将该值存放在 JPEG 文件的 Exif 头部。其中，参数 processing_method 表示获得位置的处理方法。

（10）setGpsTimestamp（long timestamp）：设置 GPS 的时间戳，以便于将该值存放在 JPEG 文件的 Exif 头部。参数 timestamp 表示时间戳，为自 UTC 时间 1970 年 1 月 1 日 0 时 0 分 0 秒起流逝的秒数。

（11）setJpegQuality（int quality）：设置 JPEG 图像的质量。其中，参数 quality 表示质量值，取值范围是 1～100。数值越大，则质量越好。

（12）setJpegThumbnailQuality（int quality）：设置 JPEG 缩略图的质量。其中，参数 quality 表示质量值，取值范围是 1～100。数值越大，则质量越好。

（13）setJpegThumbnailSize（int width，int height）：设置 JPEG 缩略图的尺寸。其中，参数 width 表示缩略图的宽度，参数 height 表示缩略图的高度。

（14）setPictureFormat（int pixel_format）：设置图像的格式。其中，参数 pixel_format 表示想要的图像格式，取值有 ImageFormat.NV21、ImageFormat.RGB_565 或 ImageFormat.JPEG 等。

（15）setPictureSize（int width，int height）：设置照片的尺寸。其中，参数 width 表示照片的宽度，单位为像素；参数 height 表示照片的高度，单位也为像素。

（16）setPreviewFormat（int pixel_format）：设置预览照片的图像格式。如果该方法没有被调用，则默认为 NV21 编码格式。可以使用 getSupportedPreviewFormats()方法来获得可用的预览图像格式列表。建议使用 NV21 或者 YV12 编码格式，所有的相机设备都支持这两种格式。其中，参数 pixel format 表示图像格式，取值已在 ImageFormat 中定义，例如，ImageFormat.NV21 或者 ImageFormat.YV12。

（17）setPreviewFrameRate（int fps）：设置预览帧接收的速度。其中，参数 fps 表示每秒接收的帧数。

（18）setPreviewSize（int width，int height）：设置预览照片的尺寸。其中，参数 width 表示照片的宽度，单位为像素；参数 height 表示照片的高度，单位也为像素。

（19）setRotation（int rotation）：设置相对于相机方位顺时针旋转的角度。参数 rotation 的取值为 0、90、180 或者 270。

（20）setSceneMode（String value）：设置场景模式。场景模式的设置有可能覆盖其他参数的设置，如闪光模式、聚焦模式。因此在设置场景模式之后，应用程序可以再次查询是否有参数被改变。参数 value 表示场景模式，取值可以为 SCENE_MODE_AUTO、SCENE_MODE_ACTION、SCENE_MODE_PORTRAIT、SCENE_MODE_LANDSCAPE、SCENE_MODE_NIGHT、SCENE_MODE_SPORTS、SCENE_MODE_THEATRE、SCENE_MODE_BEACH、SCENE_MODE_SNOW、SCENE_MODE_SUNSET、SCENE_MODE_FIREWORKS、SCENE_MODE_PARTY、SCENE_MODE_STEADYPHOTO、SCENE_MODE_NIGHT_PORTRAIT 或 SCENE_MODE_CANDLELIGHT。

（21）setWhiteBalance（String value）：设置白平衡。其中，参数 value 表示白平衡的取值，可以为 WHITE_ BALANCE_AUTO、WHITE_BALANCE_INCANDESCENT、WHITE_BALANCE_DAYLIGHT、WHITE_BALANCE_ CLOUDY_DAYLIGHT、WHITE_BALANCE_TWILIGHT、WHITE_BALANCE_SHADE 等。

（22）setZoom（int value）：设置变焦。其中，参数 value 的取值范围从 0 ~ getMaxZoom()。

3．实现拍照程序的步骤

上面只介绍了拍照的相关概念和方法，下面结合拍照过程中用到的 API 对拍照流程进行简要的描述。

（1）定义一个 Activity 类实现 Callback 接口，在 OnCreate 方法中获得 SurfaceView 和 SurfaceHolder 对象，并设置 SurfaceHolder.Callback 回调对象，具体如下。

```
// R.id.camera 为布局文件中的 SurfaceView 控件的 ID
SurfaceView mSurfaceView =(SurfaceView)this.findViewById(R.id.camera);
SurfaceHolder mSurfaceHolder=mSurfaceView.getHolder();
mSurfaceHolder.addCallback(this);
mSurfaceHolder.setType(SurfaceHolder.SURFACE_TYPE_PUSH_BUFFERS);
```

（2）在 SurfaceHolder.Callback 接口的 surfaceCreated 方法中，使用 Camera 的 Open()方法打开相机，获得 Camera 对象。

```
mCamera=Camera.open();
```

（3）成功开启相机后，在 SurfaceHolder.Callback 接口的 surfaceCreated 方法中调用 getParameters 方法获得相机的默认设置。如果有需要可以对其返回的 Camera.Parameters 对象的参数进行修改，还可调用 setDisplayOrientation 方法设置图像预览显示的角度。

```
Parameters params=mCamera.getParameters();
//拍照时自动对焦
params.setFocusMode(Parameters.FOCUS_MODE_AUTO);
//设置预览帧速率
params.setPreviewFrameRate(3);
```

（4）在 surfaceChanged 方法中，通过 setPreviewDisplay 方法为相机设置 SurfaceHolder 对象。该方法将 Camera 和 SurfaceHolder 对象关联在一起。没有 Surface，相机就不能进行预览。设置成功后调用 startPreview 方法开启预览功能，在拍照之前，必须先进行预览。

```
public void surfaceChanged(SurfaceHolder holder,int format,int w,int h)
{
  try {
         //将 SurfaceHolder 设置为相机的预览显示
         mCamera.setPreviewDisplay(holder);
  } catch(IOException exception){
         mCamera.release();
         mCamera= null;
  }
   //开始预览
   mCamera.startPreview();
```

```
}
```

（5）假设需要自动对焦功能，在上述 surfaceChanged 调用完 startPreview 方法后，可以调用 Camera.autoFocus 方法来设置自动对焦回调方法。该步为可选操作，参考代码如下。

```
// 自动对焦
camera.autoFocus(new AutoFocusCallback()
{
   @Override
   public void onAutoFocus(boolean success,Camera camera)
   {
     if(success)
     {
        // success 为 true 表示对焦成功,改变对焦状态图像
        ivFocus.setImageResource(R.drawable.focus2);
     }
   }
});
```

（6）调用 takePicture（Camera.ShutterCallback，Camera.PictureCallback，Camera.PictureCallback，Camera.PictureCallback）方法进行拍照，等待回调完成后获得实际的图像数据。

（7）每次调用 takePicture 获取图像之后，相机会停止预览。假如需要继续拍照，则在上面的 PictureCallback 接口的 onPictureTaken()方法中再次调用 startPreview 方法。

（8）在不需要拍照的时候，应主动调用 stopPreview()方法停止预览功能，并且调用 release()方法释放相机，以便让其他应用程序能够调用相机。建议在 onPause()方法中释放相机，在 onResume()方法中重新打开相机。

如果需要进入相机的视频拍摄模式，除了需要使用 Camera 类之外，还需要使用 MediaRecorder 类，并使用一组方法按照一定的步骤进行视频拍摄，在此不再做详细介绍。

三、任务实施

（1）设置相机参数。首先使用 Camera 类的静态方法 open()获得相机对象，然后通过相机对象的 getParameters 方法获得其参数对象，最后利用相机参数对象调用相关方法来设置相机参数。

```
public  void  setCameraParams()
{
    if(mCamera != null)
    {
        return ;
    }
    //创建并打开相机
    mCamera=Camera.open();
    //设置相机参数
    Parameters params=mCamera.getParameters();
    //拍照时自动对焦
    params.setFocusMode(Parameters.FOCUS_MODE_AUTO);
    //设置预览帧速率
    //params.setPreviewFrameRate(3);
    //设置预览格式
    params.setPreviewFormat(PixelFormat.YCbCr_422_SP);
    //设置图片质量百分比
    params.set("jpeg-quality",85);
    //获取相机支持的图片分辨率
    List<Size> list=params.getSupportedPictureSizes();
    Size size=list.get(0);
    int w=size.width;
    int h=size.height;
    //设置图片大小
    params.setPictureSize(w,h);
    //设置自动闪光灯
```

```
    params.setFlashMode(Camera.Parameters.FLASH_MODE_AUTO);
}
```

（2）实现 Surface 创建时的回调方法。

```
public void surfaceCreated(SurfaceHolder holder){
        setCameraParams();
}
```

（3）实现 Surface 大小发生改变时的回调方法。设置预览的 SurfaceHolder，并在屏幕上捕捉和画出预览帧。

```
public void surfaceChanged(SurfaceHolder holder,int format,int width, int height)
{
     try{
         //判断相机是否在预览，若是则先停止
         if(mPreviewRunning){
              mCamera.stopPreview();
         }
         //设置用于相机预览显示的 SurfaceHolder
         mCamera.setPreviewDisplay(holder);
          //开始在屏幕上捕捉和画出预览帧
         mCamera.startPreview();
         mPreviewRunning=true;
    } catch(Exception e){
         e.printStackTrace();
    }
}
```

（4）处理按钮拍照事件，设置自动对焦并进行拍照。

```
public void onClick(View v){
     //判断是否可以进行拍照
     if(mPreviewRunning){
          shutter.setEnabled(false);
          //设置自动对焦
          mCamera.autoFocus(new AutoFocusCallback() {
               @Override
               public void onAutoFocus(boolean success,Camera camera){
                //对焦后进行拍照
                mCamera.takePicture(mShutterCallback,null,mPictureCallback);
          }});
     }
}
```

（5）定义相机捕捉图片的回调接口，可以从中得到图片数据，将其保存为图片和显示图片的操作将在 5.3 节中介绍。

```
PictureCallback mPictureCallback=new PictureCallback() {
        public void onPictureTaken(byte[] data,Camera camera1){
                //断定照片数据是否不为空
                if(data != null){
                        //在 5.3 节的任务实施中完成照相数据保存功能
                         saveAndShow(data);
                }
        }
};
```

（6）定义照片被捕捉瞬间的回调接口，即快照。

```
ShutterCallback mShutterCallback=new ShutterCallback() {
        public void onShutter() {
            System.out.println("快照回调函数……");
        }
};
```

（7）实现 Surface 被销毁时的回调方法。停止相机预览，并回收相机资源。

```
public void surfaceDestroyed(SurfaceHolder holder){
        if(mCamera!=null)
        {
            //停止相机预览
            mCamera.stopPreview();
            mPreviewRunning=false;
```

```
            //回收相机资源
            mCamera.release();
            mCamera=null;
        }
}
```

5.3 照片保存和预览

一、任务分析

本任务需要实现的效果如图 5-5 所示。

要对拍摄的照片进行保存，首先应把照片数据从相机捕捉图片的回调接口中读取出来，然后将其保存到手机 SD 卡的相应目录下，最后将其显示在手机上。要完成本次任务，需要思考如下两个问题。

（1）如何访问 Android 系统下的 SD 卡目录?

（2）如何从 SD 卡指定目录上读取图片数据并显示到用户界面?

图 5-5 保存预览

二、相关知识

1．BitmapFactory 类

BitmapFactory 类的作用是根据文件、流、字节数据等不同的来源创建 Bitmap（BMP）图片。BitmapFactory 类提供的都是静态方法，用于创建 Bitmap 文件。主要写法有：

① Bitmap decodeFile（String pathName）。

② Bitmap decodeFile（String pathName，BitmapFactory.Options opts）。

上面两种方法都是根据给出的文件路径，将文件解码转换成 BMP 格式。其中，参数 pathName 表示完整路径的文件名；参数 opts 表示 Bitmap 的参数设置。

③ Bitmap decodeFileDescriptor（FileDescriptor fd）：根据文件描述符，编码转换成 Bitmap 文件。参数 fd 表示包含编码数据的文件描述符。

④ Bitmap decodeStream（InputStream is）：从输入流中解码转换为 BMP 文件。其中，参数 is 为用于编码的数据流。

BMP 文件通常是不压缩的，因此往往比同一幅图像的压缩图像文件要大很多。一个 800 像素× 600 像素的 24 位 BMP 文件大约占用 1.4 MB 的空间存储。因此 BMP 文件通常不适合在互联网、其他低速或者有容量限制的媒介上进行传输。

在 Android 3.0 版本之前，Dalvik 虚拟器为一个应用提供 16 MB 的内存，一般处理超过 8MB 的图片将会出现“OutOfMemoryError”异常。BMP 文件会将图片的所有像素（即长 × 宽）加载到内存中，如果图片分辨率过大，会直接导致内存溢出（java.lang.OutOfMemoryError）。因此 BitmapFactory 加载图片时需使用 BitmapFactory.Options 对相关参数进行配置以减少加载的像素。下面介绍其主要属性。

① boolean inJustDecodeBounds：若取值为 true，解码器返回空，即不生成文件，但可计算出原始图片的 outwidth 和 outheight 两个属性。有了这两个参数，再通过一定的算法，即可计算出一个恰当的 inSampleSize 数值，用于裁剪原始的图像。

② int inSampleSize：为了节省内存，请求编码器裁剪原始的图像，以便于返回一个更小的图像，取值应大于 1。图像大小根据整体取值进行等比例缩小，如 inSampleSize=4，则表示

返回图像的宽度和高度都相当于原来图像的 1/4，也就相当于原来的图像大小的 1/16。如果取值小于等于 1，则相当于没有对原来的图像做任何改变。

③ byte[] inTempStorage：用于解码的临时控件，建议取值为 16 × 1024。

④ int outHeight：BMP 文件的高度。

⑤ int outWidth：BMP 文件的宽度。

下面介绍如何根据上述属性，创建一个 BMP 图片的缩略文件。

（1）获得图片的 Options 对象。

```
Bitmap bitmap;
BitmapFactory.Options options=new BitmapFactory.Options();
options.inJustDecodeBounds=true;
bitmap=BitmapFactory.decodeFile(pathName,options); //并没有真正生成 Bitmap 文件
```

（2）定义图片缩略的目标宽度，假设为 200，然后等比例计算出相应的目标高度。

```
Int scaleWidth=200;
int height=(options.outHeight/options.outWidth)*scaleWidth;
```

将新的高度和宽度设置到属性中：

```
options.outHeight=height;
options.outWidth=scaleWidth;
```

（3）将新的参数属性作为参数，解码生成新的 BMP 文件。

```
options.inJustDecodeBounds=false;
//读取 Bitmap 照片
bitmap=BitmapFactory.decodeFile(pathName,options);
```

上述代码实现了图片的缩放，但在执行 BitmapFactory.decodeFile（path，options）代码时，并不能够节约内存。要想真正节约内存，需要使用 inSampleSize 属性，即用下面的代码替换掉第（2）步的对 BMP 文件的高度和宽度赋值代码。

```
options.inSampleSize=options.outWidth/scaleWidth;
```

三、任务实施

（1）保存和显示图片，通过 5.2 节的任务获得了照片的数据 byte[]。这里要将数据保存到指定的 SDK 目录“/com.demo.pr5/”文件夹中，然后将图片数据重新加载到 ImageView 控件并显示出来。

```
public void  saveAndShow(byte[] data)
{
    try
    {
        //图片 ID
        String imageId=System.currentTimeMillis()+"";
       //照片保存路径
       String pathName= "/sdcard/com.demo.pr5/";
       //创建文件夹
       File file=new File(pathName);
        if(!file.exists())
        {
           file.mkdirs();
        }
        //创建文件
        pathName+=imageId+".jpeg";
        file=new File(pathName);
        if(!file.exists()){
           file.createNewFile();//若文件不存在则新建
       }
       FileOutputStream fos=new FileOutputStream(file);
       fos.write(data);
       fos.close();
       //读取照片,并对其进行缩放
```

```
        BitmapFactory.Options options=new BitmapFactory.Options();
        options.inJustDecodeBounds=true;
        //此时返回的 bitmap 为空
      bitmap=BitmapFactory.decodeFile(pathName,options);
      //获取屏幕的宽度
      WindowManager manage=getWindowManager();
      Display display=manage.getDefaultDisplay();
      //将 Bitmap 的显示宽度设置为手机屏幕的宽度
      int screenWidth=display.getWidth();
      options.inSampleSize=options.outWidth/screenWidth;
      //将 inJustDecodeBounds 设置为 false,以便可以解码为 Bitmap 文件
      options.inJustDecodeBounds=false;
      //读取照片 Bitmap
      bitmap=BitmapFactory.decodeFile(pathName,options);
      //将图片设置在控件上显示出来
      mImageView.setImageBitmap(bitmap);
      mImageView.setVisibility(View.VISIBLE);
      mSurfaceView.setVisibility(View.GONE);
      //停止相机浏览
      if(mPreviewRunning){
        mCamera.stopPreview();
        mPreviewRunning=false;
      }
      shutter.setEnabled(true);
   }catch(Exception e)
   {
      e.printStackTrace();
   }
}
```

（2）创建重拍菜单，并实现重拍功能。重新拍照是通过改变屏幕方向以使 SurfaceView 重新构建，从而实现相机的初始化。

```
public boolean onCreateOptionsMenu(Menu menu){
    menu.add(0,MENU_START,0, "重拍");
    return super.onCreateOptionsMenu(menu);
}

public boolean onOptionsItemSelected(MenuItem item){
    if(item.getItemId() == MENU_START){
        //重启相机拍照
        setRequestedOrientation(ActivityInfo.
                  SCREEN_ORIENTATION_PORTRAIT); //手机屏幕竖排显示
        setRequestedOrientation(ActivityInfo.
                  SCREEN_ORIENTATION_LANDSCAPE); //手机屏幕横排显示
        return true;
    }
  }
```

5.4 照片浏览

一、任务分析

本任务需要实现的效果如图 5-6 所示。

本任务是采用传感器技术来帮助用户更方便地对照片进行预览。具体的做法是：用户将手机向左晃动，则显示下一张图片；将手机向右晃动，则显示上一张照片。

二、相关知识

1. SensorManager 类

传感器是一种物理装置，能够探测和感受外界的信号、物理条件（如光、热、湿度）或化学成分（如烟雾），并将探知的信息传递给其他装置。Android 平台支持较为丰富的传感器

功能，现支持的传感器类型共有 11 种：加速度（accelerometer）、磁场（magnetic field）、方位角（orientation）、陀螺仪（gyroscope）、光线（light）、压力（pressure）、温度（temperature）、周围物体感应（proximity）、重力（gravity）、线性加速度（linear acceleration）、旋转矢量（rotation vector）。需要注意的是，并不是每部手机都装置了这些传感器。编写传感器应用的难点在于获取传感数据后如何处理，以实现所需要的效果，这涉及很多数学物理方面的知识。

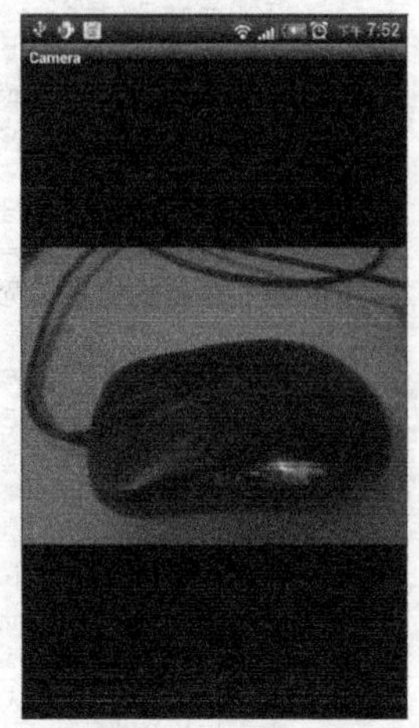

图 5-6　照片浏览界面

SensorManager 类可以让程序员方便地访问手机上的各种传感器，提供传感器的种类、采样率、精准度等方面的设置，因此可以将 SensorManager 看作是处理各种传感器的统一接口。可通过 Context.getSystemService（Context.SENSOR_SERVICE）获得 SensorManager 对象，代码如下。

```
SensorManager SensorManager=
                    (SensorManager)getSystemService(Context.SENSOR_SERVICE);
```

使用传感器时，需要通过 registerListener 函数注册传感器，使用完后应该通过 unregister Listener 取消注册。

在不需要的时候，要确保将传感器关闭。否则手机电池的电量可能很快就会耗完。这是因为即使屏幕关闭了，系统也不会自动关闭传感器。下面介绍 SensorManager 类的主要方法。

（1）Sensor getDefaultSensor（int type）：获得给定类型的默认传感器。其中，参数 type 表示所请求的传感器类型，其取值有多种选择。

① 方向传感器：Sensor.TYPE_ORIENTATION。

② 加速度传感器：Sensor.TYPE_ACCELEROMETER。

③ 重力传感器：Sensor.TYPE_GRAVITY。

④ 光线传感器：Sensor.TYPE_LIGHT。

⑤ 磁场传感器：Sensor.TYPE_MAGNETIC_FIELD。

⑥ 距离（临近性）传感器：Sensor.TYPE_PROXIMITY。

⑦ 温度传感器：Sensor.TYPE_TEMPERATURE 或 Sensor.TYPE_AMBIENT_TEMPERATURE。

⑧ 压力传感器：Sensor.TYPE_PRESSURE。

⑨ 陀螺仪传感器：Sensor.TYPE_GYROSCOPE。

（2）float getInclination（float[] I）：计算地磁传感器倾斜角的弧度。其中，参数 I 表示倾斜矩阵，可通过 getRotationMatrix 方法计算获得。

（3）float[] getOrientation（float[] R，float[] values）：计算设备的旋转方向。其中，参数 R 表示旋转矩阵，可通过 getRotationMatrix 方法计算获得；参数 values 是一个包含三个数值的数组，用于存放计算结果。其中 values[0]为绕 Z 轴旋转的方位角，取值范围为[0, 360]；values[1]是绕 X 轴旋转的倾斜角，取值范围为[−180，180]；values[2]是绕 Y 轴旋转的回转角，取值范围为[−90，90]。

（4）boolean getRotationMatrix（float[] R，float[] I，float[] gravity，float[] geomagnetic）：计算从设备坐标系到世界坐标系的倾斜矩阵和旋转矩阵。其中，参数 R 表示计算得到的旋转矩阵；参数 I 表示计算得到的倾斜矩阵；参数 gravity 是用重力矢量表示的设备坐标；参数 geomagnetic 是用地磁矢量表示的设备坐标。

（5）List<Sensor> getSensorList（int type）：得到指定类型可用的传感器列表。其中，参数 type 表示传感器类型。如果 type 取值为 Sensor.TYPE_ALL，则返回所有的传感器。

（6）boolean registerListener（SensorEventListener listener，Sensor sensor，int rate）：对某个给定的传感器注册传感事件监听器，如果注册成功则返回 true，否则则返回 false。其中，参数 listener 表示一个 SensorEventListener 对象；参数 sensor 表示需要注册的传感器；参数 rate 表示传感器事件传送的速度，可以理解为采样率。该参数的数值仅作为系统参考，感应事件的频率可以比给定的速度更快或者更慢，参数的取值有如下四种。

① 标准延迟：SENSOR_DELAY_NORMAL，是默认取值，适用于屏幕方位的改变。

② 用户界面延迟：SENSOR_DELAY_UI，适用于用户界面。

③ 游戏延迟：SENSOR_DELAY_GAME，适用于游戏。

④ 尽量少延迟：SENSOR_DELAY_FASTEST，尽可能快地获得传感器的数据，选择该参数会使手机产生较大的电力消耗。

（7）boolean registerListener（SensorEventListener listener，Sensor sensor，int rate，Handler handler）：对某个给定的传感器注册传感事件监听器。其中，参数 handler 表示传感器事件被传递给的 Handler 对象。

（8）boolean remapCoordinateSystem（float[] inR，int X，int Y，float[] outR）：将方向值转换为应用程序自定义参照系的坐标。其中，参数 inR 表示要被转换的旋转矩阵；参数 X 和 Y 表示手机在世界坐标系中 X 轴、Y 轴的方向，取值 AXIS_MINUS_X、AXIS_MINUS_Y、AXIS_X、AXIS_Y。参数 outR 表示转换后的旋转矩阵。

（9）void unregisterListener（SensorEventListener listener，Sensor sensor）：取消传感器监听器的注册。参数 listener 表示监听器；参数 sensor 表示需要取消注册的传感器。

（10）void unregisterListener（SensorEventListener listener）：取消监听器的注册，取消所有与该监听器有关的传感器的监听。

2．SensorEventListener 接口

SensorEventListener 是一个接口，当传感器的数值发生变化时，该接口被用来接收来自 SensorManager 的通知。SensorManager 的 registerListener 和 unregisterListener 方法都使用了 SensorEventListener 对象作为参数。要实现 SensorEventListener 接口，必须实现如下两个方法。

（1）onAccuracyChanged（Sensor sensor，int accuracy）：当传感器的精度发生变化时自动调用该方法。其中，参数 sensor 表示发生精度变化的传感器；参数 accuracy 表示传感器新的精度值。

（2）onSensorChanged（SensorEvent event）：当传感器的数值发生变化时自动调用该方法。其中，参数 event 表示一个 SensorEvent 对象，SensorEvent 类主要用于表示传感器的事件、类型、时间戳和精度等信息。SensorEvent 类有一个数组 values，用于存放传感器的数值。

SensorManager 类对于不同传感器的调用都使用同一个接口，这样编程时就非常简单。如果想获得某一种感应事件，只需要将 registerListener 方法的第二个参数修改为相应传感器的枚举数值，再设置一个监听器，利用监听接口 onSensorChanged 来读取具体感应的内容。编写传感器程序的步骤一般是先在 onCreate()方法中获得传感器的管理器和想要操作的传感器，然后在 onResume()方法中给该传感器添加监听，编写监听器的处理代码，最后在 onPause()方法中注销对传感器的监听。传感器的框架代码示例如下。

```
public class SensorFrameDemo extends Activity implements SensorEventListener {
private SensorManager mSensorManager;
   private Sensor mAccelerometer;
   public void onCreate(Bundle savedInstanceState){
      //获得传感器管理器
    mSensorManager =(SensorManager)getSystemService(SENSOR_SERVICE);
      //获得加速度传感器
    mAccelerometer=mSensorManager.getDefaultSensor(Sensor.TYPE_ACCELEROMETER);
   }
protected void onResume() {
      super.onResume();
      //注册传感器监听器
      mSensorManager.registerListener(this,mAccelerometer,SensorManager.
                                                SENSOR_DELAY_NORMAL);

   }

   protected void onPause() {
      super.onPause();
      //取消传感器监听器的注册
      mSensorManager.unregisterListener(this);
   }

   public void onAccuracyChanged(Sensor sensor,int accuracy){
    //在此方法中,编写当某个传感器的精度发生变化时应执行的操作
   }

   public void onSensorChanged(SensorEvent event){
    //在此方法中,编写当某个传感器的数值发生变化时应执行的操作
   }

}
```

在上述代码的基础上，可以进一步编写获得方向、加速度、光线数据的代码，实现效果如图 5-7 所示。

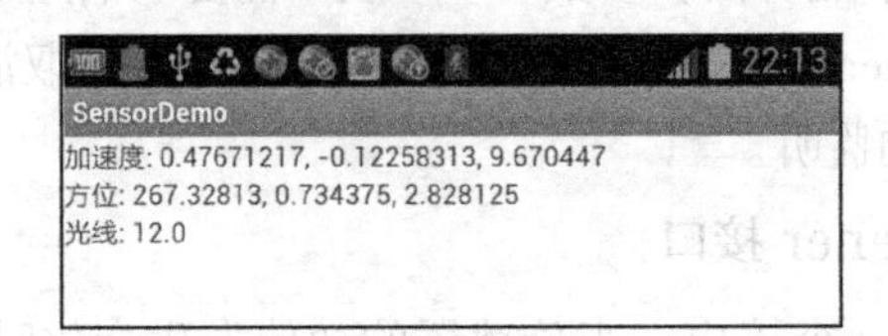

图 5-7　真机效果截图

（1）创建布局文件【sensordemo.xml】，在布局文件中设置 3 个 TextView 控件，分别显示传感器的加速度、方位和光线数值。

```
<LinearLayout xmlns:android="http://schemas.android.com/apk/res/android"
   xmlns:tools="http://schemas.android.com/tools"
   android:id="@+id/LinearLayout1"
   android:layout_width="match_parent"
   android:layout_height="match_parent"
   android:orientation="vertical" >
   <TextView
      android:id="@+id/tAccelerometer"
      android:layout_width="match_parent"
      android:layout_height="wrap_content"
      android:text="TextView" />
   <TextView
      android:id="@+id/tOrientation"
      android:layout_width="match_parent"
      android:layout_height="wrap_content"
      android:text="TextView" />
   <TextView
      android:id="@+id/tLight"
```

```
        android:layout_width="match_parent"
        android:layout_height="wrap_content"
        android:text="TextView" />
</LinearLayout>
```

（2）程序代码如下。

```
public class SensorDemo extends Activity implements SensorEventListener {
private SensorManager mSensorManager;
private Sensor mAccelerometer;
private Sensor mOrientation;
private Sensor mLight;
private TextView tAccelerometer;
private TextView tOrientation;
private TextView tLight;

@Override
public void onCreate(Bundle savedInstanceState){
    super.onCreate(savedInstanceState);
    setContentView(R.layout.sensordemo);
    tAccelerometer =(TextView)this.findViewById(R.id.tAccelerometer);
    tOrientation =(TextView)this.findViewById(R.id.tOrientation);
    tLight =(TextView)this.findViewById(R.id.tLight);
    //获得传感器管理器
    mSensorManager =(SensorManager)getSystemService(SENSOR_SERVICE);
    //获得加速度传感器
    mAccelerometer=mSensorManager.getDefaultSensor(Sensor.TYPE_ACCELEROMETER);
    //获得方向传感器
    mOrientation=mSensorManager.getDefaultSensor(Sensor.TYPE_ORIENTATION);
    //获得光线传感器
    mLight=mSensorManager.getDefaultSensor(Sensor.TYPE_LIGHT);
}

protected void onResume() {
    super.onResume();
    //对加速度传感器注册传感器监听器
    mSensorManager.registerListener(this,mAccelerometer,
            SensorManager.SENSOR_DELAY_NORMAL);
    //对方向传感器注册传感器监听器
    mSensorManager.registerListener(this,mOrientation,
            SensorManager.SENSOR_DELAY_NORMAL);
    //对光线传感器注册传感器监听器
    mSensorManager.registerListener(this,mLight,
            SensorManager.SENSOR_DELAY_NORMAL);
}

protected void onPause() {
    super.onPause();
    // 取消传感器监听器的注册
    mSensorManager.unregisterListener(this);
}

public void onAccuracyChanged(Sensor sensor,int accuracy){
    // 在此方法中,编写当某个传感器的精度发生变化时应执行的操作
}

public void onSensorChanged(SensorEvent event){
    // 在此方法中,编写当某个传感器的数值发生变化时应执行的操作
    // 得到方向的值
    float x=event.values[SensorManager.DATA_X];
    float y=event.values[SensorManager.DATA_Y];
    float z=event.values[SensorManager.DATA_Z];
    if(event.sensor.getType() == Sensor.TYPE_ORIENTATION){
        tOrientation.setText("方位: "+x+","+y+","+z);
    }
    // 得到加速度的值
```

```
    else if(event.sensor.getType() == Sensor.TYPE_ACCELEROMETER){
        tAccelerometer.setText("加速度: "+x+","+y+","+z);
    } else if(event.sensor.getType() == Sensor.TYPE_LIGHT){
        tLight.setText("光线: "+event.values[0]);
    }
}
    }
```

3．视图切换

实现屏幕切换的一种方法是对 Activity 进行切换，另外一种方法是对 View 进行切换。前面的章节已经介绍了通过 Intent 来实现 Activity 的切换，下面介绍如何在一个 Activity 内切换不同的 View。下面先介绍实现技术中相关的几个概念。

前面已经介绍过 Android 有五大布局对象，分别为线性布局（LinearLayout）、表格布局（TableLayout）、相对布局（RelativeLayout）、单帧布局（FrameLayout）、绝对布局（AbsoluteLayout）。这里对 FrameLayout 布局进行介绍。FrameLayout 是一种简单的布局对象，占据屏幕上的特定区域，只能显示一个对象，如一张图片。建议只放置一个 View 在 FrameLayout 中，这是因为所有的子元素将会固定在屏幕的左上角，它们不能够被放在指定的位置，这样后一个子元素将直接在前一个子元素上进行覆盖填充，将前一个子元素部分或全部遮掩（除非后一个子元素是透明的）。

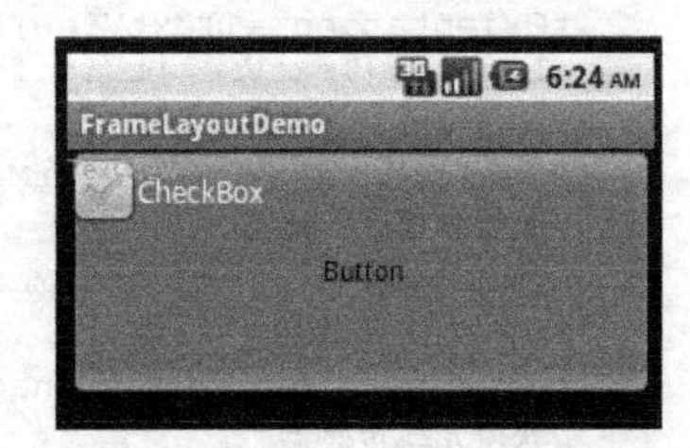

图 5-8　FrameLayout 布局示例

下面举例说明 FrameLayout 布局的使用方法，实现效果如图 5-8 所示。

```
<FrameLayout xmlns:android="http://schemas.android.com/apk/res/android"
   xmlns:tools="http://schemas.android.com/tools"
   android:id="@+id/FrameLayout1"
   android:layout_width="match_parent"
   android:layout_height="match_parent" >
   <Button
      android:id="@+id/button1"
      android:layout_width="match_parent"
      android:layout_height="134dp"
      android:text="Button" />
   <CheckBox
      android:id="@+id/checkBox1"
      android:layout_width="wrap_content"
      android:layout_height="wrap_content"
      android:text="CheckBox" />
   <TextView
      android:id="@+id/textView1"
      android:layout_width="wrap_content"
      android:layout_height="wrap_content"
      android:text="TextView" />
</FrameLayout>
```

ViewAnimator 是 FrameLayout 的子类，其作用是以动画的方式切换 FrameLayout 容器中的视图。该类的主要方法介绍如下。

（1）void addView（View child，int index，ViewGroup.LayoutParams params）。

（2）void addView（View child，ViewGroup.LayoutParams params）。

前两种方法都使用指定的布局参数集来添加一个子视图。其中，参数 child 表示需要添加的子视图；参数 index 表示子视图存放的位置；参数 params 表示布局参数集。

（3）View getCurrentView()：获得当前显示的子视图。

（4）int getDisplayedChild()：获得当前显示的子视图索引。

（5）Animation getInAnimation()：返回用于视图进入屏幕时的动画。

（6）Animation getOutAnimation()：返回用于视图退出屏幕时的动画。

（7）void removeAllViews()：清除 ViewGroup 中所有的子视图。

（8）void removeView（View view）：清除参数 view 所代表的视图。

（9）void removeViewAt（int index）：清除参数 index 所指定的视图。

（10）void setAnimateFirstView（boolean animate）：指出当前视图在首次加载时是否以动画显示。

（11）void setDisplayedChild（int whichChild）：设置将被显示的子视图。其中，参数 whichChild 表示要显示的视图索引。

（12）void setInAnimation（Context context，int resourceID）：指定视图进入屏幕时用到的动画。其中，参数 context 表示上下文；参数 resourceID 表示动画的资源 ID。

（13）void setInAnimation（Animation inAnimation）：指定视图进入屏幕时用到的动画。其中，其中，参数 inAnimation 表示动画对象。

（14）void setOutAnimation（Animation outAnimation）：指定视图退出屏幕时用到的动画。其中，参数 outAnimation 表示动画对象。

（15）void setOutAnimation（Context context，int resourceID）：指定视图退出屏幕时用到的动画。其中，参数 context 表示上下文；参数 resourceID 表示动画的资源 ID。

（16）void showNext ()：手动显示下一个子视图。

（17）void showPrevious ()：手动显示上一个子视图。

LayoutParams 类的作用是告诉父视图，子视图希望如何在父视图中摆放。主要的取值与 XML 布局属性取值类似，如视图宽度、高度取值可以为 WRAP_CONTENT、FILL_PARENT、MATCH_PARENT。LayoutParams 类有以下 3 种构造方法。

（1）LayoutParams（Context c，AttributeSet attrs）：根据给定的属性值，创建一个新的布局参数集。

（2）LayoutParams（int width，int height）：根据给定的高度和宽度值，创建一个新的布局参数集。

（3）LayoutParams（LayoutParams source）：复制一个已有的布局参数集来构造一个新的布局参数集。

一般情况下不直接使用 ViewAnimator，而是使用其子类 ViewFlipper 或 ViewSwitcher，或间接子类 ImageSwitcher 和 TextSwitcher。下面对 ViewFlipper 类进行介绍。

ViewFlipper 类继承于 FrameLayout，是一个视图 View 容器类，在 Layout 里面可以放置多个 View。该类可以用于实现视图页面的切换，也可以设定时间间隔，让视图自动播放。注意：尽量选择将压缩过的小图片放在视图内，否则可能会因为图片过大而无法显示出图片。ViewFlipper 类额外提供了如下几个方法。

（1）boolean isAutoStart ()：如果视图显示到窗口上时会自动调用 startFlipping()方法，则返回 true。

（2）boolean isFlipping ()：如果子视图正在切换，则返回 true。

（3）void setAutoStart（boolean autoStart）：设置视图显示到窗口上时是否会自动调用 startFlipping()方法。

（4）void setFlipInterval（int milliseconds）：设置视图间切换的时间间隔。其中，参数 milliseconds 表示毫秒数。

（5）void startFlipping ()：开始定时循环切换子视图。

（6）void stopFlipping ()：停止子视图的切换。

如果希望能够通过手势的左右滑动来切换视图，那么就要用到 GestureDetector 类。该类通过使用 MotionEvent 来检测不同的手势事件，主要有下面 3 种构造方法。

（1）GestureDetector（Context context，GestureDetector.OnGestureListener listener）。

（2）GestureDetector（Context context，GestureDetector.OnGestureListener listener，Handler handler）。

（3）GestureDetector（Context context，GestureDetector.OnGestureListener listener，Handler handler，boolean ignoreMultitouch）。其中，参数 context 表示上下文；参数 listener 表示 GestureDetector.OnGestureListener 监听器；参数 handler 表示 Handler 对象；参数 ignoreMultitouch 表示是否忽略多点触控。

GestureDetector 类的一个重要监听器是 GestureDetector.OnGestureListener，可用来通知单击、长按、滑动、拖曳等手势事件。该接口需要实现 6 个抽象回调函数。

（1）boolean onDown（MotionEvent e）：用户轻触触摸屏。

（2）boolean onFling（MotionEvent e1，MotionEvent e2，float velocityX，float velocityY）：用户按下触摸屏、快速移动后松开。其中，参数 e1 表示首先按下的事件对象；参数 e2 表示触发当前滑动事件的对象；参数 velocityX 表示 X 轴上的移动速度，单位为像素/秒；参数 velocityY 表示 Y 轴上的移动速度，单位为像素/秒。

（3）void onLongPress（MotionEvent e）：用户长按触摸屏。

（4）boolean onScroll（MotionEvent e1，MotionEvent e2，float distanceX，float distanceY）：用户按下触摸屏并拖曳。其中，参数 e1 表示刚开始拖曳的事件对象；参数 e2 表示触发当前拖曳的事件对象；参数 distanceX 表示在上次调用 onScroll 之后沿 X 轴拖曳的距离；参数 distanceY 表示在上次调用 onScroll 之后沿 Y 轴拖曳的距离。

（5）void onShowPress（MotionEvent e）：用户轻触触摸屏，尚未松开或拖曳。与 onDown() 的区别是强调没有松开或者拖曳的状态。

（6）boolean onSingleTapUp（MotionEvent e）：用户轻触触摸屏后松开。

需要注意的是，调用 GestureDetector 回调函数的方法主要是在 Activity 的 onTouchEvent() 事件或者 OnTouchListener 的 onTouch()事件中，将捕捉到的 MotionEvent 交给 GestureDetector 对象来分析是否有合适的 Callback 回调函数来处理用户的手势。例如，

```
public boolean onTouchEvent(MotionEvent event){
   …
   return gestureDetector.onTouchEvent(event);
}
或者
public boolean onTouch(View v,MotionEvent event){
    …
    return gestureDetector.onTouchEvent(event);
}
```

当两个视图切换时，可以使用动画技术以使切换过渡更为自然。在 Android 系统中，Animation 效果的实现可以通过两种方式：一种是渐变动画，对 View 的内容进行一系列的图形变换，包括平移、缩放、旋转、改变透明等；另一种是逐帧动画，将一帧一帧的内容连起来播放。

可以先在 Android 项目的 anim 目录下使用 XML 文件定义动画效果，然后在程序中使用 AnimationUtils.loadAnimation(Context context，int ResourcesId)载入，以生成 Animation 对象。在 View 对象需要显示动画效果时，将 Animation 对象作为参数传递给相关的方法，例如 start Animation 方法。而 ViewFlipper 则是使用 setInAnimation 和 setOutAnimation 方法分别设置切换前和切换后的动画效果。下面介绍使用 XML 来定义动画效果的元素和属性。

渐变动画有以下两种类型。

① alpha：渐变透明度动画效果。

② scale：渐变尺寸伸缩动画效果。

逐帧动画有以下两种类型。

① translate：画面转换位置移动动画效果。

② rotate：画面转移旋转动画效果。

translate 的属性取值有以下多种选择。

① android:duration：动画持续时间，单位为毫秒（ms）。

② android:fromXDelta：动画起始时，X 坐标上的位置。

③ android:toXDelta：动画结束时，X 坐标上的位置。

④ android:fromYDelta：动画起始时，Y 坐标上的位置。

⑤ android:toYDelta：动画结束时，Y 坐标上的位置。

这些属性取值可以是数值、百分数、百分数 p。数字的取值范围都是-100～100，表示的是相对值如 50、50%、50%p。当为数值 50 时，表示在当前控件的左上角，即原点处加上 50px，做为起始点；如果是 50%，表示在当前控件的左上角加上自己宽度的 50%做为起始点；如果是 50%p，那么就是表示在当前控件的左上角加上父控件（例如，屏幕）宽度的 50%做为起始点，例如，

```
android:toXDelta="100%"  表示以 View 自己的宽度作为参照
android:toXDelta="80%p"  表示父层 View 的 80%，是以父层 View 的 0.8 倍宽度作为参照
android:fromXDelta="-100%p" android:toXDelta="0"  表示图片从左进入，状态从不可见到可见
android:fromXDelta="0" android:toXDelta="100%p"  表示图片从右滑出，状态从可见到不可见
```

alpha 的属性取值也有多种选择。

① android:duration：动画持续时间，单位为毫秒（ms）。

② android:fromAlpha：动画开始时的透明度。取值范围为[0.0，1.0]，其中 1.0 表示不透明，0.0 表示完全透明。

③ android:toAlpha：动画结束时的透明度。取值范围为[0.0，1.0]，其中 1.0 表示不透明，0.0 表示完全透明。

例如，

```
android:fromAlpha="1.0"  android:toAlpha="0.1"  表示图片状态从不透明到透明
android:fromAlpha="0.1"  android:toAlpha="1.0"  表示图片状态从透明到不透明
```

根据前面介绍的知识，下面示例实现图片自动播放和滑动播放效果，分别如图 5-9 和图 5-10 所示。

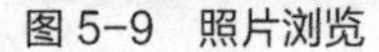

图 5-9 照片浏览

图 5-10 两张照片切换

（1）创建一个名为【view_flipper_demo.xml】的布局文件，其中只有一个 ViewFlipper 控件。

```
<?xml version="1.0" encoding="utf-8"?>
<LinearLayout xmlns:android="http://schemas.android.com/apk/res/android"
          android:orientation="vertical"
          android:layout_width="fill_parent"
          android:layout_height="fill_parent"
          >
   <ViewFlipper android:id="@+id/ViewFlipper01"
            android:layout_width="fill_parent"
            android:layout_height="fill_parent">
   </ViewFlipper>
</LinearLayout>
```

（2）动画定义：当手势从左向右滑动时，新图片从左侧进入，旧图片从右侧滑出，当手势从右向左滑动时，新图片从右侧进入，旧图片从左侧滑出。因此应定义 4 个动画效果文件。

① 定义左进的动画效果文件【push_left_in.xml】。

```
<?xml version="1.0" encoding="utf-8"?>
<set xmlns:android="http://schemas.android.com/apk/res/android" >
   <translate
      android:duration="1500"
      android:fromXDelta="-100%p"
      android:toXDelta="0" />
   <alpha
      android:duration="1500"
      android:fromAlpha="0.1"
      android:toAlpha="1.0" />
</set>
```

② 定义左出的动画渐变效果文件【push_left_out.xml】。

```
 <?xml version="1.0" encoding="utf-8"?>
<set xmlns:android="http://schemas.android.com/apk/res/android" >
   <translate
      android:duration="1500"
      android:fromXDelta="0"
      android:toXDelta="-100%p" />
   <alpha
      android:duration="1500"
      android:fromAlpha="1.0"
      android:toAlpha="0.1" />
</set>
```

③ 定义右进的动画效果文件【push_right_in.xml】。

```
  <?xml version="1.0" encoding="utf-8"?>
```

```
<set xmlns:android="http://schemas.android.com/apk/res/android" >
    <translate
        android:duration="1500"
        android:fromXDelta="100%p"
        android:toXDelta="0" />
    <alpha
        android:duration="1500"
        android:fromAlpha="0.1"
        android:toAlpha="1.0" />
</set>
```

④ 定义右出的动画渐变效果文件【push_right_out.xml】。

```
<?xml version="1.0" encoding="utf-8"?>
<set xmlns:android="http://schemas.android.com/apk/res/android" >
    <translate
        android:duration="1500"
        android:fromXDelta="0"
        android:toXDelta="100%p" />

    <alpha
        android:duration="1500"
        android:fromAlpha="1.0"
        android:toAlpha="0.1" />
</set>
```

（3）主程序代码如下。

```
public class ViewFlipperDemo extends Activity implements android.view.
GestureDetector.OnGestureListener {
    private int[] images={ R.drawable.image1,R.drawable.image2,
            R.drawable.image3,R.drawable.image4,R.drawable.image5 };

private GestureDetector gestureDetector=null;
private ViewFlipper viewFlipper=null;
private static final int FLING_MIN_DISTANCE=100;
private static final int FLING_MIN_VELOCITY=200;
private Activity mActivity=null;
private Animation rInAnim,rOutAnim,lInAnim,lOutAnim;
    @Override
    public void onCreate(Bundle savedInstanceState){
        super.onCreate(savedInstanceState);
        setContentView(R.layout.view_flipper_demo);
        mActivity=this;
        viewFlipper =(ViewFlipper)findViewById(R.id.ViewFlipper01);
        gestureDetector=new GestureDetector(this,this);    // 声明检测手势事件
        //AnimationUtils 类用于定义常用的动画处理功能，下面分别对四个动画文件进行加载
        //loadAnimation(Context context,int id)用于从资源文件中加载一个 Animation 对象
        rInAnim=AnimationUtils.loadAnimation(mActivity,R.anim.push_right_in);
        rOutAnim=AnimationUtils.loadAnimation(mActivity,R.anim.push_right_out);
        lInAnim=AnimationUtils.loadAnimation(mActivity,R.anim.push_left_in);
        lOutAnim=AnimationUtils.loadAnimation(mActivity,R.anim.push_left_out);
        for(int i=0; i < images.length; i++){      // 添加图片源，建议选择小图片
          ImageView iv=new ImageView(this);
          iv.setImageResource(images[i]);
          iv.setScaleType(ImageView.ScaleType.FIT_XY);
          viewFlipper.addView(iv,i,new LayoutParams(LayoutParams.
                               FILL_PARENT,LayoutParams.FILL_PARENT));
        }
        viewFlipper.setAutoStart(true);  //设置自动播放功能(在用户单击屏幕前,会自动播放)
        viewFlipper.setFlipInterval(2000);
        viewFlipper.startFlipping();
    }
    @Override
    public boolean onTouchEvent(MotionEvent event){
        viewFlipper.stopFlipping(); //在用户点击屏幕触发单击事件后,停止自动播放
        viewFlipper.setAutoStart(false);
        //调用 GestureDetector 的 onTouchEvent()方法,将捕捉到的 MotionEvent
```

```
        //交给 GestureDetector 分析是否有合适的 callback 函数来处理用户的手势
        return gestureDetector.onTouchEvent(event);
    }
    @Override
    public boolean onDown(MotionEvent e){
        // TODO Auto-generated method stub
        return false;
    }
    @Override
    public boolean onFling(MotionEvent e1,MotionEvent e2,float velocityX,
            float velocityY){
        // 从左向右滑动(左进右出),X 轴的坐标位移大于 FLING_MIN_DISTANCE,
        //且移动速度大于 FLING_MIN_VELOCITY 像素/秒
        if(e2.getX() - e1.getX() > FLING_MIN_DISTANCE && Math.abs(velocityX)>
            FLING_MIN_VELOCITY){
                viewFlipper.setInAnimation(lInAnim);
                viewFlipper.setOutAnimation(rOutAnim);
                viewFlipper.showPrevious();
                setTitle("相片编号: "+viewFlipper.getDisplayedChild());
                return true;
        } else if(e1.getX() - e2.getX()> FLING_MIN_DISTANCE && Math.
            abs(velocityX)> FLING_MIN_VELOCITY){//从右向左滑动(右进左出)
                viewFlipper.setInAnimation(rInAnim);
                viewFlipper.setOutAnimation(lOutAnim);
                viewFlipper.showNext();
                setTitle("相片编号: "+viewFlipper.getDisplayedChild());
                return true;
        }
        return true;
    }
    @Override
    public void onLongPress(MotionEvent e){
        // TODO Auto-generated method stub

    }

    @Override
    public boolean onScroll(MotionEvent e1,MotionEvent e2,float distanceX,
            float distanceY){
        return false;
    }
    @Override
    public void onShowPress(MotionEvent e){
        // TODO Auto-generated method stub

    }
    @Override
    public boolean onSingleTapUp(MotionEvent e){
        return false;
    }
}
```

三、任务实施

（1）定义照片浏览界面布局文件【album.xml】，在布局中设置一个 ViewFlipper 控件。

```
<?xml version="1.0" encoding="utf-8"?>
<LinearLayout xmlns:android="http://schemas.android.com/apk/res/android"
        android:orientation="vertical"
        android:layout_width="fill_parent"
        android:layout_height="fill_parent"
        >
    <ViewFlipper android:id="@+id/ViewFlipper01"
            android:layout_width="fill_parent"
            android:layout_height="fill_parent">
    </ViewFlipper>
</LinearLayout>
```

（2）定义 XML 动画。当手势从左向右甩动时，新图片左进，旧图片右出。当手势从右向左甩动时，新图片右进，旧图片左出。因此相应地定义四个动画效果文件 push_left_in.xml、push_left_out.xml、push_right_in.xml 和 push_right_out.xml，它们已经在相关知识中进行定义。

（3）定义一个 Activity 类 AlbumActivity，用于实现照片的加载、重力监听、甩动照片进行前后浏览。为了避免传感器的感应过于灵敏，不仅要检测甩动的幅度，还要检测甩动的频率。为了使照片随着手势的左右晃动切换得更为合理，读者可以对本部分代码做进一步的优化，例如，延长晃动的频率或者幅度，对上一次晃动方向的判断等。

```
public class AlbumActivity extends Activity {
   private ViewFlipper   flipper;
   private Bitmap[]      mBgList;//需要预览的图片列表
   private long         startTime=0;
   private SensorManager sm;    //重力感应硬件控制器
   private SensorEventListener sel;//重力感应侦听
   @Override
    //加载指定目录下的照片
   private  String[] loadAlbum()
   {
      //从 5.3 节任务照片保存的路径读取照片
String pathName=android.os.Environment.getExternalStorageDirectory().getPath()
              +"/com.demo.pr5";
      File file=new File(pathName);
      Vector<Bitmap> fileName=new Vector<Bitmap>();
      if(file.exists() && file.isDirectory())
      {
        String[] str=file.list();
        for(String s : str)
        {
             if(new File(pathName+"/"+s).isFile())
             {
              fileName.addElement(loadImage(pathName+"/"+s));
             }
        }
        mBgList=fileName.toArray(new Bitmap[]{});
    }
    return null;
   }
   //读取图片
   public Bitmap loadImage(String pathName){

      //读取照片,并对照片进行缩放
      BitmapFactory.Options options=new BitmapFactory.Options();
      options.inJustDecodeBounds=true;
      //此时返回的 bitmap 为空
          Bitmap  bitmap=BitmapFactory.decodeFile(pathName,options);
      //获取屏幕的宽度
      WindowManager manage=getWindowManager();
      Display display=manage.getDefaultDisplay();
      //将 Bitmap 的显示宽度设置为手机屏幕的宽度
      int screenWidth=display.getWidth();
      //计算 Bitmap 的高度等比变化数值
      options.inSampleSize=options.outWidth/screenWidth;
      //将 inJustDecodeBounds 设置为 false,以便于解码为 Bitmap 文件
      options.inJustDecodeBounds=false;
      //读取照片 Bitmap
      bitmap=BitmapFactory.decodeFile(pathName,options);
      return bitmap;
   }
      //添加显示图片 View
   private View addImage(Bitmap  bitmap){
      ImageView img=new ImageView(this);
      img.setImageBitmap(bitmap);
```

```
        return img;
    }
    public void onCreate(Bundle savedInstanceState){
        super.onCreate(savedInstanceState);
        setContentView(R.layout.album);
        flipper =(ViewFlipper)this.findViewById(R.id.ViewFlipper01);
          //加载照片
        loadAlbum();
        if(mBgList==null)
        {
           Toast.makeText(this,"相册无照片",Toast.LENGTH_SHORT).show();
           finish();
           return ;
        }else{
           for(int i=0;i<=mBgList.length-1;i++){
               flipper.addView(addImage(mBgList[i]),i,new
ViewGroup.LayoutParams(ViewGroup.LayoutParams.FILL_PARENT,
ViewGroup.LayoutParams.FILL_PARENT));
           }
             }
        //获得重力感应硬件控制器
        sm=(SensorManager)this.getSystemService(SENSOR_SERVICE);
        Sensor sensor=sm.getDefaultSensor(Sensor.TYPE_ACCELEROMETER);
        //添加重力感应侦听
        sel=new SensorEventListener(){
            public void onSensorChanged(SensorEvent se){
               float x=se.values[SensorManager.DATA_X];
               //控制甩动的幅度和甩动的频率（在 1s 内只有一次）
               if(x>8&&System.currentTimeMillis()>startTime+1000)
               {
                     flipper.setInAnimation(AnimationUtils.
               loadAnimation(AlbumActivity.this,R.anim.push_right_in));
               flipper.setOutAnimation(AnimationUtils.
              loadAnimation(AlbumActivity.this,R.anim.push_right_out));
               flipper.showPrevious();
               //记录右甩动开始时间
                startTime=System.currentTimeMillis();
               }else if(x<-8&&System.currentTimeMillis()>startTime+1000)
               {
                flipper.setInAnimation(AnimationUtils.
                loadAnimation(AlbumActivity.this,R.anim.push_left_in));
                flipper.setOutAnimation(AnimationUtils.
               loadAnimation(AlbumActivity.this,R.anim.push_left_out));
                flipper.showNext();
               //记录左甩动开始时间
                startTime=System.currentTimeMillis();
               }
            }
            public void onAccuracyChanged(Sensor arg0,int arg1){
            }
        };
     //注册 Listener，参数 SENSOR_DELAY_GAME 为检测的精确度
     sm.registerListener(sel,sensor,SensorManager.SENSOR_DELAY_GAME);
   }
   @Override
   protected void onDestroy() {
       // TODO Auto-generated method stub
       super.onDestroy();
       //注销重力感应侦听
       sm.unregisterListener(sel);
   }
}
```

（4）修改 CameraActivity 类的 onCreateOptionsMenu（Menu menu）方法，增加打开相册菜单功能；修改 onOptionsItemSelected（MenuItem item）方法，增加对打开相册菜单的处理。

```
public boolean onCreateOptionsMenu(Menu menu){
```

```
    …
    menu.add(0,MENU_SENSOR,0,"打开相册");
    …
}

public boolean onOptionsItemSelected(MenuItem item){
    …
if(item.getItemId() == MENU_SENSOR)
    {
       Intent intent=new Intent(this,AlbumActivity.class);
        startActivity(intent);
    }
    return super.onOptionsItemSelected(item);
}
```

5.5 完整项目实施

手机相机程序由多个类和 XML 布局文件组成，下面分别对每个类的实现进行详细介绍。

（1）手机相机程序运行的 AndroidManifest 描述：【AndroidManifest.xml】。

```
<?xml version="1.0" encoding="utf-8"?>
<manifest xmlns:android="http://schemas.android.com/apk/res/android"
     package="com.demo.pr5"
     android:versionCode="1"
     android:versionName="1.0">
   <uses-sdk android:minSdkVersion="8" />
   <application android:icon="@drawable/icon"
             android:label="@string/app_name">
      <activity android:name=".CameraActivity"
              android:screenOrientation="landscape"
              android:label="@string/app_name">
         <intent-filter>
           <action android:name="android.intent.action.MAIN" />
           <category
                 android:name="android.intent.category.LAUNCHER"/>
         </intent-filter>
      </activity>
       <activity android:name=".AlbumActivity"></activity>
   </application>
   <!--相机拍照权限 -->
   <uses-permission android:name="android.permission.CAMERA">
   </uses-permission>
   <uses-feature android:name="android.hardware.camera.autofocus"
      android:glEsVersion="4"></uses-feature>
   <uses-feature android:name="android.hardware.camera"
      android:glEsVersion="4"></uses-feature>
   <!--读写文件权限 -->
   <uses-permission
   android:name="android.permission.WRITE_EXTERNAL_STORAGE"/>
   <uses-permission
   android:name="android.permission.MOUNT_UNMOUNT_FILESYSTEMS"/>
</manifest>
```

（2）手机相机程序拍照界面 XML 布局：【cameralayout.xml】。

```
<?xml version="1.0" encoding="utf-8"?>
<LinearLayout
  xmlns:android="http://schemas.android.com/apk/res/android"
   android:orientation="horizontal"
   android:layout_width="fill_parent"
   android:layout_height="fill_parent"
   android:background="@drawable/camera_background"
   >
   <!--相机拍照预览窗口-->
   <SurfaceView android:id="@+id/camera"
      android:layout_width="fill_parent"
```

```
        android:layout_height="fill_parent"
        android:layout_weight="1.0"
        android:layout_marginLeft="-20.0dip"
    />
    <!--照片-->
    <ImageView android:id="@+id/image"
        android:layout_width="fill_parent"
        android:layout_weight="1.0"
        android:layout_height="fill_parent"/>
    <!--快照-->
     <LinearLayout
    android:orientation="horizontal"
    android:layout_width="72.0dip"
    android:layout_height="fill_parent"
    android:layout_gravity="center"
    android:background="@drawable/bg_camera_pattern"
    >
    <ImageView android:id="@+id/shutter"
        android:layout_width="wrap_content"
        android:layout_gravity="center"
        android:src="@drawable/btn_shutter"
        android:clickable="true"
        android:layout_height="wrap_content"/>
   </LinearLayout>
</LinearLayout>
```

（3）手机相机程序拍照处理类：【CameraActivity.java】。

```
package com.demo.pr5;
import java.io.File;
import java.io.FileOutputStream;
import java.util.List;
import android.app.Activity;
import android.content.Intent;
import android.content.pm.ActivityInfo;
import android.graphics.Bitmap;
import android.graphics.BitmapFactory;
import android.graphics.PixelFormat;
import android.hardware.Camera;
import android.hardware.Camera.AutoFocusCallback;
import android.hardware.Camera.Parameters;
import android.hardware.Camera.PictureCallback;
import android.hardware.Camera.ShutterCallback;
import android.hardware.Camera.Size;
import android.os.Bundle;
import android.view.Display;
import android.view.Menu;
import android.view.MenuItem;
import android.view.SurfaceHolder;
import android.view.SurfaceView;
import android.view.View;
import android.view.Window;
import android.view.WindowManager;
import android.view.SurfaceHolder.Callback;
import android.view.View.OnClickListener;
import android.widget.ImageView;

public class CameraActivity extends Activity implements OnClickListener,Callback {

    private SurfaceView   mSurfaceView;//相机视频浏览
    private ImageView     mImageView;  //照片
    private SurfaceHolder mSurfaceHolder;
    private ImageView     shutter;    //快照按钮
    private Camera        mCamera=null;    //相机
    private boolean       mPreviewRunning;//运行相机浏览
    private static final int MENU_START=1;
```

```
private static final int MENU_SENSOR=2;
private Bitmap bitmap;          //照片 Bitmap
@Override
protected void onCreate(Bundle savedInstanceState){
    super.onCreate(savedInstanceState);
    //隐藏标题
    requestWindowFeature(Window.FEATURE_NO_TITLE);
  getWindow().setFlags(WindowManager.LayoutParams.FLAG_FULLSCREEN,
    WindowManager.LayoutParams.FLAG_FULLSCREEN);
    //设置布局文件
    setContentView(R.layout.cameralayout);
    mSurfaceView =(SurfaceView)findViewById(R.id.camera);
    mImageView =(ImageView)findViewById(R.id.image);
    shutter=(ImageView)findViewById(R.id.shutter);
    //设置快照按钮事件
    shutter.setOnClickListener(this);
    mImageView.setVisibility(View.GONE);
    mSurfaceHolder=mSurfaceView.getHolder();
    //设置 SurfaceHolder 回调事件
    mSurfaceHolder.addCallback(this);
    mSurfaceHolder.setType(SurfaceHolder.SURFACE_TYPE_PUSH_BUFFERS);
}
//快照按钮拍照事件
public void onClick(View v){
  //判断是否可以进行拍照
  if(mPreviewRunning){
        shutter.setEnabled(false);
        //设置自动对焦
        mCamera.autoFocus(new AutoFocusCallback() {
            @Override
            public void onAutoFocus(boolean success,Camera camera){
                //聚焦后进行拍照
            mCamera.takePicture(mShutterCallback,null,mPictureCallback);
        }});
    }
}
//相机图片拍照回调函数
PictureCallback mPictureCallback=new PictureCallback() {
    public void onPictureTaken(byte[] data,Camera camera1){
            //判断照片数据是否为空
            if(data != null){
                saveAndShow(data);
            }
    }
};
//快照回调函数
ShutterCallback mShutterCallback=new ShutterCallback() {
    public void onShutter() {
        System.out.println("快照回调函数……");
    }
};
//SurfaceView 改变时调用
public void surfaceChanged(SurfaceHolder holder,int format,int width,
        int height){
   try{
        //判断相机是否在运行,若运行则停止
        if(mPreviewRunning){
            mCamera.stopPreview();
        }
        //启动相机
        mCamera.setPreviewDisplay(holder);
        mCamera.startPreview();
        mPreviewRunning=true;
    } catch(Exception e){
        e.printStackTrace();
    }
```

```
    }
    //Surface 创建时调用
    public void surfaceCreated(SurfaceHolder holder){
        setCameraParams();
    }
    //设置 Camera 参数
    public  void  setCameraParams()
    {
        if(mCamera != null)
        {
            return ;
        }
        //创建并打开相机
        mCamera=Camera.open();
        //设置相机参数
        Parameters params=mCamera.getParameters();
        //拍照时自动对焦
        params.setFocusMode(Parameters.FOCUS_MODE_AUTO);
        //设置预览帧的速率
         params.setPreviewFrameRate(3);
        //设置预览格式
        params.setPreviewFormat(PixelFormat.YCbCr_422_SP);
        //设置图片质量百分比
        params.set("jpeg-quality",85);
        //获取相机支持的图片分辨率
        List<Size> list=params.getSupportedPictureSizes();
        Size size=list.get(0);
        int w=size.width;
        int h=size.height;
        //设置图片大小
        params.setPictureSize(w,h);
        //设置自动闪光灯
        params.setFlashMode(Camera.Parameters.FLASH_MODE_AUTO);
    }

    //SurfaceView 销毁时调用
    public void surfaceDestroyed(SurfaceHolder holder){
        if(mCamera!=null)
        {
            //停止相机预览
            mCamera.stopPreview();
            mPreviewRunning=false;
            //回收相机
            mCamera.release();
            mCamera=null;
        }
    }
    //创建菜单
    public boolean onCreateOptionsMenu(Menu menu){
        menu.add(0,MENU_START,0, "重拍");
        menu.add(0,MENU_SENSOR,0,"打开相册");
        return super.onCreateOptionsMenu(menu);
    }
    //菜单事件
    public boolean onOptionsItemSelected(MenuItem item){
        if(item.getItemId() == MENU_START){
            //重启相机拍照
            setRequestedOrientation(ActivityInfo.SCREEN_ORIENTATION_PORTRAIT);
            setRequestedOrientation(ActivityInfo.SCREEN_ORIENTATION_LANDSCAPE);
            return true;
        }else if(item.getItemId() == MENU_SENSOR)
        {
            Intent intent=new Intent(this,AlbumActivity.class);
               startActivity(intent);
        }
```

```
        return super.onOptionsItemSelected(item);
    }
    //保存和显示图片
    public void  saveAndShow(byte[] data)
    {
        try
        {
            //图片 ID
            String imageId=System.currentTimeMillis()+"";
            //照片保存路径
            String pathName=android.os.Environment.
            getExternalStorageDirectory().getPath()
            +"/com.demo.pr5";
            //创建文件夹
            File file=new File(pathName);
            if(!file.isDirectory())
            {
                file.exists();
            }
            //创建文件
            pathName+="/"+imageId+".jpeg";
            file=new File(pathName);
            if(!file.exists()){
                file.createNewFile();//文件不存在则新建
            }
            FileOutputStream fos=new FileOutputStream(file);
            fos.write(data);
            fos.close();
            AlbumActivity album=new AlbumActivity();
            //读取照片 Bitmap
            bitmap= album.loadImage(pathName);
            //将图片设置在控件上显示
            mImageView.setImageBitmap(bitmap);
            mImageView.setVisibility(View.VISIBLE);
            mSurfaceView.setVisibility(View.GONE);
            //停止相机浏览
            if(mPreviewRunning){
               mCamera.stopPreview();
               mPreviewRunning=false;
            }
            shutter.setEnabled(true);
        }catch(Exception e)
        {
            e.printStackTrace();
        }
    }
}
```

（4）手机相机程序图像浏览布局 XML：【album.xml】。

```
<?xml version="1.0" encoding="utf-8"?>
<LinearLayout
  xmlns:android="http://schemas.android.com/apk/res/android"
            android:orientation="vertical"
            android:layout_width="fill_parent"
            android:layout_height="fill_parent"
            >
    <ViewFlipper android:id="@+id/ViewFlipper01"
              android:layout_width="fill_parent"
              android:layout_height="fill_parent">
    </ViewFlipper>
</LinearLayout>
```

（5）手机相机程序动画定义 XML。

① 定义左进的动画效果文件【/res/anim/push_left_in.xml】。

```
<?xml version="1.0" encoding="utf-8"?>
<set xmlns:android="http://schemas.android.com/apk/res/android">
```

```
    <translate android:fromXDelta="-100%p" android:toXDelta="0"
        android:duration="500" />
    <alpha android:fromAlpha="1.0" android:toAlpha="0.1"
        android:duration="500" />
</set>
```

② 定义左出的动画渐变效果文件【/res/anim/push_left_out.xml】。

```
<?xml version="1.0" encoding="utf-8"?>
<set xmlns:android="http://schemas.android.com/apk/res/android">
    <translate android:fromXDelta="0" android:toXDelta="-100%p"
        android:duration="500" />
    <alpha android:fromAlpha="1.0" android:toAlpha="0.1"
        android:duration="500" />
</set>
```

③ 定义右进的动画效果文件【/res/anim/push_right_in.xml】。

```
<?xml version="1.0" encoding="utf-8"?>
<set xmlns:android="http://schemas.android.com/apk/res/android">
    <translate android:fromXDelta="100%p" android:toXDelta="0"
        android:duration="500" />
    <alpha android:fromAlpha="0.1" android:toAlpha="1.0"
        android:duration="500" />
</set>
```

④ 定义右出的动画渐变效果文件【/res/anim/push_right_out.xml】。

```
<?xml version="1.0" encoding="utf-8"?>
<set xmlns:android="http://schemas.android.com/apk/res/android">
    <translate android:fromXDelta="0" android:toXDelta="100%p"
        android:duration="500" />
    <alpha android:fromAlpha="1.0" android:toAlpha="0.1"
        android:duration="500" />
</set>
```

（6）手机相机程序图像浏览实现类：【AlbumActivity.java】。

```
package com.demo.pr5;
import java.io.File;
import java.util.Vector;
import android.app.Activity;
import android.graphics.Bitmap;
import android.graphics.BitmapFactory;
import android.hardware.Sensor;
import android.hardware.SensorEvent;
import android.hardware.SensorEventListener;
import android.hardware.SensorManager;
import android.os.Bundle;
import android.view.Display;
import android.view.View;
import android.view.ViewGroup;
import android.view.WindowManager;
import android.view.animation.AnimationUtils;
import android.widget.ImageView;
import android.widget.Toast;
import android.widget.ViewFlipper;

public class AlbumActivity extends Activity {
    private ViewFlipper   flipper;
    private Bitmap[]      mBgList;//需要预览的图片列表
    private long          startTime=0;
    private SensorManager sm;    //重力感应硬件控制器
    private SensorEventListener sel;//重力感应侦听
    @Override
    public void onCreate(Bundle savedInstanceState){
        super.onCreate(savedInstanceState);
        setContentView(R.layout.album);
        flipper =(ViewFlipper)this.findViewById(R.id.ViewFlipper01);
        //加载照片
        loadAlbum();
        if(mBgList==null)
```

```
        {
          Toast.makeText(this,"相册无照片",Toast.LENGTH_SHORT).show();
          finish();
          return ;
        }else{
              for(int i=0;i<=mBgList.length-1;i++){
                  flipper.addView(addImage(mBgList[i]),i,new
ViewGroup.LayoutParams(ViewGroup.LayoutParams.FILL_PARENT,
ViewGroup.LayoutParams.FILL_PARENT));
          }
           }
        //获得重力感应硬件控制器
        sm=(SensorManager)this.getSystemService(SENSOR_SERVICE);
        Sensor sensor=sm.getDefaultSensor(Sensor.TYPE_ACCELEROMETER);
        //添加重力感应侦听
        sel=new SensorEventListener(){
            public void onSensorChanged(SensorEvent se){
               float x=se.values[SensorManager.DATA_X];
               //控制甩动的幅度和甩动的频率（在 1s 内只有一次）
               if(x>8&&System.currentTimeMillis()>startTime+1000)
               {
                flipper.setInAnimation(AnimationUtils.
               loadAnimation(AlbumActivity.this,R.anim.push_right_in));
               flipper.setOutAnimation(AnimationUtils.
              loadAnimation(AlbumActivity.this,R.anim.push_left_out));
               flipper.showPrevious();
                //记录右甩动开始时间
               startTime=System.currentTimeMillis();
               }else if(x<-8&&System.currentTimeMillis()>startTime+1000)
               {
              flipper.setInAnimation(AnimationUtils.
                loadAnimation(AlbumActivity.this,R.anim.push_left_in));
                flipper.setOutAnimation(AnimationUtils.
               loadAnimation(AlbumActivity.this,R.anim.push_right_out));
                flipper.showNext();
               //记录左甩动开始时间
             startTime=System.currentTimeMillis();
               }
            }
            public void onAccuracyChanged(Sensor arg0,int arg1){
            }
        };
     //注册 Listener,参数 SENSOR_DELAY_GAME 为检测的精确度
     sm.registerListener(sel,sensor,SensorManager.SENSOR_DELAY_GAME);
    }
    @Override
     protected void onDestroy() {
        super.onDestroy();
        //注销重力感应侦听
        sm.unregisterListener(sel);
     }
     //加载相册
     private  String[] loadAlbum()
     {
     //从本节中照片保存的路径读取照片
        String pathName=android.os.Environment.
        getExternalStorageDirectory().getPath()
        +"/com.demo.pr5";
        File file=new File(pathName);
        Vector<Bitmap> fileName=new Vector<Bitmap>();
        if(file.exists() && file.isDirectory())
        {
            String[] str=file.list();
            for(String s : str)
            {
                if(new File(pathName+"/"+s).isFile())
```

```
            {
             fileName.addElement(loadImage(pathName+"/"+s));
            }
        }
        mBgList=fileName.toArray(new Bitmap[]{});
    }
    return null;
}
//读取照片
public Bitmap loadImage(String pathName){
 //读取照片,并对照片进行缩放
    BitmapFactory.Options options=new BitmapFactory.Options();
    options.inJustDecodeBounds=true;
    //此时返回的bitmap为空
     Bitmap  bitmap=BitmapFactory.decodeFile(pathName,options);
    //获取屏幕的宽度
    WindowManager manage=getWindowManager();
    Display display=manage.getDefaultDisplay();
    //将Bitmap的显示宽度设置为手机屏幕的宽度
    int screenWidth=display.getWidth();
    //计算Bitmap的高度等比变化数值
    options.inSampleSize=options.outWidth/screenWidth;
    //将inJustDecodeBounds设置为false,以便于可以解码为Bitmap文件
    options.inJustDecodeBounds=false;
    //读取照片Bitmap
    bitmap=BitmapFactory.decodeFile(pathName,options);
    return bitmap;
}
    //添加显示图片View
private View addImage(Bitmap  bitmap){
    ImageView img=new ImageView(this);
    img.setImageBitmap(bitmap);
    return img;
}
}
```

5.6 实训项目

一、实现手机录像功能

1. 实训目的与要求

学会利用Camera、MediaRecorder、SurfaceView和SurfaceHoler等类实现手机的录像功能。

2. 实训内容

（1）程序运行后打开录像预览界面，单击【录制】按钮，开始录制视频；单击【暂停】按钮，暂停录制，并保存前面的视频。

（2）在视频保存目录下，单击视频，可以播放视频文件。播放界面上有【暂停】,【向前快进】,【向后快进】按钮。

3. 思考

（1）程序实现手机拍照和手机录像的不同点有哪些？

（2）程序实现视频播放和音乐播放的不同点有哪些？

（3）查阅资料，探索如何使用xml selector 来设置按钮的外观和点击效果？

二、对手机晃动进行检测

1．实训目的与要求

学会利用传感器实现检测手机的晃动程度。

2．实训内容

（1）手机一旦开始晃动，则开始播放一首音乐。

（2）手机晃动的幅度越大，则播放的声音越大。反之，手机晃动的幅度越小，则播放的声音越小。

（3）手机停止晃动，则音乐也随之暂停播放。

3．思考

（1）传感器数据的变化如何体现出手机的晃动程度？

（2）如何编写其他类型的传感器程序？

PART 6 项目六 开发地图应用

本项目工作情景的目标是让学生掌握利用百度地图的主要开发技术。主要的工作任务划分为① 显示百度地图；② 地图的基础应用；③ 实现 POI 查询；④ 实现定位。本项目主要涉及的关键技术包括：百度地图的显示、获取地图上某个点的经纬度、地图覆盖层的显示、POI 搜索、定位等。在本项目中，将大量使用百度地图的 SDK。需要注意的是，本项目的案例需要打开手机的网络连接功能，通过网络访问百度服务器，获得其返回的地理数据，但未必每次都会很快从服务器中得到响应。因此在关于地图的应用上，除了需要更多一些耐心之外，还需要多考虑对各种可能的异常情况的处理。

6.1 显示百度地图

一、任务分析

本任务需要实现的效果如图 6-1 所示。

本任务要求打开地图，并在地图上显示出当前的位置。地图上的数据需要通过网络来获取，通常需要花费一定的时间。要完成本次任务，需要思考如下两个问题。

（1）如何在 Android 上配置百度地图 API？

（2）如何在 Android 中打开一幅百度地图？

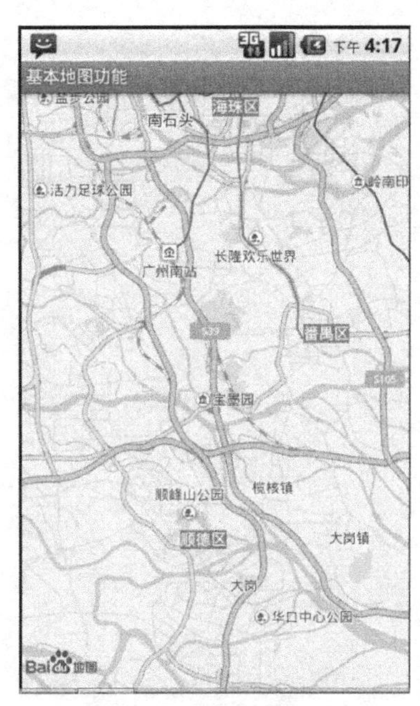

图 6-1　百度地图效果图

二、相关知识

1．百度地图 Android SDK 的下载

开发者可在百度地图 Android SDK 的下载页面下载最新版的地图 SDK，下载地址为。http://developer.baidu.com/map/sdkandev- download.htm（见图 6-2），选择全部下载。

2．开发密钥

在使用百度地图 SDK 之前，开发者需要获取百度地图移动版的开发密钥，该密钥与开发者的百度账户相关联。因此，必须先有百度帐户（如果没有，则在 http://passport.baidu.com/网址上注册），才能获得开发密钥，具体流程如下。

（1）单击图 6-2 页面右上角的【API 控制台】，进入应用列表页面（见图 6-3）。

图 6-2　百度地图 Android SDK 的下载页面

图 6-3　开发者创建的百度应用列表

（2）单击【创建应用】按钮，输入应用名称、选择应用类型为 Android SDK，以及输入安全码（见图 6-4）。

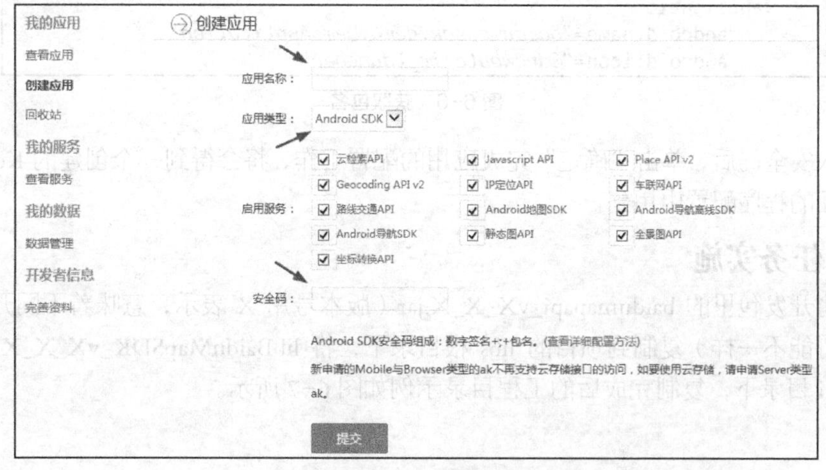

图 6-4　创建百度应用

其中，安全码的组成规则为 Android 签名证书的 SHA1 值+“;”+ packagename，即数字签名+分号+包名。下面分别介绍如何获取 SHA1 值和包名。

①在 Eclipse 编辑器的【Windows】→【Preferences】→【Android】→【Build】下查看 SHA1，如图 6-5 所示。

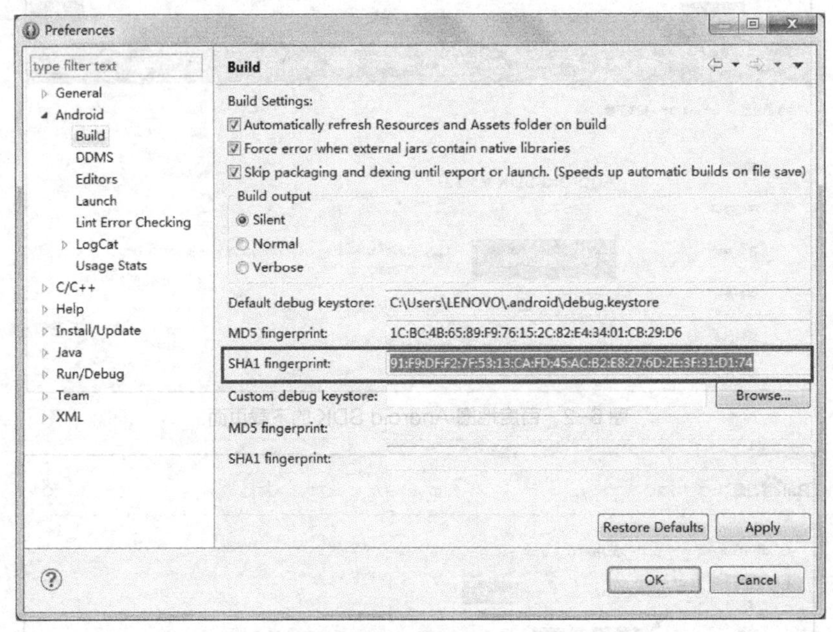

图 6-5　获取 SHA1

②在 Android 应用程序的 AndroidManifest.xml 中获得项目定义的包名，如图 6-6 所示。

```
<?xml version="1.0" encoding="utf-8"?>
<manifest xmlns:android="http://schemas.android.com/apk/res/android"
    package="baidumapsdk.demo"
    android:versionCode="1"
    android:versionName="1.0" >

    <uses-sdk
        android:minSdkVersion="7"
        android:targetSdkVersion="20" />
    <application
        android:name="baidumapsdk.demo.DemoApplication"
        android:icon="@drawable/ic_launcher"
```

图 6-6　获取包名

在输入安全码后，单击“确定”完成应用的配置工作，将会得到一个创建的 Key，该 Key 将会在后面的程序配置中用到。

三、任务实施

（1）将开发包里的 baidumapapi_vX_X_X.jar（版本号用 X 表示，意味着不同开发者下载的版本号可能不一样）复制到项目的 libs 根目录下，将 libBaiduMapSDK_vX_X_X.so 复制到 libs\armeabi 目录下，复制完成后的工程目录示例如图 6-7 所示。

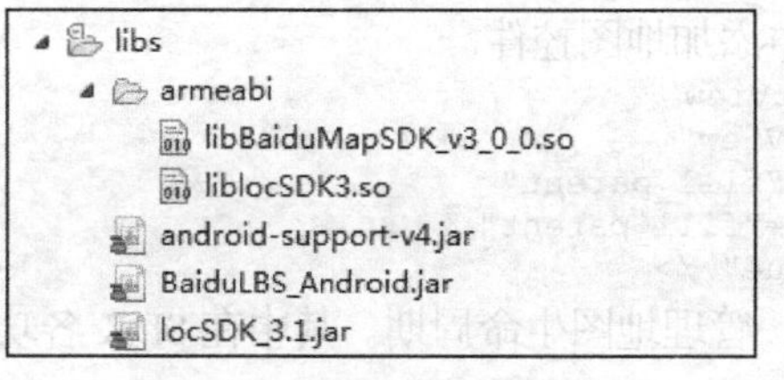

图 6-7 百度地图开发库配置

（2）选中项目，单击右键，选择 Properties，在 Java Build Path->Libraries 中选择“Add JARs...”，选定上一步的 baidumapapi_vX_X_X.jar，确定后返回，如图 6-8 所示。

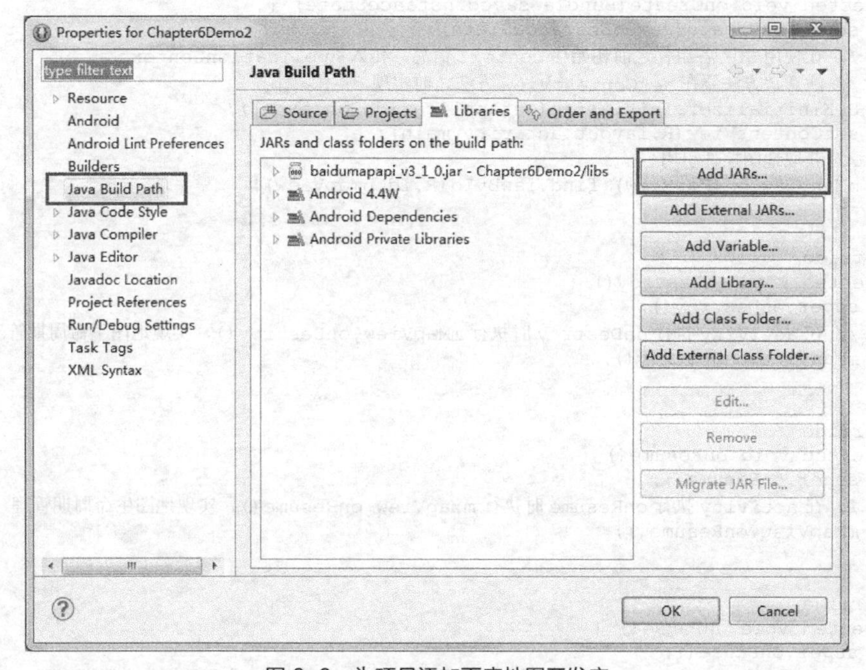

图 6-8 为项目添加百度地图开发库

（3）在 AndroidManifest 中添加开发密钥。注意：<meta-data>元素是作为<application>的子元素。

```
<application>
   <meta-data
      android:name="com.baidu.lbsapi.API_KEY"
      android:value="开发者 key" />
</application>
```

（4）在 AndroidManifest 中添加所需权限。注意：<uses-permission>元素与<application>是平行的，即在 AndroidManifest 文件中的位置是同一个层次的。

```
<uses-permission android:name="android.permission.GET_ACCOUNTS" />
<uses-permission android:name="android.permission.USE_CREDENTIALS" />
<uses-permission android:name="android.permission.MANAGE_ACCOUNTS" />
<uses-permission android:name="android.permission.AUTHENTICATE_ACCOUNTS" />
<uses-permission android:name="android.permission.ACCESS_NETWORK_STATE" />
<uses-permission android:name="android.permission.INTERNET" />
<uses-permission android:name="com.android.launcher.permission.READ_SETTINGS" />
<uses-permission android:name="android.permission.CHANGE_WIFI_STATE" />
<uses-permission android:name="android.permission.ACCESS_WIFI_STATE" />
<uses-permission android:name="android.permission.READ_PHONE_STATE" />
<uses-permission android:name="android.permission.WRITE_EXTERNAL_STORAGE" />
<uses-permission android:name="android.permission.BROADCAST_STICKY" />
<uses-permission android:name="android.permission.WRITE_SETTINGS" />
```

（5）在布局 xml 文件中添加地图控件。

```
<com.baidu.mapapi.map.MapView
    android:id="@+id/bmapView"
    android:layout_width="fill_parent"
    android:layout_height="fill_parent"
    android:clickable="true" />
```

（6）创建地图 Activity，管理地图生命周期。其中在 SDK 各功能组件使用之前都需要调用 SDKInitializer.initialize(Context context)方法。

```
public class MainActivity extends ActionBarActivity {
MapView mMapView;

@Override
protected void onCreate(Bundle savedInstanceState) {
     super.onCreate(savedInstanceState);
     // 在使用 SDK 各组件之前初始化 context 信息，传入 ApplicationContext
     // 注意该方法要在 setContentView 方法之前实现
     SDKInitializer.initialize(getApplicationContext());
     setContentView(R.layout.activity_main);
     // 获取地图控件引用
     mMapView = (MapView) findViewById(R.id.bmapView);
}

@Override
protected void onDestroy() {
     super.onDestroy();
     // 在 activity 执行 onDestroy 时执行 mMapView.onDestroy()，实现地图生命周期管理
     mMapView.onDestroy();
}

@Override
protected void onResume() {
     super.onResume();
     // 在 activity 执行 onResume 时执行 mMapView.onResume()，实现地图生命周期管理
     mMapView.onResume();
}

@Override
protected void onPause() {
     super.onPause();
     // 在 activity 执行 onPause 时执行 mMapView.onPause()，实现地图生命周期管理
     mMapView.onPause();
}
}
```

6.2 地图基础应用

一、任务分析

本任务需要实现的效果如图 6-9 所示。

百度地图提供了丰富的地图操作功能。本任务将实现基础的地图操作，例如设置地图的类型、打开指定经纬度的地图、获取地图上某个点的经纬度、在地图上描绘文字、标注覆盖物等。此外，由于百度地图的应用需要依托网络的支持，所以需要提供检测手机数据网络是否打开的功能。

图 6-9 地图基础应用

二、相关知识

1．系统服务

服务（Service）是 Android 系统中的重要组件，服务可

以在不显示界面的情况下在后台运行指定的任务或者在两个不同进程之间进行通信。

Android 在后台运行着许多服务，这些服务在系统启动时被开启，支持系统的正常工作。例如，来电显示服务。在编程时通过调用系统服务可以便捷地实现系统功能，提高编程的效率。

getSystemService（String name）是 Activity 的一个重要方法，根据传入的服务名称来获得系统级的服务。系统服务可以被看作一个对象，getSystemService 方法返回的是一个 Object 对象类型，在实际使用中需要进行强制类型转换。参数 name 表示系统服务的 ID，在 android.content.Context 类中定义了常用的 Android 系统服务如表 6-1 所示。

表 6-1 Android 系统服务

name	返回的对象类型	说 明
ACCESSIBILITY_SERVICE	AccessibilityManager	可访问服务
ACCOUNT_SERVICE	AccountManager	账户服务
ACTIVITY_SERVICE	ActivityManager	活动服务，管理应用程序的系统状态
ALARM_SERVICE	AlarmManager	闹钟服务
AUDIO_SERVICE	AudioManager	音频服务
CLIPBOARD_SERVICE	ClipboardManager	剪切板服务
CONNECTIVITY_SERVICE	ConnectivityManager	网络连接服务
DEVICE_POLICY_SERVICE	DevicePolicyManager	设备管理策略服务
DROPBOX_SERVICE	DropBoxManager	记录诊断日志服务
INPUT_METHOD_SERVICE	InputMethodManager	输入法服务
KEYGUARD_SERVICE	KeyguardManager	键盘保护服务
WINDOW_SERVICE	WindowManager	窗口服务
LAYOUT_INFLATER_SERVICE	LayoutInflater	布局映射服务，取得 XML 布局文件
LOCATION_SERVICE	LocationManager	位置服务，如 GPS
NOTIFICATION_SERVICE	NotificationManager	状态栏的通知服务
POWER_SERVICE	PowerManager	管理控制电源
SEARCH_SERVICE	SearchManager	搜索服务
SENSOR_SERVICE	SensorManager	传感器服务
TELEPHONY_SERVICE	TelephonyManager	电话服务
UI_MODE_SERVICE	UiModeManager	UI 模式服务
VIBRATOR_SERVICE	Vibrator	手机震动服务
WALLPAPER_SERVICE	WallpaperService	墙纸服务
WIFI_SERVICE	WifiManager	Wi-Fi 服务
DOWNLOAD_SERVICE	DownloadManager	下载服务

要掌握系统服务的详细使用方法需要先了解服务对象所提供的变量和方法，下面举例说明。

（1）剪切板服务。在下面的例子中，Android 系统中的所有文本输入框都可以通过系统的剪切板获得"设置剪切板中的信息"这段文本内容。

```
ClipboardManager
clipboardManager=(ClipboardManager)getSystemService(Context.CLIPBOARD_ SERVICE);
clipboardManager.setText("设置剪切板中的信息");
```

（2）窗口服务。在下面的例子中，窗口的标题栏将输出当前窗口的宽度和高度。

```
WindowManager windowManager=(WindowManager)getSystemService(Context.WINDOW_SERVICE);
setTitle(String.valueOf(windowManager.getDefaultDisplay().getWidth())+"*"+String.valu
eOf(windowManager.getDefaultDisplay().getHeight()));
```

（3）布局映射服务。下面的代码将打开布局文件，获得布局文件中的控件。

```
LayoutInflater inflater =(LayoutInflater)getSystemService(LAYOUT_INFLATER_SERVICE);
View layout=inflater.inflate(R.layout.main,null);
    EditText et =(EditText)layout.findViewById(R.id.edittext);
```

（4）移动数据网络服务。下面的代码将诊断移动数据网络是否已经打开，其中 bis ConnFlag 变量的值表示移动数据网络是否打开。

```
boolean bisConnFlag=false;
    ConnectivityManager conManager =
    (ConnectivityManager)getSystemService(Context.CONNECTIVITY_SERVICE);
NetworkInfo network = conManager.getActiveNetworkInfo();
if(network!=null){
    bisConnFlag=conManager.getActiveNetworkInfo().isAvailable();
}
```

2. Point 类

Point 类是 Android 图像包中的一个类，表示包含两个整数 x，y 的点坐标，可用于表示屏幕或者空间上某个点的位置。

3. MapView 类

MapView 类是一个显示百度地图的视图，它负责从服务端获取地图数据。它将会捕捉屏幕触控手势事件。与 Activity 类似，该类也具有生命周期。可以将它的 onResume()、onPause()、onDestroy()等方法放在 Activity 的相应方法中进行调用。在使用地图组件之前请确保已经调用了 SDKInitializer.initialize(Context) 函数以提供全局 Context 信息。百度提供了在布局 xml 文件中添加 MapView 地图控件的方法，请参考 6.1 节任务实施部分的第 5 步。在程序中可通过 findViewById()方法来获取 MapView 对象。

4. BaiduMap 类

BaiduMap 类定义百度地图对象的操作方法与接口，可以理解为地图控制器。可以通过 MapView 类的 getMap()方法获取 BaiduMap 对象。

下面介绍 BaiduMap 类的主要监听接口。

（1）OnMapClickListener：地图单击事件监听接口。设置地图单击事件监听者的方法是：setOnMapClickListener(BaiduMap.OnMapClickListener listener)。

（2）OnMapDoubleClickListener：地图双击事件监听接口。设置地图双击事件监听者的方法是：setOnMapDoubleClickListener(BaiduMap.OnMapDoubleClickListener listener)。

（3）OnMapLoadedCallback：地图加载完成回调接口。设置地图加载完成回调的方法是：setOnMapLoadedCallback(BaiduMap.OnMapLoadedCallback callback)。

（4）OnMapLongClickListener：地图长按事件监听接口。设置地图长按事件监听者的方法是：setOnMapLongClickListener(BaiduMap.OnMapLongClickListener listener)。

（5）OnMapStatusChangeListener：地图状态改变相关接口。设置地图状态监听者的方法是：setOnMapStatusChangeListener(BaiduMap.OnMapStatusChangeListener listener)。

（6）OnMarkerClickListener：地图 Marker 覆盖物点击事件监听接口。设置该点击事件监听者的方法是：setOnMarkerClickListener(BaiduMap.OnMarkerClickListener listener)。

（7）OnMarkerDragListener：地图 Marker 覆盖物拖拽事件监听接口。设置 Marker 拖拽事件监听者的方法是：setOnMarkerDragListener(BaiduMap.OnMarkerDragListener listener)。

（8）OnMyLocationClickListener：地图定位图标点击事件监听接口。设置该点击事件监听者的方法是：setOnMyLocationClickListener(BaiduMap.OnMyLocationClickListener listener)。

下面介绍 BaiduMap 类的主要方法。

（1）Overlay addOverlay(OverlayOptions options)：向地图添加一个 Overlay。

（2）void animateMapStatus(MapStatusUpdate update)：以动画方式更新地图状态。

（3）void clear()：清空地图所有的 Overlay 覆盖物以及 InfoWindow。

（4）MyLocationConfiguration getLocationConfigeration()：获取定位图层配置信息。

（5）MyLocationData getLocationData()：获取定位数据。

（6）MapStatus getMapStatus()：获取地图的当前状态。

（7）int getMapType()：获取地图当前的模式，有普通地图或者卫星图两种。

（8）float getMaxZoomLevel()：获取地图最大缩放级别。

（9）float getMinZoomLevel()：获取地图最小缩放级别。

（10）Projection getProjection()：获取地图投影坐标转换器。当地图初始化完成之前返回 null，在 OnMapLoadedCallback.onMapLoaded() 之后才能正常。

（11）UiSettings getUiSettings()：获取地图 UI 控制器。

（12）void hideInfoWindow()：隐藏当前 InfoWindow。

（13）void setBuildingsEnabled(boolean enabled)：设置是否允许楼块效果。

（14）void setMapStatus(MapStatusUpdate update)：改变地图状态。

（15）void setMapType(int type)：设置地图类型。参数 MAP_TYPE_NORMAL 为普通图；参数 MAP_TYPE_SATELLITE 为卫星图。

（16）void setMaxAndMinZoomLevel(float max, float min)：设置地图最大以及最小缩放级别，地图支持的最大最小级别分别为[3-19]。

（17）void setMyLocationConfigeration(MyLocationConfiguration configeration)：设置定位图层配置信息，只有先允许定位图层后设置定位图层配置信息才会生效，参见 setMyLocationEnabled(boolean)。

（18）void setMyLocationData(MyLocationData data)：设置定位数据，只有先允许定位图层后设置数据才会生效，参见 setMyLocationEnabled(boolean)。

（19）void setMyLocationEnabled(boolean enabled)：设置是否允许定位图层。

5．LatLng 类

LatLng 类是表示地理坐标的基本数据结构。它是一个不可变类，用于表示一对经纬度值。不可变类是指获得这个类的一个实例引用时，不可以改变实例的内容。即实例一旦创建，其内在成员变量的值就不能再被修改。

LatLng 类的构造方法是：LatLng(double latitude, double longitude)。其中，参数 latitude 和 longitude 分别表示纬度和经度坐标。

地图上某个点的经纬度，可以通过百度地图的拾取系统查询地理坐标，网址是：http://api.map.baidu.com/lbsapi/getpoint/index.html。

6．Marker 类

Marker 类是抽象类 Overlay 的子类，用来定义地图的覆盖物，即标注一个点位置的可见元素。可以通过 BaiduMap 类的 addOverlay ()方法生成 Marker 对象。每个标注自身都包含地理信息。比如在西单商场位置添加一个标注，不论地图移动、缩放，标注都会跟随一起移动，保证其始终指向正确的地理位置，如图 6-10 所示。

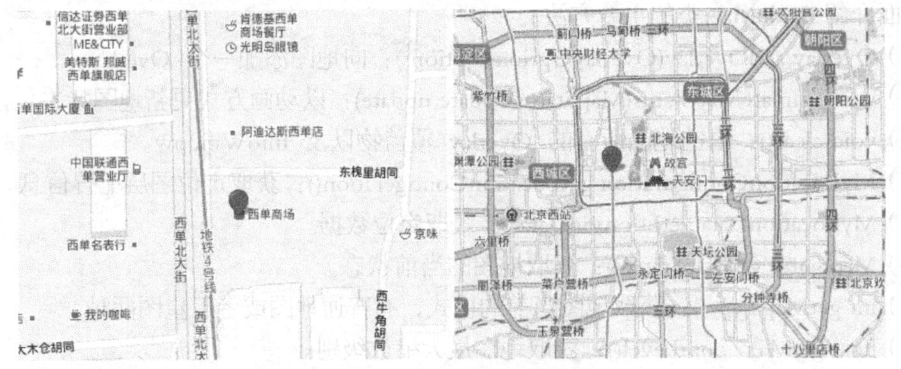

图 6-10　Marker 效果示例

7. MarkerOptions

MarkerOptions 是抽象类 OverlayOptions 的子类，用来表示 Marker 覆盖物的属性，例如，覆盖物用的图标、覆盖物的位置等。

下面介绍 MarkerOptions 类的主要方法。

（1）BitmapDescriptor getIcon()：获取 Marker 覆盖物的图标。

（2）LatLng getPosition()：获取 Marker 覆盖物的位置坐标。

（3）String getTitle()：获取 Marker 覆盖物的标题。

（4）MarkerOptions icon(BitmapDescriptor icon)：设置 Marker 覆盖物的图标，相同图案的 icon 的 Marker 最好使用同一个 BitmapDescriptor 对象以节省内存空间。

（5）MarkerOptions position(LatLng position)：设置 Marker 覆盖物的位置坐标。

（6）MarkerOptions title(java.lang.String title)：设置 Marker 覆盖物的标题。

（7）MarkerOptions visible(boolean visible)：设置 Marker 覆盖物的可见性。

8. InfoWindow

InfoWindow 类是在地图中显示一个信息窗口。可以设置一个 View 作为该窗口的内容，也可以设置一个 BitmapDescriptor 作为该窗口的内容。它有名为 OnInfoWindowClickListener 的信息窗口点击事件监听接口，可用于实现用户点击信息窗口的响应。

它有两种构造方法，格式分别如下。

①public InfoWindow(View view, LatLng position, OnInfoWindowClickListener listener)。其中 view 表示 InfoWindow 展示的控件；position 表示 InfoWindow 显示的地理位置；listener 表示 InfoWindow 点击监听者。

②public InfoWindow(BitmapDescriptor bd, LatLng position, OnInfoWindowClickListener listener)。其中参数 bd 表示 InfoWindow 展示的 bitmap，另外两个参数和第一种构造方法一样。

三、任务实施

接着 6.1 节的任务实施代码来实现更改地图类型、更改地图中心、标注覆盖物、标注文字和获取地图上某个点的经纬度、检测手机数据网络是否打开等功能。

（1）定义 5 个功能菜单项。对于获取地图上某个点经纬度的功能，单击屏幕即可以实现，所有不需要专门设置一个菜单项来响应。

```
public boolean onCreateOptionsMenu(Menu menu) {
        menu.add(1, 1, 1, "打开网络配置界面");
        menu.add(1, 2, 1, "更改地图类型");
        menu.add(1, 3, 1, "标注覆盖物");
        menu.add(1, 4, 1, "标注文字");
        menu.add(1, 5, 1, "更改地图中心");
        return true;
    }
```

（2）在菜单项上定义响应的方法。

```
public boolean onOptionsItemSelected(MenuItem item) {
        int id = item.getItemId();
        switch (id) {
        case 1:
            showNetwork(); //显示配置网络界面
            break;
        case 2:
            setMapType(); // 更改地图类型
            break;
        case 3:
            setOverlay(); // 标注覆盖物
            break;
        case 4:
            setMapText(); // 标注文字
            break;
        case 5:
            setCenter(); // 更改地图中心
            break
        }
        return super.onOptionsItemSelected(item);
    }
```

（3）定义显示配置网络界面的方法。

```
//显示配置网络界面
public void showNetwork(){
    if (!isConn()) {
            openSetting();
        } else {
            Toast.makeText(this, "网络已经打开", Toast.LENGTH_LONG).show();
        }
    }
    //检测网络是否打开
    public boolean isConn() {
        boolean bisConnFlag = false;
        ConnectivityManager         conManager          =           (ConnectivityManager)
getSystemService(Context.CONNECTIVITY_SERVICE);
        NetworkInfo network = conManager.getActiveNetworkInfo();
        if (network != null) {
            bisConnFlag = conManager.getActiveNetworkInfo().isAvailable();
        }
        return bisConnFlag;
    }
    //打开网络配置界面
    public void openSetting() {
        Intent intent = null;
        // 判断手机系统的版本 其中API大于10 就是3.0或以上版本
        if (android.os.Build.VERSION.SDK_INT > 10) {
            intent = new Intent(
                    android.provider.Settings.ACTION_WIRELESS_SETTINGS);
        } else {
            intent = new Intent();
            ComponentName component = new ComponentName("com.android.settings",
                    "com.android.settings.WirelessSettings");
            intent.setComponent(component);
            intent.setAction("android.intent.action.VIEW");
```

```
        }
        startActivity(intent);
    }
```

（4）定义更改地图类型的方法。

```
public void setMapType() {
        if (mBaiduMap.getMapType() == BaiduMap.MAP_TYPE_SATELLITE) {
            // 普通地图
            mBaiduMap.setMapType(BaiduMap.MAP_TYPE_NORMAL);
        } else {
            // 卫星地图
            mBaiduMap.setMapType(BaiduMap.MAP_TYPE_SATELLITE);
        }
    }
```

（5）定义标注覆盖物的方法。

```
public void setOverlay() {
        // 定义 Maker 坐标点
        LatLng point = new LatLng(39.963175, 116.400244);
        // 构建 Marker 图标
        BitmapDescriptor bitmap = BitmapDescriptorFactory
                .fromResource(R.drawable.ic_launcher);
        // 构建 MarkerOption，用于在地图上添加 Marker
        OverlayOptions option = new MarkerOptions().position(point).icon(bitmap);
        // 在地图上添加 Marker，并显示
        mMarkerA = (Marker) mBaiduMap.addOverlay(option);
    }
```

（6）为覆盖物添加监听，即当点击标注覆盖物时，会显示提示，若点击提示可以改变标注覆盖物的位置。

```
mBaiduMap.setOnMarkerClickListener(new OnMarkerClickListener() {
            public boolean onMarkerClick(final Marker marker) {
                Button button = new Button(getApplicationContext());
                button.setBackgroundResource(R.drawable.popup);
                final LatLng ll = marker.getPosition();
                Point p = mBaiduMap.getProjection().toScreenLocation(ll);
                p.y -= 47;
                LatLng llInfo = mBaiduMap.getProjection().fromScreenLocation(p);
                OnInfoWindowClickListener listener = null;
                if (marker == mMarkerA) {
                    button.setText("更改位置");
                    button.setTextColor(Color.BLACK);
                    listener = new OnInfoWindowClickListener() {
                        public void onInfoWindowClick() {
                            LatLng llNew = new LatLng(ll.latitude + 0.005,
                                    ll.longitude + 0.005);
                            marker.setPosition(llNew);
                            mBaiduMap.hideInfoWindow();
                        }
                    };
                }
                mInfoWindow = new InfoWindow(button, llInfo, listener);
                mBaiduMap.showInfoWindow(mInfoWindow);
                return true;
            }
        });
```

（7）定义标注文字的方法。

```
public void setMapText() {
        // 定义文字所显示的坐标点
        LatLng llText = new LatLng(39.86923, 116.397428);
        // 构建文字 Option 对象，用于在地图上添加文字
        OverlayOptions textOption = new TextOptions().bgColor(0xAAFFFF00)
                .fontSize(24).fontColor(0xFFFF00FF).text("百度地图 SDK").rotate(-30)
                .position(llText);
        // 在地图上添加该文字对象并显示
        mBaiduMap.addOverlay(textOption);
    }
```

（8）定义更改地图中心的方法。

```
public void setCenter() {
        LatLng cenpt = new LatLng(22.906, 113.310);
        // 定义地图状态
        MapStatus mMapStatus = new MapStatus.Builder().target(cenpt).zoom(18)
                .build();
        // 定义 MapStatusUpdate 对象，以便描述地图状态将要发生的变化
        MapStatusUpdate mMapStatusUpdate = MapStatusUpdateFactory
                .newMapStatus(mMapStatus);
        // 改变地图状态
        mBaiduMap.setMapStatus(mMapStatusUpdate);
    }
```

（9）定义点击获取地图上某个点获取其经纬度的方法。

```
mBaiduMap.setOnMapClickListener(new OnMapClickListener() {
            @Override
            public void onMapClick(LatLng point) {
                // 这里的 point 就是点击屏幕所获得的经纬度
                // 下面两个函数可以将经纬度转换为屏幕的点坐标，另一个是将屏幕点转换为经纬度
                // mBaiduMap.getProjection().fromScreenLocation(point);
                // mBaiduMap.getProjection().toScreenLocation(latlng);
                String hint = "纬度：" + String.valueOf(point.latitude) + ",经度"
                        + String.valueOf(point.longitude);
                Toast.makeText(MainActivity.this, hint, Toast.LENGTH_LONG).show();
             }

            @Override
            public boolean onMapPoiClick(MapPoi arg0) {
                return false;
            }
        });
```

6.3 实现 POI 查询

一、任务分析

本任务需要实现的效果如图 6-11 和图 6-12 所示。

程序可以搜索出某个城市特定类型的地点，例如餐厅、酒店、医院，用图标分别标注在地图上的相应位置。用户单击图标还可以进一步查看地点的详细信息。要完成本次任务，需要思考如下两个问题。

（1）如何在百度地图中查找感兴趣的地点？

（2）如何查看感兴趣地点的详细信息？

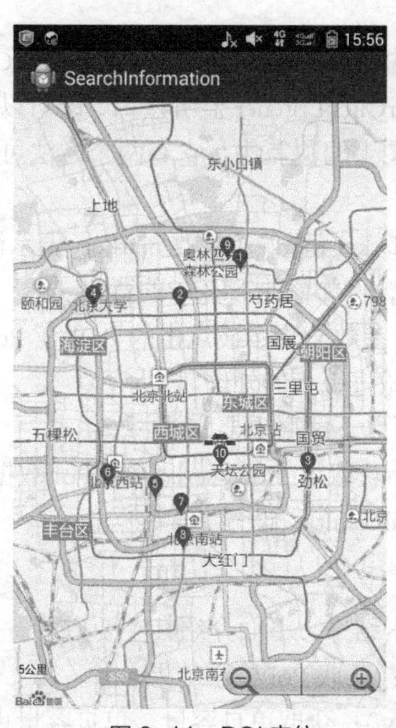

图 6-11 POI 定位

图 6-12 POI 详情

二、相关知识

1. POI

POI（Point of Interest），中文可以翻译为“兴趣点”。在地理信息系统中，一个 POI 可以是一栋房子、一个商铺、一个邮筒、一个公交站等。 每个 POI 包含四方面信息：名称、类别、经度纬度、附近的酒店饭店商铺等信息。我们可以叫它为“导航地图信息”，导航地图数据是整个导航产业的基石。

百度地图 SDK 提供三种类型的 POI 检索：周边检索、区域检索和城市内检索。

2. FragmentActivity

FragmentActivity 由 android-support-v4.jar 包（该包提供了 API 库函数的 JAR 文件，使得在旧版本的 Android 能使用一些新版本的 APIs）提供。它是 Activity 的子类，提供对片段（Fragment）的支持，提供操作 Fragment 的方法。当需要动态多界面切换时，传统的方法是通过在各种 Activity 中进行跳转来实现多界面的跳转和单个界面动态改变。在 Android 3.0 或以上系统中使用 Fragment 类来达到这个效果。在 FragmentActivity 中可以包括多个 Fragment。Fragment 支持在不同的 Activity 中使用并且可以处理自己的输入事件以及生命周期方法等，可以将它看做是一个子 Activity。

本次任务的一个需求是在地图上搜索出兴趣点后，用户点击兴趣点，可以打开新的页面来进一步显示它的有关信息，所以可以应用 Fragment 技术。

百度地图开发包提供了 com.baidu.mapapi.map.SupportMapFragment 类，用于管理百度地图上的 Fragment，开发者在程序中直接使用即可。它提供了一个非常有用的方法 getBaiduMap() 来获取百度地图 BaiduMap 对象。

3. PoiOverlay 类

PoiOverlay 类用于显示 POI 的 Overly。它的构造方法是：PoiOverlay(BaiduMap baiduMap)。

它实现了 BaiduMap.OnMarkerClickListener 监听接口，即可以把 PoiOverlay 对象作为监听器，用于监听用户是否点击了地图上的覆盖物。它有两个方法可以用于处理点击，差别在参数的含义。

（1）boolean onMarkerClick(Marker marker):参数 marker 表示被点击的 Marker 对象。在该方法内可以对 Marker 对象做操作，例如修改 Maker 的背景。

（2）boolean onPoiClick(int i)：参数 i 表示被点击的 POI 在 PoiResult.getAllPoi()中的索引。在该方法内可以获取 POI 数据，做进一步的处理。

此外，PoiOverlay 类还提供了以下两个重要方法。

（1）PoiResult getPoiResult()：获取该 PoiOverlay 的 Poi 数据。

（2）void setData(PoiResult poiResult)：设置 Poi 数据。

4. PoiResult 类

PoiResult 类用来表示 Poi 搜索结果。下面介绍其主要方法。

（1）List<PoiInfo> getAllPoi()：获取 Poi 检索结果。

（2）int getCurrentPageCapacity()：获取单页容量。单页容量可以通过检索参数指定。

（3）int getCurrentPageNum()：获取当前分页编号。

（4）int getSuggestCityList()：返回建议城市列表。

（5）int getTotalPageNum()：获取总分页数。

（6）int getTotalPoiNum()：获取 Poi 总数。

5．PoiSearch 类

PoiSearch 类是百度地图 POI 的检索接口。下面介绍其主要方法。

（1）static PoiSearch newInstance()：创建 PoiSearch 实例。

（2）boolean searchInBound(PoiBoundSearchOption option)：范围内检索。

（3）boolean searchInCity(PoiCitySearchOption option)：城市内检索。

（4）boolean searchNearby(PoiNearbySearchOption option)：周边检索。

（5）boolean searchPoiDetail(PoiDetailSearchOption option)：对 POI 进行详情检索。若在代码中使用详情检索，需要在 AndroidManifest.xml 文件中声明 PlaceCaterActivity。

（6）void setOnGetPoiSearchResultListener(OnGetPoiSearchResultListener listener)：设置 POI 检索监听者。

5．PoiCitySearchOption 类

PoiCitySearchOption 类用于设置 POI 城市内检索参数。主要方法如下。

（1）PoiCitySearchOption city(String city)：指定检索的城市。

（2）PoiCitySearchOption keyword(String key)：搜索关键字。

（3）PoiCitySearchOption pageCapacity(int pageCapacity)：设置每页容量，默认为每页 10 条。

（4）PoiCitySearchOption pageNum(int pageNum)：查询第几页。pageNum 表示分页编号。

三、任务实施

实现 POI 查询，首先在布局文件中使用 Fragment 元素，然后定义一个 FragmentActivity 的子类，在子类中实现 POI 的搜索和显示。在 AndroidManifest 中添加百度提供的用于支持点击 POI，可以打开详细信息页面的 Activity。具体实现如下。

（1）定义布局文件【activity_poisearch.xml】，在布局文件中定义一个<fragment>元素，类名为"com.baidu.mapapi.map.SupportMapFragment"。

```
 <LinearLayout xmlns:android="http://schemas.android.com/apk/res/android"
   android:layout_width="fill_parent"
   android:layout_height="fill_parent"
   android:orientation="vertical" >
   <fragment
      android:id="@+id/map"
      android:layout_width="match_parent"
      android:layout_height="match_parent"
      class="com.baidu.mapapi.map.SupportMapFragment" />
</LinearLayout>
```

（2）定义一个 FragmentActivity 的子类，并实现 OnGetPoiSearchResultListener 监听接口。通过 getSupportFragmentManager().findFragmentById(R.id.map))).getBaiduMap()获得地图对象，这里的 R.id.map 是第（1）步布局文件中的<fragment>元素的 ID。

```
public class POISearch extends FragmentActivity implements
        OnGetPoiSearchResultListener {
    private PoiSearch mPoiSearch = null;
    BaiduMap mBaiduMap;
    @Override
    protected void onCreate(Bundle savedInstanceState) {
        super.onCreate(savedInstanceState);
        // 在使用 SDK 各组件之前初始化 context 信息，传入 ApplicationContext
```

```
        // 注意该方法要在 setContentView 方法之前实现
        SDKInitializer.initialize(getApplicationContext());
        setContentView(R.layout.activity_poisearch);
        // 获取地图控件引用
        mBaiduMap = ((SupportMapFragment) (getSupportFragmentManager()
                .findFragmentById(R.id.map))).getBaiduMap();
        // 初始化搜索模块，注册搜索事件监听
        mPoiSearch = PoiSearch.newInstance();
        mPoiSearch.setOnGetPoiSearchResultListener(this);
    }

    @Override
    protected void onDestroy() {
        mPoiSearch.destroy();
        super.onDestroy();
    }
    }
```

（3）定义菜单项和菜单点击操作。

```
@Override
    public boolean onCreateOptionsMenu(Menu menu) {
        menu.add(1, 1, 1, "查找");
        return true;
    }
@Override
    public boolean onOptionsItemSelected(MenuItem item) {
        mPoiSearch.searchInCity((new PoiCitySearchOption()).city("北京")
                .keyword("餐厅").pageNum(0));
        return super.onOptionsItemSelected(item);
    }
```

（4）定义 PoiOverlay 的子类，用于响应用户点击 Overlay 的操作。

```
private class MyPoiOverlay extends PoiOverlay {
        public MyPoiOverlay(BaiduMap baiduMap) {
            super(baiduMap);
        }
        @Override
        public boolean onPoiClick(int index) {
            super.onPoiClick(index);
            PoiInfo poi = getPoiResult().getAllPoi().get(index);
            if (poi.hasCaterDetails) {
                mPoiSearch.searchPoiDetail((new PoiDetailSearchOption())
                        .poiUid(poi.uid));
            }
            return true;
        }
    }
```

（5）在 OnGetPoiSearchResultListener 接口，实现在获得 POI 信息后的处理方法。

```
public void onGetPoiResult(PoiResult result) {
        if (result == null
                || result.error == SearchResult.ERRORNO.RESULT_NOT_FOUND) {
            return;
        }
        if (result.error == SearchResult.ERRORNO.NO_ERROR) {
            mBaiduMap.clear();
            PoiOverlay overlay = new MyPoiOverlay(mBaiduMap);
            mBaiduMap.setOnMarkerClickListener(overlay);
            overlay.setData(result);
            overlay.addToMap();
            overlay.zoomToSpan();
            return;
        }
        if (result.error == SearchResult.ERRORNO.AMBIGUOUS_KEYWORD) {
            // 当输入关键字在本市没有找到，但在其他城市找到时，返回包含该关键字信息的城市列表
            String strInfo = "在";
            for (CityInfo cityInfo : result.getSuggestCityList()) {
                strInfo += cityInfo.city;
                strInfo += ",";
```

```
            }
            strInfo += "找到结果";
            Toast.makeText(POISearch.this, strInfo, Toast.LENGTH_LONG).show();
        }
}
```

（6）在 OnGetPoiSearchResultListener 接口，实现在获得 POI 详细信息后的处理方法。

```
public void onGetPoiDetailResult(PoiDetailResult result) {
    if (result.error != SearchResult.ERRORNO.NO_ERROR) {
            Toast.makeText(POISearch.this, "抱歉，未找到结果", Toast.LENGTH_SHORT)
                    .show();
    } else {
            Toast.makeText(POISearch.this, "成功，查看详情页面", Toast.LENGTH_SHORT)
                    .show();
    }
}
```

（7）在 AndroidManifest 中添加百度提供的用于支持点击 POI，可以打开详细信息页面的 PlaceCaterActivity。若没有该 activity 声明，点击地图上的 POI，将不会有响应。

```
<activity
    android:name="com.baidu.mapapi.search.poi.PlaceCaterActivity"
    android:configChanges="orientation|keyboardHidden"
    android:label="@string/demo_name_poi"
    android:screenOrientation="sensor" >
</activity>
```

6.4　实现定位

一、任务分析

本任务需要实现的效果如图 6-13 所示。

程序可以将用户当前所在的位置定位出来，并显示在地图上。要完成本次任务，需要思考如下两个问题。

（1）百度地图提供了怎样的定位技术？

（2）现有各种定位技术的差异？

图 6-13　定位

二、相关知识

百度定位 SDK 提供了 GPS+基站+WIFI+IP 混合定位功能，传感器辅助定位，定位方式可自由切换，自动给出精度最好的定位结果。基站定位根据运营商的覆盖情况，精度达到 100 米～300 米；WI-FI 定位则能实现 30 米～200 米的精度；GPS 定位能实现 15 米～20 米的精度。下面简要介绍 GPS 定位和基站定位的主要原理，以及关键的实现类。

1．GPS 定位原理

GPS（Global Position System），即全球定位系统，是 20 世纪 70 年代由美国陆海空三军联合研制的新一代空间卫星导航定位系统。

24 颗 GPS 卫星在离地面 22 000km 的高空上，以 12h 为周期环绕地球运行，使得在任意时刻，地面的任意一点都可以同时观测到 4 颗以上的卫星，在任何天气情况下随时随地获取可靠的位置信息。

GPS 定位易受周围环境的影响，并非一下就能定位成功，因此需要在程序代码中做好各种防范措施。建议在室外空旷的地方测试 GPS 是否定位成功，避免在有遮挡的地方测试，在室内很有可能无法定位成功。

2．基站原理

GPS 定位虽然比较准确，但不能在室内获取数据，在这种情况下，可以借助于移动通信网络的基站进行定位。基站定位的主要原理是：手机在开机状态下，会向一个附近合适的移动基站注册。每一个移动基站都有自己地理位置信息，可以将其粗略看作手机的地理位置信息。国内大城市市区的基站密度可以达到 500 m 以下或者更低，所以能够大体上确定用户的位置。但如果在农村地区，基站的密度较疏，可能达到几千米远，这样精确度就会比较低。

3．LocationClient 类

LocationClient 类是定位服务的客户端，是百度地图提供定位服务的一个重要类。它的构造方法有两种。

（1）LocationClient(Context context)

（2）LocationClient(Context context, LocationClientOption locOption)

下面介绍它的主要方法。

（1）BDLocation getLastKnownLocation()：返回最近一次定位结果。

（2）void registerLocationListener(BDLocationListener listener)：注册定位监听，通过该监听可以获得最新的当前位置信息。

（3）void setLocOption(LocationClientOption locOption)：设置 LocationClientOption。

（4）void start()：开始定位。

（5）void stop()：停止定位。

4．LocationClientOption 类

LocationClientOption 类用于设置定位的相关参数。它的属性 LocationMode 表示定位模式，分为高精度定位模式、低功耗定位模式、仅设备定位模式。其中：

- 高精度定位模式会同时使用网络定位和 GPS 定位，优先返回最高精度的定位结果；
- 低功耗定位模式不会使用 GPS，只会使用网络定位（Wi-Fi 和基站定位）；
- 仅用设备定位模式不需要连接网络，只使用 GPS 进行定位，这种模式下不支持室内环境的定位。

LocationClientOption 类有两种构造方法。

（1）LocationClientOption()。

（2）LocationClientOption(LocationClientOption lOption)。

下面介绍它的主要方法。

（1）void setCoorType(String coorType)：设置坐标类型。参数 coorType 取值有 3 个，分别是：gcj02（国测局经纬度坐标系）；bd09（百度墨卡托坐标系）；bd09ll（百度经纬度坐标系）。

（2）void setLocationMode(LocationClientOption.LocationMode mode)：设置定位模式。参数 mode 的取值有 3 个：Battery_Saving（低功耗模式）；Device_Sensors（仅 GPS 设备模式）；Hight_Accuracy（高精度模式）。

（3）void setOpenGps(boolean openGps)：是否打开 GPS 进行定位。

（4）setScanSpan(int scanSpan)：设置扫描间隔，单位是毫秒。即设置定位的时间间隔，能够设置的最小间隔为 1000 毫秒。

5．BDLocationListener 接口

BDLocationListener 接口用于进行定位监听，提供定位请求回调方法。onReceiveLocation (BDLocation location)。

6．MyLocationData 类

MyLocationData 类表示设置当前用户的定位信息，其包含的信息要比 BDLocation 要少。下面介绍其主要方法。

（1）MyLocationData.Builder()：构造器。

（2）MyLocationData.Builder accuracy(float accuracy)：设置定位数据的精度信息，单位：米。

（3）MyLocationData build()：构建生成定位数据对象。

（4）MyLocationData.Builder direction(float direction)：设置定位数据的方向信息

（5）MyLocationData.Builder latitude(double lat)：设置定位数据的纬度。

（6）MyLocationData.Builder longitude(double lng): 设置定位数据的经度。

7．BDLocation 类

BDLocation 类用于表示获取或者设置定位信息。下面列出其主要获取定位信息的方法。

（1）String getAddrStr()：获取详细地址信息。

（2）String getCity()：获取城市。

（3）String getDistrict()：获取区/县信息。

（4）double getLatitude()：获取纬度坐标。

（5）double getLongitude()：获取经度坐标。

（6）int getOperators()：获取运营商信息。

（7）String getProvince()：获取省份。

（8）float getRadius()：获取定位精度。

三、任务实施

（1）定义一个 Activity 的子类，在 onCreate 方法初始化定位服务的有关客户端参数，开启客户端定位。

```
public class MyLocationActivity extends Activity {
    MapView mMapView;
    BaiduMap mBaiduMap;
    LocationClient mLocClient;
    public MyLocationListenner myListener = new MyLocationListenner();
    boolean isFirstLoc = true;//变量表示是否首次定位
    @Override
    protected void onCreate(Bundle savedInstanceState) {
        super.onCreate(savedInstanceState);
        // 注意该方法要在 setContentView 方法之前实现
        SDKInitializer.initialize(getApplicationContext());
        setContentView(R.layout.activity_main);
        // 获取地图控件引用
        mMapView = (MapView) findViewById(R.id.bmapView);
        mBaiduMap = mMapView.getMap();
        // 开启定位图层
        mBaiduMap.setMyLocationEnabled(true);
        // 定位初始化
        mLocClient = new LocationClient(this);
        mLocClient.registerLocationListener(myListener);
        LocationClientOption option = new LocationClientOption();
        option.setOpenGps(true);// 打开 gps
```

```
        option.setCoorType("bd09ll"); // 设置坐标类型
        option.setScanSpan(1000);
        mLocClient.setLocOption(option);
        mLocClient.start();
    }
    @Override
    protected void onDestroy() {
        // 退出时销毁定位
        mLocClient.stop();
        // 关闭定位图层
        mBaiduMap.setMyLocationEnabled(false);
        mMapView.onDestroy();
        mMapView = null;
        super.onDestroy();
    }
}
```

（2）定义 BDLocationListener 监听器，用于定位监听。

```
public class MyLocationListenner implements BDLocationListener {
    @Override
    public void onReceiveLocation(BDLocation location) {
        // map view 销毁后不在处理新接收的位置
        if (location == null || mMapView == null)
            return;
    //定义定位数据
        MyLocationData locData = new MyLocationData.Builder()
                .accuracy(location.getRadius())
                // 此处设置开发者获取到的方向信息，顺时针 0-360
                .direction(100).latitude(location.getLatitude())
                .longitude(location.getLongitude()).build();
        //在地图上设置定位数据，只有先允许定位图层后设置数据才会生效
        mBaiduMap.setMyLocationData(locData);
        if (isFirstLoc) {
            isFirstLoc = false;
            LatLng ll = new LatLng(location.getLatitude(),
                    location.getLongitude());
        //MapStatusUpdate 类用于描述地图状态将要发生的变化。
            MapStatusUpdate u = MapStatusUpdateFactory.newLatLng(ll);
            mBaiduMap.animateMapStatus(u);
        }
    }
    public void onReceivePoi(BDLocation poiLocation) {
    }
}
```

（3）在 AndroidManifest 中添加百度提供"com.baidu.location.f"服务。百度地图 API 规定如果要使用定位服务，就需要在 Manifest 之中注册这个 Service。

```
<service
    android:name="com.baidu.location.f"
    android:enabled="true"
    android:process=":remote" >
</service>
```

（4）在 AndroidManifest 中添加定位的相关权限。

```
<uses-permission android:name="android.permission.WAKE_LOCK"/>
<permission android:name="android.permission.BAIDU_LOCATION_SERVICE"/>
<uses-permission android:name="android.permission.BAIDU_LOCATION_SERVICE"/>
<uses-permission android:name="android.permission.ACCESS_COARSE_LOCATION"/>
<uses-permission android:name="android.permission.ACCESS_MOCK_LOCATION"/>
<uses-permission android:name="android.permission.ACCESS_FINE_LOCATION"/>
<uses-permission android:name="android.permission.ACCESS_GPS" />
<uses-permission android:name="android.permission.READ_CONTACTS" />
<uses-permission android:name="android.permission.CALL_PHONE" />
<uses-permission android:name="android.permission.READ_SMS" />
<uses-permission android:name="android.permission.SEND_SMS" />
```

6.5 实训项目

一、完善百度地图应用程序的开发

1. 实训目的与要求

深入掌握百度地图应用程序的相关开发技术。

2. 实训内容

(1)完善 POI 查找功能,用户在用户界面上输入感兴趣的信息,点击搜索可以查询出结果。

(2)在界面上输入查找地址,单击查找按钮,地图将定位到相应地址的位置。

(3)实现公交路线的查询功能。

(4)实现导航功能。

(5)实现地名和经纬度的转换。

3. 思考

(1)位置服务(LBS,Location Based Services)的应用场景有哪些?

(2)其他类型的地图开发方法,例如高德地图、谷歌地图?

二、GPS 定位器

1. 实训目的与要求

通过查阅资料,学会利用 Android 自带的 Location Provider、Location、LocationManager、LocationListener 等类和接口来进行 GPS 定位。

2. 实训内容

要求:在程序运行后,界面上能够显示用户的经纬度信息以及对应的地址信息,并且位置信息能够随着用户的移动而变化,效果如图 6-14 所示。

3. 思考

(1)如何使用百度地图 SDK 技术来实现同样的效果?

(2)如何根据地点的经纬度,获得地点的详细地址信息?

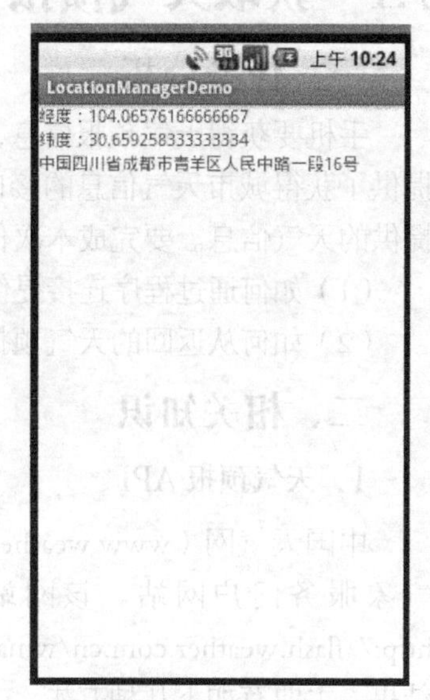

图 6-14 GPS 定位示例

PART 7 项目七 开发天气预报程序

本项目工作情景的目标是让学生掌握利用 Android 的网络技术获取数据。主要的工作任务划分为① 获取天气预报信息；② 下载天气图片；③ 显示天气预报。本项目主要涉及的关键技术包括：网络连接、网络数据和图片读取、XML 数据解析、JSON 数据解析等。

7.1 获取天气预报信息

一、任务分析

手机要获得天气预报信息，首先要获得天气预报信息的来源。目前有一些网站为程序员提供了获得城市天气信息的接口。本任务是通过网络连接访问外部的天气预报网站，获取它提供的天气信息。要完成本次任务，需要思考如下两个问题。

（1）如何通过程序连接提供天气预报接口的网站？

（2）如何从返回的天气预报信息中抽取程序所需的数据？

二、相关知识

1．天气预报 API

中国天气网（www.weather.com.cn）是中国气象局面向社会和公众、以公益性为基础的气象服务门户网站。该网站提供了天气预报信息调用接口，只需要在 URL 网址 http://flash.weather.com.cn/wmaps/xml/填写相应的查询条件，即可返回 XML 格式的天气预报结果。主要有如下几种方法。

（1）列出全国所有省的省会天气信息。

http://flash.weather.com.cn/wmaps/xml/china.xml

（2）以地方所在城市或者省会名称的汉语拼音作为查询条件。

查找北京市各区的天气：

http://flash.weather.com.cn/wmaps/xml/beijing.xml

查找广东省各市的天气：

http://flash.weather.com.cn/wmaps/xml/guangdong.xml

同时，中国天气网提供了天气信息图片的接口：

http://m.weather.com.cn/img/c0.gif

http://m.weather.com.cn/img/b0.gif

http://www.weather.com.cn/m/i/weatherpic/29x20/d0.gif

http://www.weather.com.cn/m2/i/icon_weather/29x20/n0.gif

天气预报结果 XML 中使用属性 state1 和 state2 表示天气情况，若取值为 0 表示天晴，其他天气情况的取值依此类推。以字母 c 开头的图片大小是 20 像素 × 20 像素，以字母 b 开头的图片大小是 50 像素 × 46 像素，以字母 d 开头的反白图片大小是 29 × 20 像素大小，以字母 n 开头的夜间反白图标大小是 29 × 20 像素。

2．XML 知识

下面以中国天气网返回的北京天气预报 XML 数据为例进行说明。

```
<beijing dn="nay">
    <city cityX="264.5" cityY="300.3" cityname="海淀" centername="海淀" fontColor="FFFFFF" pyName="haidian" state1="0" state2="1" stateDetailed="晴转多云" tem1="19" tem2="27" temNow="22" windState="微风" windDir="南风" windPower="1 级" humidity="81%" time="21:45" url="101010200" />
    <city cityX="344.3" cityY="317.65" cityname="朝阳" centername="朝阳" fontColor="FFFFFF" pyName="chaoyang" state1="0" state2="1" stateDetailed="晴转多云" tem1="19" tem2="27" temNow="22" windState="微风" windDir="南风" windPower="1 级" humidity="81%" time="21:45" url="101010300" />
    <city cityX="255" cityY="341.5" cityname="石景山" centername="石景山" fontColor="FFFFFF" pyName="shijingshan" state1="0" state2="1" stateDetailed="晴转多云" tem1="19" tem2="27" temNow="22" windState="微风" windDir="南风" windPower="1 级" humidity="81%" time="21:45" url="101011000" />
    <city cityX="310.05" cityY="339.3" cityname="市中心" centername="市中心" fontColor="FFFF00" pyName="shizhongxin" state1="0" state2="1" stateDetailed="晴转多云" tem1="19" tem2="27" temNow="22" windState="微风" windDir="南风" windPower="1 级" humidity="81%" time="21:45" url="101010100" />
  ......//北京其他区天气信息
  </beijing>
```

XML 文档的第一行是天气预报的城市，beijing 是文档的根元素。

从 XML 文档的第二行开始表示每个区的天气预报信息，cityX、citY、cityname、stateDetailed 等都是 city 元素的属性。下面对程序常用的属性进行举例说明。

- state1="0"：天气图标 1。
- state2="1"：天气图标 2。
- stateDetailed="晴转多云"：天气说明。
- tem1="19"：最高气温。
- tem2="27"：最低气温。
- temNow="22"：当前实况气温。
- windState="微风"：风向风速预报。
- windDir="南风"：当前实况风向。
- windPower="1 级"：当前实况风速。
- humidity="81%"：湿度。
- time="21:45"：天气预报发布时间。

3．XmlPullParser

为了更好地获得 XML 文档中标签的内容，可以借助两种编程模型对 XML 文档进行分析：一种是流模式，另一种是文档对象模型（DOM）方式。其中，DOM 方式将所有的 XML 数据以树状结构加载到内存中，这种方式适用于随机访问数据的操作，但不适用于大型文件；

流模式是基于事件的方式，即当需要数据时即开始分析，并不要求读取所有数据。流模式较适用于对速度要求高或者硬件资源条件低的场合。流模式处理又分 Pull 和 Push 两种，其中 Pull 依靠应用程序主动请求获得所需要的数据，以 Pull 方式通过读取 XML 回调相应的事件；而 Push 则是依靠解析器来发送数据。

XML Pull 解析器是一个定义解析功能的接口，基于事件对 XML 文档进行分析。

- 读取到 XML 文档的开始标签将返回：START_DOCUMENT 事件。
- 读取到 XML 文档的结束标签将返回：END_DOCUMENT 事件。
- 读取到 XML 元素的开始标签将返回：START_TAG 事件。
- 读取到 XML 元素的结束标签将返回：END_TAG 事件。
- 读取到 XML 的元素文本内容将返回：TEXT 事件。

例如，

```
<title ****></title>
```

其中，可将<titile>理解为 START_TAG，将</title>理解为 END_TAG。

定义一个 XmlPullParser 对象有两种方法。

（1）通过 XmlPullParserFactory 工厂构建 XmlPullParser 对象。

```
XmlPullParserFactory factory=XmlPullParserFactory.newInstance();
factory.setNamespaceAware(true);
XmlPullParser xpp=factory.newPullParser();
```

（2）通过 XML 类的 newPullParser 方法生成一个 XmlPullParser 对象。

```
XmlPullParser xpp=Xml.newPullParser();
```

XmlPullParser 接口主要的方法如下。

（1）int getAttributeCount()：返回元素属性的数量，如果当前事件类型不是 START_TAG，则返回-1。

（2）String getAttributeValue（int index）：根据索引 index 返回元素相应的属性值。参数 index 从 0 开始。如果当前事件类型不是 START_TAG 或者索引越界，将会抛出异常。

（3）String getAttributeValue（String namespace，String name）：根据给定的命名空间和属性名称，返回相应的属性值。其中，参数 namespace 表示属性的命名空间，如果不存在，则参数设置为 null；参数 name 表示属性名称。

（4）int getEventType ()：返回当前事件的类型，例如 START_DOCUMENT、END_DOCUMENT、START_TAG、END_TAG、TEXT 等。

（5）String getName()：如果是 START_TAG 或者 END_TAG 事件，获取当前元素的名字。

（6）String getText ()：返回当前事件的文本内容。返回的内容依赖于当前事件的类型，如对于 TEXT 事件，则返回元素的文本内容。

（7）boolean isWhitespace()：检查当前 TEXT 事件是否只包含空白字符。

（8）void setInput（Reader in）：设置解析器将要处理的输入流。执行该方法，将会重置解析器的状态，并且设置事件类型为 START_DOCUMENT。参数 in 表示读取 XML 文档的输入流。

（9）void setInput（InputStream inputStream，String inputEncoding）：设置解析器将要处理的输入流。执行该方法，将会重置解析器的状态，并且设置事件类型为 START_DOCUMENT。其中，参数 inputStream 表示读取 XML 文档的输入流；参数 inputEncoding 表示输入流的编码。

（10）int next()：获得下一个解析事件，会产生 START_TAG、END_DOCUMENT、START_TAG、END_TAG、TEXT 等事件。

（11）void require（int type，String namespace，name）：用于测试 XML 解析器的当前事件是否与所声明的类型不一致，若不一致将会报错。该方法有助于程序员检查 XML 文档结构是否与设想的一致。

（12）int nextTag()：调用 next()方法并返回事件，但要求返回事件的必须是 START_TAG 或者 END_TAG，否则会抛出异常。

（13）String nextText()：如果当前事件是 START_TAG，则返回当前元素的文本内容；若下一个事件是 END_TAG，则返回空字符串，否则会抛出异常。

Pull 解析过程就是一个文档遍历过程，每次调用 next()、nextTag()、nextToken()和 nextText()都会向前推进文档（注意：不能倒退），产生 Event，并使解析器停留在某些事件，然后再处理 Event。

下面举例说明如何使用上述方法来完成对 XML 文档的遍历。

```
public class XmlPullParserDemo extends Activity {
    @Override
    public void onCreate(Bundle savedInstanceState){
        super.onCreate(savedInstanceState);
        try {
          parse();
      } catch(XmlPullParserException e){
          e.printStackTrace();
      } catch(IOException e){
          e.printStackTrace();
      }
    }
    public  void parse ()
            throws XmlPullParserException,IOException
        {
        /*
            XmlPullParserFactory factory=XmlPullParserFactory.newInstance();
            factory.setNamespaceAware(true);
            XmlPullParser xpp=factory.newPullParser();
            */
         XmlPullParser xpp=Xml.newPullParser();//获得 XmlPullParser 解析器
            xpp.setInput(new StringReader("<title>Hello World!</title>"));
            int eventType=xpp.getEventType();
            while(eventType != XmlPullParser.END_DOCUMENT){
             if(eventType == XmlPullParser.START_DOCUMENT){
                 System.out.println("Start document");
             } else if(eventType == XmlPullParser.START_TAG){
                 System.out.println("Start tag "+xpp.getName());
             } else if(eventType == XmlPullParser.END_TAG){
                 System.out.println("End tag "+xpp.getName());
             } else if(eventType == XmlPullParser.TEXT){
                 System.out.println("Text "+xpp.getText());
             }
             eventType=xpp.next();
            }
            System.out.println("End document");
        }
}
```

执行上述程序将会输出：

```
Start document
 Start tag title
 Text Hello World!
 End tag title
 End document
```

从上面的代码可以看出，基于事件对 XML 文档进行分析，通常是在一个 while 循环内，使用 XmlPullParser 解析器的 next()来对 XML 文档进行遍历，通过 switch 或者 if 语句来对当前的事件进行判断，然后做相应的处理。

4．输入输出流

Android 网络应用涉及数据的发送和接收，将会使用到 Java 的 I/O 技术。下面简单介绍 Java 中 I/O 流的重要概念和相关类。

Java 用流的概念来管理输入输出（I/O）。当程序需要读取数据的时候，就会开启一个通向数据源的流，这个数据源可以是文件、内存或网络连接。类似地，当程序需要写入数据的时候，就会开启一个通向目的地的流。这时候可以想象数据好像在其中“流”动一样。

数据若在输入输出过程中以字节（byte，8 位单字节）为单位，则称为字节流；若在输入输出过程中以字符（char，16 位双字节）为单位，则称为字符流。

不管是对字节流还是字符流，Java 提供了多种多样的类来进行管理，提供了读入或者写入流的方法。其中 InputSteream 和 OutpurStream 分别用于基于字节输入和输出的父类（见图 7-1（a））；Reader 和 Writer 则分别用于所有读写字符数据流的父类（见图 7-1（b））。

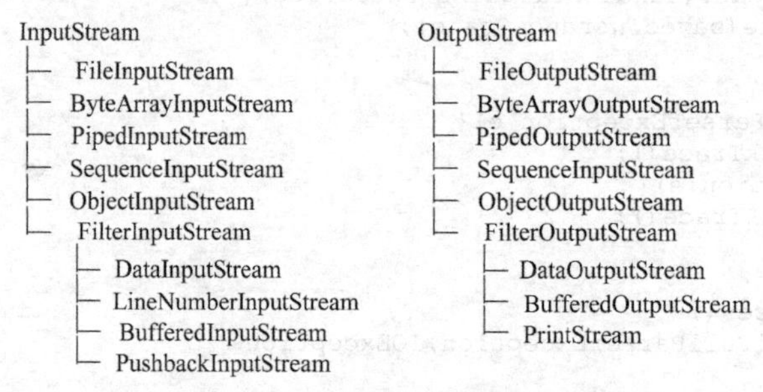

（a） 基于字节输入和输出的父类

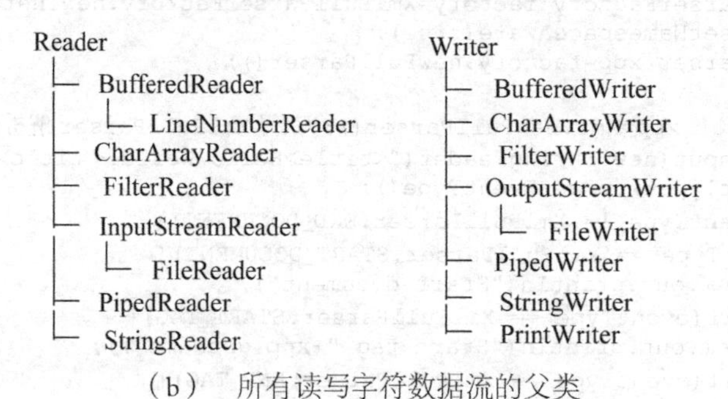

（b） 所有读写字符数据流的父类

图 7-1 字节流、字符流输入与输出类层次图

下面选取其中的一些类进行举例说明。

ByteArrayOutputStream 类的作用是，使用 write()方法以流的形式将需要输出的数据写入到字节数组当中，可以通过使用 toByteArray()方法返回其中的字节数组。

ByteArrayInputStream 类的作用是，创建一个字节类型的数组缓冲区，通过 read()方法以流的形式从字节数组中读取字节，即将内容读进内存中。定义 ByteArrayInputStream 对象，先在构造方法中传入一个字节数组，然后通过该输入流逐个或者多个字节读取出需要的信息。

```
int a=1;
int b=2;
int c=3;
int r;
ByteArrayOutputStream bout=new ByteArrayOutputStream();
bout.write(a);
bout.write(b);
```

```
bout.write(c);
byte[] buff=bout.toByteArray();
Log.v("提示","以byte数组形式输出");
for(int i=0; i<buff.length; i++){
  Log.v("buf["+i+"]",String.valueOf(buff[i]));
}
Log.v("提示","通过ByteArrayInputStream读取输出");
ByteArrayInputStream bin=new ByteArrayInputStream(buff);
while((r=bin.read())!=-1){
      Log.v("r=",String.valueOf(r));
}
```

DataOutputStream 类允许写入不同类型的数据（例如，整型 int 是 4 个字节，UTF 的每个字符是 2 个字节），并可以相应地转换成合适类型的字节，提供写入的方法有 writeBoolean()、writeByte()、writeChar()、writeChars()、writeDouble()、writeFloat()、writeInt()、writeLong()、writeShort()、writeUTF()等。

DataInputStream 类允许读入不同类型的数据，与 DataOutputStream 类相对应，提供的读取方法有 readBoolean()、readByte()、readChar()、readDouble()、readFloat()、readInt()、readLong()、readShort()、readUnsignedByte()、readUTF()等。

需要注意的是，使用 DataOutputStream 类进行数据写入与使用 DataInputStream 类进行数据读取一般是一一对应的，即如果使用 writeInt()方法写入整型数据，那么应该采用 readInt()方法进行读取。

FileInputStream 和 FileOutputStream 类分别用于与文件有关的输入和输出字节流操作。而 FileReader 和 FileWriter 类则分别用于与文件有关的输入和输出字符流操作。ObjectInputStream 和 ObjectOutputStream 类则是用于读取或写入 Java 对象的字节流。

ObjectInputStream 和 ObjectInputStream 类用于对象数据流的处理，称为对象输入流和对象输出流，它们所读写的对象必须实现 Serializable 接口。在网络传输中，通常需要将一个对象从一方传输到另外一方，这时可以使用 ObjectInputStream 和 ObjectInputStream 类。

字节流和字符流都有关于缓冲处理的类，为了提高性能，可以将其他输入输出流对象作为参数放入缓冲类中。缓冲对于读取或者写入大块的数据比较方便，尤其是在读取或者写入文件的时候。其中，BufferedInputStream 是有关读入字节流的处理类，提供 read()方法读取一个或者多个字节；BufferedOutputStream 是有关输出字节流的处理类，提供 write()方法输出一个或者多个字节。BufferedReader 是有关读入字符流的处理类，除了提供 read()方法读取一个或者多个字符之外，还提供 readLine()方法，能方便地从文件中读取一行数据；BufferedWriter 是有关输出字符流的处理类，提供 write()方法输出一个或者多个字符。

InputStreamReader 就像是把字节流转换为字符流的一座桥，它读取字节流，并使用指定的字符集将读取的字节流转换为字符流。常见的字符集有：GBK、ISO-8859-1、UTF-8。OutputStreamWriter 类则是字符流通向字节流的桥梁，使用指定的字符集将向其写入的字符编码为字节。举例说明如下。

```
BufferedReader in=new BufferedReader(new InputStreamReader(System.in));
String strLine;
while(strLine=in.readline()!=null)
{
  Log.v("输出",strLine);
}
in.close();
```

上述代码的含义为将标准输入流转换为字符流，并缓存到 BufferedReader 的缓冲区中，再将缓冲区中的数据显示在 Android 的日志上。

5. 安装 Tomcat 服务器

Tomcat 服务器是一款开源的 Web 应用服务器，由 Apache 软件基金会负责管理。可登录网站 http://tomcat.apache.org/进行下载。主页与下载页面分别如图 7-2 和图 7-3 所示。

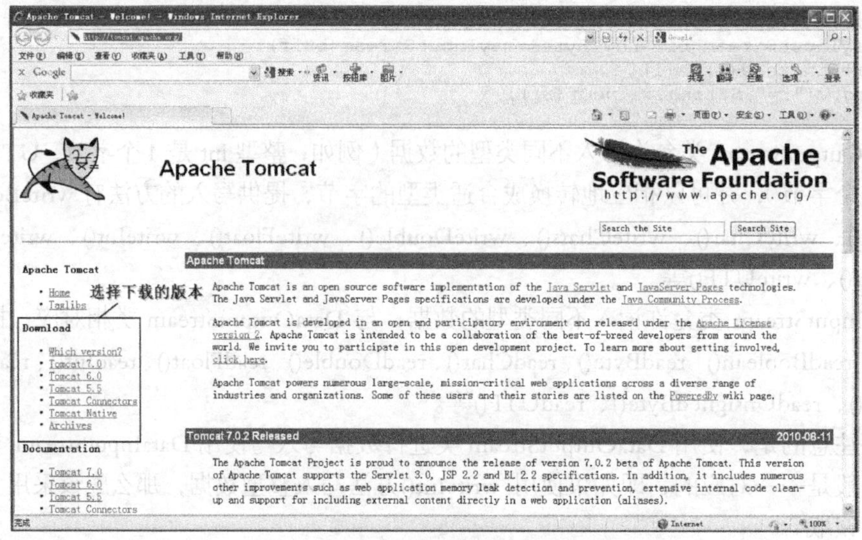

图 7-2　Tomcat 主页

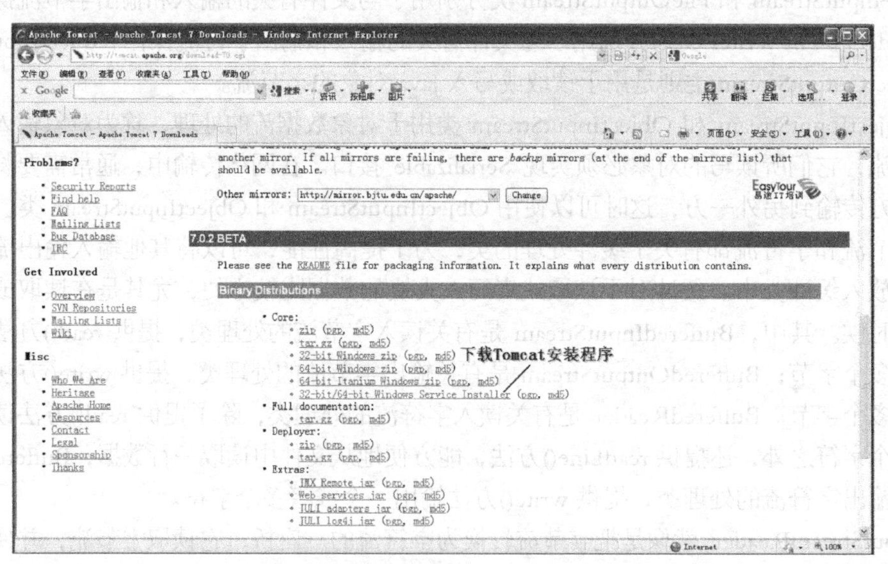

图 7-3　Tomcat 下载页面

下载完成后，直接双击安装文件即可进入 Tomcat 服务器安装界面，如图 7-4 所示。

需要注意是，在安装成功后启动 Tomcat 服务器可能会报 8080 端口被占用的错误。这是因为 Tomcat 服务器默认的端口为 8080，但如果事先已经有程序占用了该端口，而相同的端口号不允许被不同的程序使用，因此会报错使得 Tomcat 无法正常使用。解决方法有两种：一种是将启动 8080 端口的进程关闭；另一种是修改 Tomcat conf 目录下的【service.xml】文件，将 <Connector port="8080" protocol="HTTP/1.1"..../> 中的 8080 修改为其他端口号，如 8088。

打开 Tomcat 服务器文件的方法是：http://domain:端口号/目录，其中 domain 为 Tomcat 服务器的域名，一般是将需要发布的 Web 文件目录作为 Tomcat 服务器 webapps 下的子目录。

下面介绍测试 Tomcat 是否安装成功的方法。

（1）在 IE 浏览器地址栏输入：http://localhost:8080/，如果能够获得 Apache 的帮助，则表示已成功安装 Tomcat 服务器。

（2）验证是否能够正常显示 JSP 程序。

① 在 Tomcat 的安装目录 webapps 下，有 ROOT、examples、examples 等 Tomcat 自带的目录。在该目录下新建一个目录，命名为 myapp。

② 在 myapp 下新建一个测试用的 JSP 页面，文件名为【index.jsp】，文件内容如下：

图 7-4 安装 Tomcat 服务器

```
<html><body><center>
The time is: <%=new java.util.Date()%>
</center></body></html>
```

③ 打开浏览器，在地址栏输入 http://localhost:8080/myapp/index.jsp，若看到当前系统时间则说明 Tomcat 已安装成功。

6．连接网站

Android的网络连接大致可分为两种：一种是基于Socket的网络连接；另一种是基于HTTP协议的网络连接。Socket 通信方式连接一般适用于实时要求比较高的应用，例如聊天、即时战斗的网游等；HTTP 通信方式较为普遍，服务器端有较多现成的应用可以使用，开发起来速度较快。不管使用哪种方式，均需要在配置文件中添加网络访问权限，否则程序会报错。

```
<uses-permission android:name="android.permission.INTERNET">
</uses-permission>
```

（1）基于 Socket 的用法。

Socket 由 IP 地址和端口号两部分组成。IP 地址用来定位设备，端口号用来定位应用程序或者进程，比如常见的是运行在 80 端口上的 HTTP 协议。建立 Socket 连接需要一对套接字，其中一个运行客户端，另一个运行服务器端。套接字之间的连接过程分为服务器监听、客户端请求、连接确认、数据传送和接收等步骤。注意：在程序最后应关闭流和 Socket 连接。

```
服务器端框架代码如下：
//启动 Socket 服务器端，使用 ServerSocket 类，监听端口号 8989
ServerSocket svr=new ServerSocket(8989);
//监听客户端的连接请求
Socket sk= svr.accept();
//分别取得输入和输出流
DataInputStream  dis=new DataInputStream(sk.getInputStream());
DataOutputStream dos=new DataOutputStream(sk.getOutputStream());
```

Socket 的交互通过字节流来完成，因此任何文件都可以进行传送使用 DataInputStream/Data OutputStream 类来实现对基本数据类型的读写功能，如 readInt()、writeInt()、readUTF()、writeUTF()等。

```
客户端框架代码如下：
//发起一个 Socket 连接，使用 Socket 类，指定连接服务器的 IP 地址和端口号
Socket sk=new Socket("192.168.1.200",8989);
//分别取得输入和输出流
DataInputStream dis=new DataInputStream(sk.getInputStream());
DataOutputStream dos=new DataOutputStream(sk.getOutputStream());
```

（2）基于 HTTP 协议的用法。

HTTP（超文本传输协议）是一个基于请求与响应模式的、无状态的、应用层的协议。基于 TCP 的连接方式，HTTP1.1 版本给出了一种持续连接的机制。绝大多数的 Web 开发都是构建在 HTTP 协议之上的。

HttpURLConnection 类是 URLConnection 的子类，用于通过 Web 发送和接收数据，在手机应用程序和 Web 资源之间建立了通信的桥梁。该类的数据可以是任何类型和长度的数据，可以用于发送和接收长度事先不确定的流数据。HttpURLConnection 类进行网络数据连接的代码需要放在一个线程中执行，它的主要处理步骤介绍如下。

① 定义一个 URL 对象，调用其 openConnection()方法，并进行强制类型转换以获得一个 HttpURLConnection 对象。

② 准备请求。请求的主要属性是 URI，请求的头部（header）可能还包括验证凭证、首选的内容类型、会话 cookie 等元数据。

③ 可以选择上传请求正文（body）。如果包括请求正文，实例必须通过 setDoOutput（true）配置，并通过 getOutputStream()写入流的方式传送数据。

④ 读取远程发送回来的响应。响应头通常包括响应正文的内容类型和长度、修改日期和会话 cookie 等元数据。可以通过 getInputStream()方法以流的方式读取响应正文。如果响应没有包含正文，getInputStream()方法将返回一个空流。

⑤ 断开连接。一旦响应的正文已经被读取，HttpURLConnection 应调用 disconnect()关闭连接，这样能够释放连接资源，以便于重用。

客户向服务器发送的请求包括请求方法、头和正文。其中请求方法用于确定如何将数据发送给远程服务器。在 Android 系统中可以使用 GET、POST 和 HEAD 等方法。使用 GET 方法，会把客户端数据作为 URL 的一部分一起发送，不需要程序员做太多的工作；使用 POST 方法，客户端数据通过建立的连接请求不同的的流进行发送，即 POST 参数并没有存放在 URL 字串中，而是存放在 HTTP 请求的正文内，使用该方法通常需要设置较多的连接属性；使用 HEAD 方法，请求不向服务器发送任何数据，而是向服务器请求关于远程资源的首部信息。

本项目的任务是通过手机客户端利用 HTTP 协议访问 Google 服务器获取天气信息。下面简单举例说明如何使用 GET 方法打开一个 HTTP 连接。

```
String httpUrl=" http://flash.weather.com.cn/wmaps/xml/beijing.xml
try{
    URL url=new URL(httpUrl);
    HttpURLConnection urlConnection =(HttpURLConnection)url.openConnection();
    urlConnection.setRequestMethod("GET");//默认是 GET，该语句可以不需要
    InputStream in=new BufferedInputStream(urlConnection.getInputStream());
    …
} catch(Exception ex){
    Log.v("exception",ex.getMessage());
} finally {
    if(urlConnection != null){
    urlConnection.disconnect();
    }
}
```

HttpURLConnection 是基于 HTTP 协议的，其底层通过 Socket 实现通信。如果没有超时设置，在网络异常的情况下，可能会导致程序僵死而无法继续往下执行。可以通过以下两个语句来设置相应的超时：

```
System.setProperty("sun.net.client.defaultConnectTimeout",超时毫秒数字符串);
System.setProperty("sun.net.client.defaultReadTimeout",超时毫秒数字符串);
```

其中，参数 sun.net.client.defaultConnectTimeout 用于设置连接主机的超时时间，单位为ms；参数 sun.net.client. defaultReadTimeout 用于设置从主机读取数据的超时时间，单位为 ms。

例如，

```
System.setProperty("sun.net.client.defaultConnectTimeout","30000");
System.setProperty("sun.net.client.defaultReadTimeout","50000");
```

下面对 HttpURLConnection 类的主要方法进行介绍。

（1）void disconnect()：释放连接，使得资源能够被重用或者关闭。

（2）String getContentEncoding ()：返回数据的内容编码格式。如果内容没有被编码，则返回值为 null 或者 identity。如果内容是被 GZIP 压缩过的，则返回 GZIP。若把被 GZIP 压缩过的内容直接当作普通的 InputStream 来读取，会出现数据读取错误，应使用 GZIPInputStream 来进行读取。

（3）InputStream getErrorStream ()：当出现请求的文件在远程服务器不存在等错误时，对从服务器中返回的错误信息输入流进行分析，以便找出出错的原因。

（4）String getRequestMethod()：得到当前的请求方法，返回值有 GET、POST、HEAD、PUT、DELETE、TRACE 或 OPTIONS。

（5）int getResponseCode()：获得 HTTP 服务器响应的代码，返回值类型为整型。如果没有响应码，则返回-1。

（6）String getResponseMessage()：获得 HTTP 服务器响应的消息，返回值类型为字符串。

（7）void setChunkedStreamingMode（int chunkLength）：设置流的块模式，将流分为多个块。参数 chunkLength 表示每块的字节数，若取值为 0 则使用系统默认的块长度。为了使上传数据的性能更高，若预先知道传送正文的长度，建议调用 setFixedLengthStreamingMode（int）方法；若预先并不知道传送正文的大小，则调用 setChunkedStreamingMode（int）方法。否则，HttpURLConnection 将在内存中先缓存所有要传送的数据，然后再发送，这样会造成堆空间的浪费，并增加延迟。

（8）void setFixedLengthStreamingMode（int contentLength）：如果 HTTP 请求正文的长度事先已经知道，将其设置为固定的长度，以便无需缓冲就能够传送流。参数 contentLength 表示 HTTP 请求正文的固定长度。

（9）void setRequestMethod（String method）：设置传送给远程 HTTP 服务器的请求命令。参数 method 的取值可以为 GET、POST、HEAD、PUT、DELETE、TRACE 或 OPTIONS。

（10）boolean usingProxy ()：判断连接是否使用了代理服务器。

（11）long getDate()：获取头部的日期信息。

（12）long getExpiration()：获取头部的过期时间。

（13）String getHeaderField（int pos）：根据位置 pos 返回头部相应的字段值。

（14）String getHeaderField（String key）：根据关键字 key 得到头部的字段值。

（15）long getLastModified()：返回最后修改的字段值。

（16）String getRequestProperty（String key）：获得连接中请求属性为 key 的值。

（17）URL getURL()：返回连接的 URL。

（18）InputStream getInputStream()：获取输入流。

（19）OutputStream getOutputStream()：获取输出流。

（20）void setRequestProperty（String field，String newValue）：设置请求属性的值。其中，参数 field 表示需要设置的头部字段；参数 newValue 用于指定属性的值。如 setRequestProperty

("Content-type","application/x-java-serialized-object")表示设定传送的内容类型是可序列化的 Java 对象。

(21) void connect(): 连接到指定的 URL 资源，执行该方法没有返回值。

下面举例说明手机如何通过 GET 和 POST 方法访问服务器端的 Servlet 程序。为了更好地帮助读者理解流技术在网络传输中的应用，本例对 GET 和 POST 的流传送采用不同的技术实现。其中，GET 传送采用 DataInputStream、DataOutputStream、InputStream 和 OutputStream 等类来实现；POST 传送则采用 BufferedReader、InputStreamReader、InputStream、BufferedWriter、OutputStreamWriter、OutputStream 等类来实现。

(1) 编写手机客户端程序。

① 创建布局文件【httpurlconnection_demo.xml】，设置一个文本框，用于提示信息以及接收服务器端返回的信息。

```
<LinearLayout xmlns:android="http://schemas.android.com/apk/res/android"
    xmlns:tools="http://schemas.android.com/tools"
    android:id="@+id/LinearLayout1"
    android:layout_width="match_parent"
    android:layout_height="match_parent"
    android:orientation="vertical" >
    <TextView
        android:id="@+id/tResponse"
        android:layout_width="match_parent"
        android:layout_height="wrap_content"
        android:text="TextView" />
</LinearLayout>
```

② 主程序部分建立两个菜单项，当用户单击【Get 连接】，将通过 GET 方式连接远程 Servlet 程序；当用户单击【Post 连接】，将通过 POST 方式连接远程 Servlet 程序。

```
public class HttpURLConnectionDemo extends Activity {
    private TextView tResponse;
    //sUrl 为 Servlet 服务器的所在地址，其中 ServletDemo 为提供服务的 Servlet 的名字
    private String sUrl="http://192.168.1.100:8080/MyServlet/servlet/ServletDemo";
    Handler handler =new Handler(){
        @Override
        public void handleMessage(Message msg) {
            super.handleMessage(msg);
            switch(msg.what){
                case 1:
                    String mymessage=(String)msg.obj;
                    tResponse.setText(mymessage);
                    break;
            }
        }
    };
    @Override
    public void onCreate(Bundle savedInstanceState) {
        super.onCreate(savedInstanceState);
        setContentView(R.layout.activity_http_urlconnection_demo);
        tResponse = (TextView) this.findViewById(R.id.tResponse);
        tResponse.setText("请点击菜单选择需要的连接");
    }
    @Override
    public boolean onCreateOptionsMenu(Menu menu) {
        menu.add(Menu.NONE, 1, Menu.NONE, "Get 连接");
    menu.add(Menu.NONE, 2, Menu.NONE, "Post 连接");
        return super.onCreateOptionsMenu(menu);
    }
    @Override
    public boolean onOptionsItemSelected(MenuItem item) {
        switch(item.getItemId()){
        case 1:
```

```
            connectServerByGet thread1=new connectServerByGet(this,sUrl);
            thread1.start();
            break;
        case 2:
            connectServerByPost thread2=new connectServerByPost(this,sUrl);
            thread2.start();
            break;
        }
        return super.onOptionsItemSelected(item);
    }
}

class connectServerByGet extends Thread{
    HttpURLConnectionDemo activity;
    String sUrl;
    public connectServerByGet(HttpURLConnectionDemo activity,String sUrl){
        this.activity=activity;
        this.sUrl=sUrl;
    }
    @Override
    public void run() {
        connectServerByGet();
    }
    public void connectServerByGet() {
        HttpURLConnection urlConnection = null;
        try {
            String severUrl=sUrl+"?name=wzq&password=123";
            URL url = new URL(severUrl);
            urlConnection = (HttpURLConnection) url.openConnection();
            urlConnection.connect();
            String lines="";
            DataInputStream reader = new DataInputStream(
                    urlConnection.getInputStream());
            lines = reader.readUTF()+"\n";
            lines=lines+reader.readUTF();
            Message message1=new Message();
            message1.what=1;
            message1.obj=lines;
            activity.handler.sendMessage(message1);
            // 关闭输入流
            reader.close();
            // 关闭连接
            urlConnection.disconnect();
        } catch (Exception ex) {
            Log.v("exception", ex.getMessage());
        } finally {
            if (urlConnection != null) {
                urlConnection.disconnect();
            }
        }
    }
}

class connectServerByPost extends Thread{
    HttpURLConnectionDemo activity;
    String sUrl;
    public connectServerByPost(HttpURLConnectionDemo activity,String sUrl){
        this.activity=activity;
        this.sUrl=sUrl;
    }

    @Override
    public void run() {
        connectServerByPost();
    }

    public void connectServerByPost() {
        HttpURLConnection urlConnection = null;
        try {
```

```
            URL url = new URL(sUrl);
            urlConnection = (HttpURLConnection) url.openConnection();
            // 参数要放正文内，所以设置为 POSt 方式请求，HttpURLConnection 默认是 GET 方式;
            urlConnection.setRequestMethod("POST");
            // 设置允许输入输出
            urlConnection.setDoOutput(true);
            urlConnection.setDoInput(true);
            // 设置输出流，向服务器发送数据。
            DataOutputStream dos = new DataOutputStream(
                    urlConnection.getOutputStream());
            // 向流写数据
            dos.writeBytes("name=" + URLEncoder.encode("wzq", "gb2312"));
            dos.writeBytes("&password=" + "123");
            // 刷新输出流，将任何字节都写入潜在的流中，即发送数据
            dos.flush();
            // 关闭输出流
            dos.close();
            //服务端返回数据
            //因为读取服务器端的汉字，所以用"UTF-8"，否则可以不需要使用该参数
            BufferedReader reader = new BufferedReader(new InputStreamReader(
                    urlConnection.getInputStream(),"UTF-8"));
            String lines="",temp;
            while ((temp = reader.readLine()) != null) {
                lines =lines+ temp + "\n";
            }
            Message message2=new Message();
            message2.what=1;
            message2.obj=lines;
            activity.handler.sendMessage(message2);
            // 关闭输入流
            reader.close();
            // 关闭连接
            urlConnection.disconnect();
        } catch (Exception ex) {
            Log.v("exception", ex.getMessage());
        } finally {
            if (urlConnection != null) {
                urlConnection.disconnect();
            }
        }
    }
}
```

（2）编写服务器端的 Servlet 程序。

① Servlet 程序部分的代码。

```
public class ServletDemo extends HttpServlet {
    public void doGet(HttpServletRequest request,HttpServletResponse response)
            throws IOException,ServletException {
        response.setContentType("text/html; charset=GBK");
        // 接收客户端传递的参数
        String sName=request.getParameter("name"); // 名字
        String sPassword=request.getParameter("password");// 密码
        String sResponse="";
        System.out.println("servlet 端接收到手机传来的姓名是: "+sName);
        // 主要起调试作用,调试结果显示在 Tomcat 启动的 DOS 窗口中
        System.out.println("servlet 端接收到手机传来的密码是: "+sPassword);
        sResponse="Get 连接成功!\n";
        if(sName.equals("wzq")&& sPassword.equals("123")){
            sResponse =sResponse+ "您好! 您的账号和密码正确,欢迎使用本系统";
        } else {
            sResponse =sResponse+ "您好! 您的账号或密码不正确! ";
        }
        // 发送处理后的参数给手机
        DataOutputStream dos=new DataOutputStream(response.getOutputStream());
        dos.writeUTF(sResponse);
```

```
        dos.writeUTF("使用本软件前,请先查看帮助");
    }

    public void doPost(HttpServletRequest request,HttpServletResponse response)
            throws ServletException,IOException {
        response.setContentType("text/html; charset=GBK");
        // 接收客户端传递的参数
        String sName=request.getParameter("name"); // 姓名
        String sPassword=request.getParameter("password"); // 密码
        String sResponse="";
        System.out.println("servlet 端接收到手机传来的姓名是: "+sName);
        // 主要起调试作用,调试结果显示在 Tomcat 启动的 DOS 窗口中
        System.out.println("servlet 端接收到手机传来的密码是: "+sPassword);
        PrintWriter out=response.getWriter();
        sResponse="Post 连接成功!\n";
        if(sName.equals("wzq")&& sPassword.equals("123")){
            sResponse=sResponse+"您好! 您的账号和密码正确,欢迎使用本系统";
        } else {
            sResponse=sResponse+"您好! 您的账号或密码不正确! ";
        }
        // 发送处理后的参数给手机,因为需要写入汉字,所以使用"UTF-8"参数,否则无需使用该参数
         BufferedWriter writer=new BufferedWriter(new OutputStreamWriter(
         response.getOutputStream(),"UTF-8"));
         writer.write(sResponse+"\n");
         writer.write("使用本软件前,请先查看帮助");
         writer.flush();
         writer.close();
    }
}
```

② 【Web.xml】配置文件。

```
<?xml version="1.0" encoding="UTF-8"?>
<web-app version="2.5"
    xmlns="http://java.sun.com/xml/ns/javaee"
    xmlns:xsi="http://www.w3.org/2001/XMLSchema-instance"
    xsi:schemaLocation="http://java.sun.com/xml/ns/javaee
    http://java.sun.com/xml/ns/javaee/web-app_2_5.xsd">
  <servlet>
    <description>This is the description of my J2EE component</description>
    <display-name>This is the display name of my J2EE component</display-name>
    <servlet-name>ServletDemo</servlet-name>
    <servlet-class>ServletDemo</servlet-class>
  </servlet>

  <servlet-mapping>
    <servlet-name>ServletDemo</servlet-name>
    <url-pattern>/ServletDemo</url-pattern>
  </servlet-mapping>
  <welcome-file-list>
    <welcome-file>index.jsp</welcome-file>
  </welcome-file-list>
</web-app>
```

代码说明如下。

（1）建议使用 MyEclipse 开发上面的 Servlet 程序，这样容易部署和调试。

（2）客户端连接服务器的代码需要放在一个线程中执行，否则会报错。

（3）对于客户端的 Get 请求，服务器的 Servlet 端应实现 doGet 方法，在 doGet 方法中编写接受请求并作出响应的代码。类似地，对于客户端的 Post 请求，服务器的 Servlet 端应实现 doPost 方法，在 doPost 方法中编写接受请求并作出响应的代码。

（4）对于 Get 请求，可以先在网页上通过输入参数的方法来进行测试，即在网址后附加参数名和参数值：

```
http://192.168.1.100:8080/AndroidServer/ServletDemo?name=wzq&password=123
```

（5）在 Servlet 中使用 System.out.println()输出客户端发送过来的消息，以便于测试。

（6）Android 系统不能识别类似 localhost 或 127.0.0.1 的环回本地地址。可以通过 cmd 指令在命令行状态下输入 ipconfig /all 来查看自己的网络 IP，使用该 IP 地址进行测试。

三、任务实施

（1）定义程序界面的 XML 布局文件【main.xml】，界面划分为两个区域。一个区域用于设置查询条件，其中使用一个 TextView 表示城市名字，EditText 表示用户输入的城市拼音名；使用一个 Button 按钮用于查询，该区域总体使用水平放置的 LinearLayout 布局。第二个区域布局天气预报查询结果，只使用一个 LinearLayout 即可，结果在程序中动态生成。

```
<?xml version="1.0" encoding="utf-8"?>
<LinearLayout xmlns:android="http://schemas.android.com/apk/res/android"
   android:orientation="vertical"
   android:layout_width="fill_parent"
   android:layout_height="fill_parent"
   >
  <LinearLayout
   android:layout_width="fill_parent"
        android:layout_height="wrap_content"
   android:orientation="horizontal"
        android:weightSum="1">
  <TextView android:layout_width="50.0dip"
   android:layout_height="wrap_content"
   android:text="城市:"
   android:layout_marginLeft="5.0dip"
   android:layout_marginRight="2.0dip"
  />
  <EditText android:id="@+id/value"
   android:layout_height="wrap_content"
   android:text="guangzhou"
   android:layout_width="185dp"/>
  <Button
   android:id="@+id/find"
   android:layout_height="wrap_content"
   android:text="查询"
  android:layout_width="match_parent"/>
  </LinearLayout>
  <LinearLayout android:id="@+id/my_body"
   android:layout_width="fill_parent"
   android:layout_height="wrap_content"
   android:layout_weight="1.0">
  </LinearLayout>
</LinearLayout>
```

（2）将需要查询的城市拼音作为参数，建立 URL 地址，通过 Connector 类的 open 方法向服务器请求天气预报信息。

```
  //city 变量表示城市名字的拼音
   String
     weatherUrl="http://flash.weather.com.cn/wmaps/xml/"+city+".xml";
   //天气情况图标的基础网址
   String weatherIcon="http://m.weather.com.cn/img/c";
   URL  url=new URL(weatherUrl);
   //建立天气预报查询连接
   httpConn =(HttpURLConnection)url.openConnection();
   //采用 GET 请求方法
   httpConn.setRequestMethod("GET");
   //打开数据输入流
   din=httpConn.getInputStream();
```

（3）利用 XmlPullParser 对象对返回的天气预报 XML 文件进行分析，循环读出所需要的天气信息，并使用集合变量 cityname、low、high、icon 和 summary 分别存放城市名、最低温度、最高温度、天气图标和天气情况信息。

```
//获得 XmlPullParser 解析器
 XmlPullParser xmlParser=Xml.newPullParser();
 xmlParser.setInput(din, "UTF-8");
 //获得解析到的事件类别，这里有开始文档，结束文档，开始标签，结束标签，文本等等事件。
 int evtType=xmlParser.getEventType();
 while(evtType!=XmlPullParser.END_DOCUMENT)//一直循环，直到文档结束
 {
    switch(evtType)
    {
      case XmlPullParser.START_TAG:
          String tag = xmlParser.getName();
          //如果是 city 标签开始，则说明需要实例化对象了
         if (tag.equalsIgnoreCase("city")){
      //城市天气预报
          cityname.addElement(xmlParser.getAttributeValue(null, "cityname")+"天气：");
      //天气情况概述
      summary.addElement(xmlParser.getAttributeValue(null, "stateDetailed"));
      //最低温度
      low.addElement("最低："+xmlParser.getAttributeValue(null, "tem2"));
      //最高温度
      high.addElement("最高："+xmlParser.getAttributeValue(null, "tem1"));
      //天气情况图标网址
 icon.addElement(weatherIcon+xmlParser.getAttributeValue(null,     "state1")+".gif");
             }
         break;
      case XmlPullParser.END_TAG:
           //标签结束
           default:break;
    }
      evtType=xmlParser.next();//如果 xml 没有结束，则导航到下一个节点
 }
```

（4）在【AndroidManifest.xml】中添加网络访问权限。

```
<uses-permission android:name="android.permission.INTERNET">
```

7.2 下载天气图片

一、任务分析

中国天气网提供了描述天气信息的 GIF 格式图片，可以将其下载到手机，这样天气信息的显示显得更为直观。要完成本次任务，需要思考如下两个问题。

（1）如何从网络下载图片文件？

（2）如何在手机上显示下载的图片？

二、任务实施

通过网络下载图片主要有两个步骤。

（1）在 7.1 节的任务实施中获得包含图片 URL 地址的 icon 集合，这样就可以通过 URL 类的 openConnection 方法向服务器提出图片获取请求。

（2）利用 BitmapFactory 类将返回的图片流转换成程序内部能够处理的图像。

主要实现代码示例如下。

```
private void downImage()
   {
   //天气情况图标获取
       int i=0;
```

```
for(i=0;i<icon.size();i++){
        try
        {
            URL  url=new URL(icon.elementAt(i));
            httpConn =(HttpURLConnection)url.openConnection();
            httpConn.setRequestMethod("GET");
            din =httpConn.getInputStream();
            //Vector<Bitmap> bitmap=new Vector<Bitmap>();
            //bitmap是专门存放Bitmap 类型的全局集合变量，上面用注释方式列出，是为了便于说明
            bitmap.addElement(BitmapFactory.decodeStream(httpConn.getInputStream()));
        }catch(Exception ex){
            ex.printStackTrace();
        }finally{
            //释放连接
            try{
                din.close();
                httpConn.disconnect();
            }catch(Exception ex){
                ex.printStackTrace();
            }
        }
    }
}
```

7.3 显示天气预报

一、任务分析

天气预报的文字信息和图片已经分别在7.1节和7.2节中实现，本节任务需要完成将天气预报的文字信息和图片统一放在界面上。要完成本次任务，需要思考如下两个问题。

（1）如何让手机在后台进行网络连接？

（2）如何进行界面数据的刷新？即用户单击查询新的城市名字后，前面旧的查询结果被清除，界面上显示新的天气预报信息。

二、相关知识

前面任务的界面一般都预先通过XML文件设置。本任务将介绍如何在一个定义好的XML布局文件中，动态地增加新的控件，并设置其布局。要在Android程序中动态创建UI控件，首先创建UI对象，设置其相应属性，然后再将其添加到布局中。

下面对一些界面UI类进行介绍，以便于在程序中调用。

（1）LayoutParams类用于告诉父视图如何对其进行布局，主要描述元素的宽度和高度，属性取值有以下几种。

① FILL_PARENT或者MATCH_PARENT：希望和父视图的尺寸相同。其中MATCH_PARENT用在Android 8以上的版本中。

② WRAP_CONTENT：视图的大小恰好能够放置内容。

③ 具体的数值。

LayoutParams类的主要构造函数是LayoutParams（int width，int height），该函数表示根据给定的宽度和高度，创建一个新的布局参数集。其中，参数width表示视图的宽度，对应于XML属性android:layout_width；参数height表示视图的高度，对应于XML属性android:layout_height。

View类的方法setLayoutParams（ViewGroup.LayoutParams params），可以对视图设置布局参数。所以在设置好LayoutParams对象后，可作为参数调用上述方法对视图进行布局设置。

（2）ViewGroup 类是特殊的视图类，用于包含其他视图，相当于容器。以下方法可以将子视图添加到 ViewGroup 中。

① void addView（View child，int index，ViewGroup.LayoutParams params）。

② void addView（View child，ViewGroup.LayoutParams params）。

③ void addView（View child，int index）。

④ void addView（View child）。

⑤ public void addView（View child，int width，int height）。

其中，参数 child 表示要添加的子视图；参数 index 表示要放置的位置；参数 params 为布局参数；参数 width 表示视图的宽度；参数 height 表示视图的高度。

相应地，ViewGroup 类提供了 removeAllViews()方法，用于从 ViewGroup 中移除掉所有的子视图。

（3）LinearLayout 类是关于线性布局的类，也比较常用。

① setOrientation（int orientation）：设置布局中是按行还是按列放置。参数 orientation 的取值为 HORIZONTAL 或者 VERTICAL，分别表示水平或者垂直排列。相当于设置 XML 属性的 android: orientation。

② void setGravity（int gravity）：设置子视图如何在布局中放置。相当于设置 XML 属性的 android:gravity。

三、任务实施

下面介绍在界面上显示天气预报文字和图片信息的关键步骤。

（1）使用 LinearLayout 类的 removeAllViews 方法清除存储原有查询结果的组件，然后为每一个城市的天气信息创建一个新的 LinearLayout 布局，设置显示天气预报信息的控件布局参数，新建显示预报信息内容的控件，向控件设置显示信息，并将其添加到 LinearLayout 布局中，最后将 LinearLayout 添加到显示天气结果的布局中。下面给出主要的实现方法。

```
//body 为布局参数中用于显示天气预报结果的顶层 LinearLayout
LinearLayout body=(LinearLayout)findViewById(R.id.my_body);
body.removeAllViews();//清除存储原有的查询结果的组件
body.setOrientation(LinearLayout.VERTICAL);
LayoutParams params=new LayoutParams(
LayoutParams.WRAP_CONTENT,LayoutParams.WRAP_CONTENT);
params.weight=80;
params.height=50;
for(int i=0;i<cityname.size();i++){
    //linerlayout 表示每一条天气信息（城市、天气描述、天气情况图片、最低和最高气温）
    LinearLayout  linerlayout=new LinearLayout(this);
    linerlayout.setOrientation(LinearLayout.HORIZONTAL);
    //城市
    TextView  cityView=new TextView(this);
    cityView.setLayoutParams(params);
    cityView.setText(cityname.elementAt(i));
    linerlayout.addView(cityView);
    //天气描述
    TextView  summaryView=new TextView(this);
    summaryView.setLayoutParams(params);
    summaryView.setText(summary.elementAt(i));
    linerlayout.addView(summaryView);
    //图标
    ImageView iconView=new ImageView(this);
    iconView.setLayoutParams(params);
    iconView.setImageBitmap(bitmap.elementAt(i));
    linerlayout.addView(iconView);
```

```
        //最低气温
        TextView  lowView=new TextView(this);
        lowView.setLayoutParams(params);
        lowView.setText(low.elementAt(i));
        linerlayout.addView(lowView);
        //最高气温
        TextView  highView=new TextView(this);
        highView.setLayoutParams(params);
        highView.setText(high.elementAt(i));
        linerlayout.addView(highView);
        //将每一条城市天气信息加入到body布局中
        body.addView(linerlayout);
}
```

（2）采用多线程技术处理和显示天气预报信息。由于获取天气预报信息受网络的影响，而且分析和显示天气预报信息也需要花费一定的时间，所以将这些操作交由线程来执行，这样可以使程序整体运行更流畅。下面给出主要的实现方法。

① 定义一个实现 Runnable 接口的 Activity 子类。

② 在 Activity 子类中定义一个 Handler 消息处理对象。

```
private final Handler handler=new Handler(){
     public void handleMessage(Message msg){
         switch(msg.what){
             case 1:
                  调用第(1)步中的代码来更新主界面的信息
                  break;
             }
             super.handleMessage(msg);
         }
};
```

③ 在 run 方法中，调用 7.1 节和 7.2 节的实现，并使用 handler 对象发送更新界面的消息。

④ 在执行查询操作的按钮函数 OnClick 中，开始线程。

```
Thread th=new Thread(this);
th.start();
```

7.4 完整项目实施

天气预报程序主界面和查询界面效果分别如图 7-5 和图 7-6 所示。

图 7-5 天气预报程序主界面

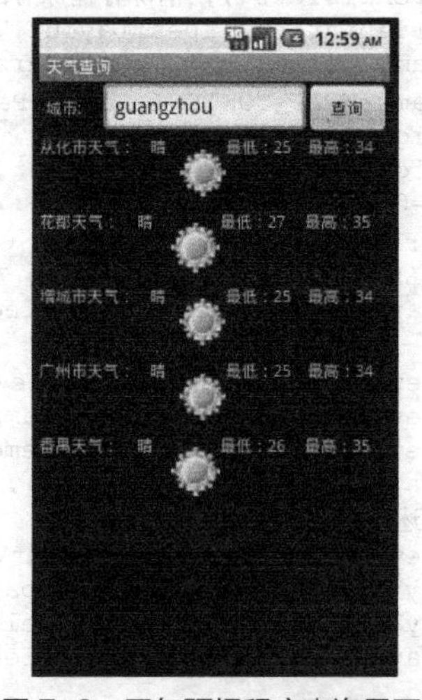

图 7-6 天气预报程序查询界面

天气预报程序由【main.xml】、【AndroidManifest.xml】布局文件及【WebWeatherActivity.

java】类组成，其完整代码如下。

（1）天气预报程序运行的 AndroidManifest 描述：【AndroidManifest.xml】。

```
<?xml version="1.0" encoding="utf-8"?>
<manifest xmlns:android="http://schemas.android.com/apk/res/android"
     package="com.demo.pr7"
     android:versionCode="1"
     android:versionName="1.0">
   <uses-sdk android:minSdkVersion="8" />
   <application android:icon="@drawable/icon"
                android:label="@string/app_name">
       <activity android:name=".WebWeatherActivity"
               android:label="@string/app_name">
           <intent-filter>
              <action android:name="android.intent.action.MAIN" />
              <category
                   android:name="android.intent.category.LAUNCHER"/>
           </intent-filter>
       </activity>
</application>
<!-网络访问权限-->
    <uses-permission android:name="android.permission.INTERNET">
    </uses-permission>
</manifest>
```

（2）天气预报程序界面布局 XML：【main.xml】。

```
<?xml version="1.0" encoding="utf-8"?>
<LinearLayout
  xmlns:android="http://schemas.android.com/apk/res/android"
    android:orientation="vertical"
    android:layout_width="fill_parent"
    android:layout_height="fill_parent"
    >
   <LinearLayout
        android:layout_width="fill_parent"
        android:layout_height="wrap_content"
        android:orientation="horizontal"
        android:weightSum="1">
       <TextView android:layout_width="50.0dip"
        android:layout_height="wrap_content"
        android:text="城市:"
        android:layout_marginLeft="5.0dip"
        android:layout_marginRight="2.0dip"
       />
       <EditText android:id="@+id/value"
        android:layout_height="wrap_content"
        android:text="guangzhou"
        android:layout_width="185dp"/>
       <Button
        android:id="@+id/find"
        android:layout_height="wrap_content"
        android:text="查询"
       android:layout_width="match_parent"/>
   </LinearLayout>
   <LinearLayout android:id="@+id/my_body"
       android:layout_width="fill_parent"
       android:layout_height="wrap_content"
       android:layout_weight="1.0">
   </LinearLayout>
</LinearLayout>
```

（3）天气预报程序功能实现类：【WebWeatherActivity.java】。

```
package com.demo.pr7;
```

```
import java.io.BufferedReader;
import java.io.ByteArrayInputStream;
import java.io.InputStream;
import java.io.InputStreamReader;
import java.net.HttpURLConnection;
import java.net.URL;
import java.util.Vector;
import org.xmlpull.v1.XmlPullParser;
import android.app.Activity;
import android.graphics.Bitmap;
import android.graphics.BitmapFactory;
import android.os.Bundle;
import android.os.Handler;
import android.os.Message;
import android.util.Xml;
import android.view.View;
import android.view.View.OnClickListener;
import android.widget.Button;
import android.widget.EditText;
import android.widget.ImageView;
import android.widget.LinearLayout;
import android.widget.TextView;
import android.widget.Toast;
import android.widget.LinearLayout.LayoutParams;

public class WebWeatherActivity extends Activity  implements Runnable{
    HttpURLConnection  httpConn = null;
    InputStream din = null;
    Vector<String> cityname=new Vector<String>();
    Vector<String> low=new Vector<String>();
    Vector<String> high=new Vector<String>();
    Vector<String> icon=new Vector<String>();
    Vector<Bitmap> bitmap=new Vector<Bitmap>();
    Vector<String> summary=new Vector<String>();
    int weatherIndex[]=new int[20];
    String city="guangzhou";
    boolean bPress=false;
    boolean bHasData=false;
    LinearLayout  body;
    Button       find;
    EditText     value;
   @Override
   public void onCreate(Bundle savedInstanceState) {
      super.onCreate(savedInstanceState);
      setContentView(R.layout.main);
      setTitle("天气查询");
      body=(LinearLayout)findViewById(R.id.my_body);
      find=(Button)findViewById(R.id.find);
      value=(EditText)findViewById(R.id.value);
      find.setOnClickListener(new OnClickListener() {
            @Override
            public void onClick(View v) {
                body.removeAllViews();//移除当前的所有结果
                city=value.getText().toString();
                Toast.makeText(WebWeatherActivity.this,
                    "正在查询天气信息...", Toast.LENGTH_LONG).show();
                Thread th=new Thread(WebWeatherActivity.this);
                th.start();
            }
        });
   }
   @Override
    public void run() {
        // TODO Auto-generated method stub
      cityname.removeAllElements();
      low.removeAllElements();
    high.removeAllElements();
    icon.removeAllElements();
```

```
    bitmap.removeAllElements();
    summary.removeAllElements();
    //获取数据
        parseData();
        //下载图片
        downImage();
        //通知UI线程显示结果
      Message message = new Message();
      message.what = 1;
      handler.sendMessage(message);
    }
   //获取数据
   public void parseData(){
    int i=0;
    String sValue;
    //city变量表示城市名字的拼音
    String weatherUrl="http://flash.weather.com.cn/wmaps/xml/"+city+".xml";        //
表示天气情况图标的基础网址
    String weatherIcon="http://m.weather.com.cn/img/c";
    try
        {
    URL  url=new URL(weatherUrl);
    //建立天气预报查询连接
    httpConn =(HttpURLConnection)url.openConnection();
    //采用GET请求方法
    httpConn.setRequestMethod("GET");
    //打开数据输入流
    din=httpConn.getInputStream();
    //获得XmlPullParser解析器
    XmlPullParser xmlParser=Xml.newPullParser();
    xmlParser.setInput(din, "UTF-8");
    //获得解析到的事件类别，这里有开始文档，结束文档，开始标签，结束标签，文本等等事件。
       int evtType=xmlParser.getEventType();
       while(evtType!=XmlPullParser.END_DOCUMENT)//一直循环，直到文档结束
       {
          switch(evtType)
          {
          case XmlPullParser.START_TAG:
              String tag = xmlParser.getName();
             //如果是city标签开始，则说明需要实例化对象了
              if (tag.equalsIgnoreCase("city"))
              {
                  //城市天气预报
       cityname.addElement(xmlParser.getAttributeValue(null, "cityname")+"天气：");
                   //天气情况概述
        summary.addElement(xmlParser.getAttributeValue(null, "stateDetailed"));
                  //最低温度
         low.addElement("最低："+xmlParser.getAttributeValue(null, "tem2"));
                  //最高温度
         high.addElement("最高："+xmlParser.getAttributeValue(null, "tem1"));
                //天气情况图标网址
                      icon.addElement(weatherIcon+xmlParser.getAttributeValue(null,
"state1")+".gif");

              }
              break;
         case XmlPullParser.END_TAG:
             //标签结束
             default:break;
         }
         //如果xml没有结束，则导航到下一个节点
         evtType=xmlParser.next();
        }

         }catch (Exception ex){
```

```
            ex.printStackTrace();
        }finally{
            //释放连接
            try{
              din.close();
              httpConn.disconnect();
            }catch(Exception ex){
              ex.printStackTrace();
            }
        }

    }
    // 下载图片
    private void downImage()
    {
     //天气情况图标获取
        int i=0;
     for(i=0;i<icon.size();i++){
         try
         {
                URL  url=new URL(icon.elementAt(i));
                System.out.println(icon.elementAt(i));
               httpConn =(HttpURLConnection)url.openConnection();
               httpConn.setRequestMethod("GET");
               din =httpConn.getInputStream();
               //图片数据 Bitmap
bitmap.addElement(BitmapFactory.decodeStream(httpConn.getInputStream()));
        }catch (Exception ex){
            ex.printStackTrace();
        }finally{
            //释放连接
            try{
              din.close();
              httpConn.disconnect();
            }catch(Exception ex){
              ex.printStackTrace();
            }
        }
     }
    }
    //显示结果 handler
     private final Handler handler = new Handler(){
       public void handleMessage(Message msg) {
         switch (msg.what) {
             case 1:
                 showData();
                 break;
             }
             super.handleMessage(msg);
         }
    };
    //显示结果
    public void showData(){
     body.removeAllViews();//清除存储旧的查询结果的组件
     body.setOrientation(LinearLayout.VERTICAL);
     LayoutParams params = new LayoutParams(
            LayoutParams.WRAP_CONTENT,LayoutParams.WRAP_CONTENT);
     params.weight=80;
     params.height=50;
     for(int i=0;i<cityname.size();i++){
         LinearLayout  linerlayout=new LinearLayout(this);
        linerlayout.setOrientation(LinearLayout.HORIZONTAL);
        //城市
        TextView  dayView=new TextView(this);
        dayView.setLayoutParams(params);
        dayView.setText(cityname.elementAt(i));
        linerlayout.addView(dayView);
```

```
    //描述
      TextView  summaryView=new TextView(this);
      summaryView.setLayoutParams(params);
      summaryView.setText(summary.elementAt(i));
      linerlayout.addView(summaryView);
      //图标
      ImageView  icon=new ImageView(this);
      icon.setLayoutParams(params);
      icon.setImageBitmap(bitmap.elementAt(i));
      linerlayout.addView(icon);
      //最低气温
      TextView  lowView=new TextView(this);
      lowView.setLayoutParams(params);
      lowView.setText(low.elementAt(i));
      linerlayout.addView(lowView);
      //最高气温
      TextView  highView=new TextView(this);
      highView.setLayoutParams(params);
      highView.setText(high.elementAt(i));
      linerlayout.addView(highView);
      body.addView(linerlayout);
    }
  }
}
```

7.5 扩展项目 JSON 格式接口调用

一、任务分析

在上面的天气预报项目中，已经详细介绍了基于 XML 格式的数据解析方法。除了 XML，JSON 是另外一种业界常用的数据交换格式。本任务是获取天气预报网站的 JSON 数据，并进行解析。要完成本次任务，需要思考如下两个问题。

（1）什么是 JSON 数据格式?

（2）Android 提供了什么方法去解析 JSON 数据格式?

二、相关知识

1. JSON 概念

JSON(JavaScript Object Notation) 是一种轻量级的数据交换格式。它是基于 JavaScript 的一个子集。JSON 采用完全独立于语言的文本格式，但是也使用了类似于 C 语言家族的习惯（包括 C, C++, C#, Java, JavaScript, Perl, Python 等）。这些特性使 JSON 成为理想的数据交换语言，易于人阅读和编写，同时也易于机器解析和网络传输。

下面来介绍 JSON 的格式。JSON 是一个“名称/值”对的集合，其中“名称/值”在集合中是无序的。一个 JSON 对象是以左括号“{”开始，右括号“}”结束。每个“名称”后跟一个冒号“:”。“‘名称/值’”对之间使用逗号“,”隔开。例如下面是一个简单的表达学生信息的 JSON 数据：

```
{"ID":"20150603","name":"张三","sex":"男","age":19,"department":"信息工程学院"}
```

“‘名称/值’”中的“名称’”是关键字，用字符串表示。“值’”是双引号括起来的字符串、数值、true、false、 null、JSON 对象或者数组，例如上面例子中的"ID"为名称，"20150603"为其对应的值。这些结构可以嵌套，从而表达出各种复杂的数据关系。其中数组是“值’”的有序集合。一个数组以左中括号“[”开始，右中括号“]”结束。值之间使用逗号“,”分隔。下面是一个较为复杂的个人信息 JSON 数据：

```
{
    "name" : "李四", //值为字符串
    "age" : 30, //值为数值
    "married" : false, //值为布尔值
    "phone" : ["12345678", "131111111"], //值为数组
    "address" : {"country" : "中国", "province" : "广东" } //值为 JSON 对象
}
```

与 XML 相对比，JSON 有以下相同和不同点：

- JSON 和 XML 的数据可读性基本相同；
- JSON 和 XML 同样拥有丰富的解析手段；
- JSON 相对于 XML 来讲，数据的体积小；
- JSON 与 JavaScript 的交互更加方便；
- JSON 对数据的描述性比 XML 差；
- JSON 的速度要远远快于 XML；

2．JSONObject

Android 的 JSON 解析部分都在包 org.json 下。JSONObject 类用于表示一个 JSON 对象。这是系统中有关 JSON 定义的基本单元，其包含一组(Key/Value)数值。Value 的数据类型可以是 JSONObjects, JSONArrays, Strings, Booleans, Integers, Longs, Doubles 或者 NULL。下面介绍 JSONObject 类的两个主要构造方法。

（1）JSONObject()：创建一个空的 JSON 对象，没有任何“名称/值’”对。在创建后需要向对象添加 Key/Value 值对。

（2）JSONObject(String json)：创建一个 JSON 对象，数据来自 JSON 字符串参数 json。

JSONObject 类提供 put 方法来添加“名称/值’”对，使用 get 方法来获取某个名称对应的值。

3．JSONArray

JSONArray 类代表一组有序的值，可以理解为用一个数组来表示 JSON 对象的值，值的形式为 [value1,value2,value3]。下面介绍 JSONArray 类的两个主要构造方法。

（1）JSONArray()：创建一个空的 JSONArray 对象，没有任何值。在创建后需要向对象添加值。

（2）JSONArray (String json)：创建一个 JSONArray 对象，数据来自 JSON 字符串参数 json。

JSONArray 类提供 put 方法来添加值，使用 get 方法来获取某个位置的值。

下面介绍创建一个 JSON 对象，点击菜单可以显示 JSON 数据的例子，其中 JSON 数据来自前面的个人信息 JSON 数据：

```
public class JsonDemo extends Activity {
    JSONObject data;

    @Override
    public void onCreate(Bundle savedInstanceState) {
        super.onCreate(savedInstanceState);
        data = createJSONObject();
    }

    @Override
    public boolean onCreateOptionsMenu(Menu menu) {
        menu.add(1, 1, 1, "读取 JSON 数据");
        return true;
    }
```

```
    @Override
    public boolean onOptionsItemSelected(MenuItem item) {
        switch (item.getItemId()) {
        case 1:
            showJSON();
            break;
        }
        return super.onOptionsItemSelected(item);
    }
//创建 JSON 对象
    public JSONObject createJSONObject() {
        JSONObject person = new JSONObject();
        try {
            // 第 1 个 value 一个字符串
            person.put("name", "李四");
            // 第 2 个 value 一个字符串数值
            person.put("age", 30);
            // 第 3 个 value 一个布尔值
            person.put("married", false);
            // 第 4 个 value 是[ ]，也就是一个 JSONArray 对象
            JSONArray phone = new JSONArray();
            phone.put("12345678");
            phone.put("131111111");
            person.put("phone", phone);
            // 第 5 个 value 一个 JSONObject 对象
            JSONObject address = new JSONObject();
            address.put("country", "中国");
            address.put("province", "广东");
            person.put("address", address);
        } catch (JSONException e) {
            e.printStackTrace();
        }
        return person;
    }
//显示 JSON 数据
    public void showJSON()

    {
        String text;
        try {
            text = "姓名为" + data.getString("name") + "\n";
            text = text + "年龄为" + data.getInt("age") + "\n";
            text = text + "电话号码为" + data.getJSONArray("phone").getString(0)
                    + "," + data.getJSONArray("phone").getString(1);
            Toast.makeText(this, text, Toast.LENGTH_LONG).show();
        } catch (Exception ex) {
          ex.printStackTrace();
        }
    }
}
```

4．JSONTokener

JSONTokener 为 JSON 数据的解析类，大部分情况下，只用到它的构造方法和 nextValue()方法。

public Object nextValue ()：从输入中返回下一个值。值的数据类型为 JSONObject, JSONArray, String, Boolean, Integer, Long, Double 或 NULL。

下面是 JSONTokener 的一个使用例子：

```
String json = "{"
      + "  \"query\": \"Pizza\", "
      + "  \"locations\": [ 94043, 90210 ] "
      + "}";
 JSONTokener jsonParser = new JSONTokener(json);
//此时还未读取任何 json 文本，直接读取就是一个 JSONObject 对象。
```

```
//如果此时的读取位置在"query" : 了，那么nextValue就是"Pizza"
JSONObject object = (JSONObject) jsonParser.nextValue();
 String query = object.getString("query");
 JSONArray locations = object.getJSONArray("locations");
```

5. JSON格式的天气预报数据源

中国天气网提供JSON格式的天气预报接口有以下三个，其中网址末尾的"101010100"表示北京的城市代码。这三个网址都能够返回城市的天气预报信息，差异在于预报信息的详细程度不同。

- http://www.weather.com.cn/data/sk/101010100.html
- http://www.weather.com.cn/data/cityinfo/101010100.html
- http://m.weather.com.cn/data/101010100.html

下面列出部分城市的代码：

```
北京:101010100、天津:101030100、上海:101020100、石家庄:101090101、郑州:101180101、
合肥:101220101、杭州:101210101、重庆:101040100、福州:101230101、兰州:101160101、
广州:101280101、南宁:101300101、贵阳:101260101、昆明:101290101、呼和浩特:101080101、
南昌:101240101、武汉:101200101、成都:101270101、银川:101170101、济南:101120101、
西安:101110101、太原:101100101、乌鲁木齐:101130101、拉萨:101140101、长春:101060101、
台北县:101340101、海口:101310101、长沙:101250101、南京:101190101、哈尔滨:101050101
```

三、项目实施

1. 以http://www.weather.com.cn/data/cityinfo/101010100.html网址进行JSON数据查询，下面数据为其JSON数据格式


```
{"weatherinfo":{"city":" 北 京 ","cityid":"101010100","temp1":"15 ℃ ","temp2":"5 ℃
","weather":"多云","img1":"d1.gif","img2":"n1.gif","ptime":"08:00"}}
```


2. 在相关知识中创建的JsonDemo类中，定义一个线程类，用于连接JSON天气预报网址，获取JSON数据，对其信息进行抽取，并显示。

```
    class connectWeatherServer extends Thread{
        JsonDemo activity;
        String sUrl;
        public connectWeatherServer(JsonDemo activity,String sUrl){
            this.activity=activity;
            this.sUrl=sUrl;
        }

        @Override
        public void run() {
            showWeatherJSON();
        }
        public void showWeatherJSON() {
            try {
                URL url = new URL(sUrl);
                // 建立天气预报查询连接
                HttpURLConnection urlConnection = (HttpURLConnection) url.openConnection();
                // 采用GET请求方法
                urlConnection.setRequestMethod("GET");
                // 打开数据输入流
                InputStream din = urlConnection.getInputStream();
                InputStreamReader    in    =    new    InputStreamReader(urlConnection.
getInputStream());
                // 为输出创建BufferedReader
                BufferedReader buffer = new BufferedReader(in);
                String inputLine =null;
                StringBuffer JsonData = new StringBuffer();;
                // 使用循环来读取获得的数据
                while ((inputLine = buffer.readLine()) != null) {
                    JsonData.append(inputLine);
                }
```

```
                String sJsonData = JsonData.toString();
                JSONObject jsonObject = new JSONObject(sJsonData);
                JSONObject cityweather = jsonObject.getJSONObject("weatherinfo");
                StringBuffer weatherInfo=new StringBuffer();
                weatherInfo.append("城市: "+cityweather.getString("city"));
                weatherInfo.append("天气情况: "+cityweather.getString("weather"));
                weatherInfo.append("最高温度: "+cityweather.getString("temp1"));
                weatherInfo.append("最低温度: "+cityweather.getString("temp2"));
                Message message=new Message();
                message.what=1;
                message.obj=weatherInfo;
                activity.handler.sendMessage(message);
                // 关闭输入流
                buffer.close();
                // 关闭连接
                urlConnection.disconnect();
            } catch (Exception ex) {
                ex.printStackTrace();
            }
        }
}
```

3．增加一个菜单项，点击执行第 2 步的线程连接。

```
public boolean onCreateOptionsMenu(Menu menu) {
    menu.add(1, 2, 1, "读取 JSON 天气预报数据");
}

public boolean onOptionsItemSelected(MenuItem item) {
    switch (item.getItemId()) {
    case 2:
        String weatherUrl =
                "http://www.weather.com.cn/data/cityinfo/101010100.html";
        connectWeatherServer thread=new connectWeatherServer(this,weatherUrl);
        thread.start();
        break;
    }
    return super.onOptionsItemSelected(item);
}
```

7.6 实训项目

一、开发天气预报程序

1．实训目的与要求

掌握在 Android 中通过网络获取数据并进行进一步处理的技能。

2．实训内容

在完成本项目的三个任务之后，实现如下功能以进一步完善天气预报程序。

（1）增加一个圆形进度条，用于提示正在查询天气信息，当有天气查询结果时，将进度条消掉，并显示查询结果。

（2）支持中文城市名的查询，即用户输入城市汉字后能够查询相应城市的天气信息。

（3）使用 ListView 来显示天气预报信息。

（4）使用 DOM 方式来解析天气预报信息的 XML 文档。

3．思考

（1）如何实现汉字与拼音的转换？

（2）基于流模式的 XML 文档解析与 DOM 解析有何区别？

二、开发手机聊天室

1. 实训目的与要求

掌握在手机上运用 Socket 进行通信的技能。

2. 实训内容

功能要求：实现多个手机客户端能够在聊天室中进行聊天。具体思路可细分如下：

（1）创建一个服务器程序，接收手机客户端的连接，能够发送信息和文件；

（2）创建一个客户端程序，向服务器发送消息或者文件，并从服务器接收信息或文件。

3. 思考

（1）服务器端如何监听客户端的连接？

（2）怎样以流的方式实现客户端和服务器端的通信？

（3）多线程技术在手机聊天室程序中有什么作用？